Studies in Computational Intelligence

Volume 938

The series "Studies in Computational Intelligence" (SCI) publishes new developments and advances in the various areas of computational intelligence—quickly and with a high quality. The intent is to cover the theory, applications, and design methods of computational intelligence, as embedded in the fields of engineering, computer science, physics and life sciences, as well as the methodologies behind them. The series contains monographs, lecture notes and edited volumes in computational intelligence spanning the areas of neural networks, connectionist systems, genetic algorithms, evolutionary computation, artificial intelligence, cellular automata, self-organizing systems, soft computing, fuzzy systems, and hybrid intelligent systems. Of particular value to both the contributors and the readership are the short publication timeframe and the world-wide distribution, which enable both wide and rapid dissemination of research output.

Indexed by SCOPUS, DBLP, WTI Frankfurt eG, zbMATH, SCImago.

All books published in the series are submitted for consideration in Web of Science.

More information about this series at http://www.springer.com/series/7092

Oliver Schütze · Carlos Hernández

Archiving Strategies for Evolutionary Multi-objective Optimization Algorithms

Springer

Oliver Schütze
Departamento de Computación
CINVESTAV-IPN
Mexico City, Mexico

Carlos Hernández
Departamento de Computación
CINVESTAV-IPN
Mexico City, Mexico

ISSN 1860-949X ISSN 1860-9503 (electronic)
Studies in Computational Intelligence
ISBN 978-3-030-63775-0 ISBN 978-3-030-63773-6 (eBook)
https://doi.org/10.1007/978-3-030-63773-6

This Springer imprint is published by the registered company Springer Nature Switzerland AG
The registered company address is: Gewerbestrasse 11, 6330 Cham, Switzerland

To my beloved wife and children Adriana, Fabián, and Luis Adrián

Oliver Schütze

To my mother Maria de la Luz and my grandmother Mónica

Carlos Hernández

Foreword

The integration of an archive into an algorithm for multi-objective optimization seems simple: You need a rule for deciding which solution is allowed to enter the archive and a rule for deciding which solution should leave the archive if the archive's capacity is exceeded.

So, why would someone write a whole book about it? Because different fields of application require differently shaped archives, and the art is to formulate conditions leading to a collection of solutions that is endowed with the desired properties.

The monograph of Oliver Schütze and Carlos Hernández provides a coherent account of archiving strategies and their algorithmic implementation. Certainly, practitioners and theoreticians alike will appreciate the fact that every strategy is supplemented with a theoretical foundation. Early archiving strategies have been developed heuristically without any guarantees regarding convergence to and spread across the Pareto front. The situation changed at the beginning of this millennium, and the development to the current state-of-the-art is largely a merit of both authors.

The book deserves a place on the shelf of anyone who designs set-based algorithms for multi-objective optimization. But don't just leave it on the shelf: Read it, use its algorithms, and get inspired to devise new archiving strategies with provable properties.

Dortmund, Germany
August 2020

Güter Rudolph

Preface

This book presents an overview of several archiving strategies we have developed over the last years dealing with approximations of the solution sets of multi-objective optimization problems[1] by stochastic search algorithms. More precisely, we will present and analyze several archiving strategies that aim for different finite size approximations either of the set of optimal solutions (Pareto set and front) as well as the set of approximate solutions of a given optimization problem. The convergence analysis will be done for a very broad framework that includes all existing stochastic search algorithms (such as evolutionary algorithms, subdivision, and cell mapping techniques) and that will only use minimal assumptions on the process to generate new candidate solutions. As will be seen, already small changes in the design of the archivers can have significant effects on the respective limit archives. It is important to note that all of the archivers presented here can be coupled with any set-based multi-objective search algorithm, and that the resulting hybrid method takes over the convergence properties of the used archiver. This book hence targets all algorithm designers and practitioners in the field of multi-objective optimization. We hope that the archivers can either be used to enhance their preferred search method or that they may be used as a starting point for the design of further archiving strategies that aim for different approximations of the solution set.

The archivers dealing with the computation of optimal solutions have first been published in (Schütze et al. 2007, 2008a, b, 2010b), and the archivers that deal with approximate solutions have first been published in (Schütze et al. 2007, 2011b, 2019). Discussions of couplings of these archivers with evolutionary algorithms can be found in (Sánchez and Schütze 2010; Schütze et al. 2011b, c; Hernández et al. 2020), and couplings with subdivision and cell mapping techniques in (Schütze et al. 2008; Hernández et al. 2013b; Sun et al. 2019). Further, the use and adaptation of these archivers to several applications have been discussed inlcuding space mission design (Schütze et al. 2008c, 2009, 2010a, 2011b, c), knapsack problems (Tantar et al. 2008), non-uniform beam design (Schütze et al. 2019), sliding mode

[1]With the exception of Chap. 9, where we make a short excursion to scalar optimization problems.

control (Qin et al. 2017; Sun et al. 2019), and the detecion of knee solutions (Schütze et al. 2008a).

Implementations of all the archivers in Matlab together with the test examples used in this book can be found in the GitHup software developement platform under

https://github.com/NumericalEvolutionaryOptimization/archivers.

The authors would like to thank the researchers, co-authors, and friends Carlos A. Coello Coello, Xavier Esquivel, Jesus Fernández, Adriana Lara, Günter Rudolph, Jian-Qiao Sun, Emilia Tantar, Massimiliano Vasile, and Fu-Rui Xiong who helped in the design of the archivers and in their integration into multi-objective search algorithms. In this regard, the first author wants to thank in particular Marco Laumanns, without whom the entire project would never have started. Finally, the authors acknowledge support from Conacyt project no. 285599 and SEP-Cinvestav project no. 231.

Mexico City, Mexico
July 2020

Oliver Schütze
Carlos Hernández

Contents

Acronyms

BOP	Bi-objective problem
$clos(A)$	Closure of A
d_H	Hausdorff distance
Δ_p	Averaged Hausdorff distance
DM	Decision maker
EA	Evolutionary algorithm
EMO	Evolutionary multi-objective optimization
$F(P_Q)$	Pareto front
$F(P_{Q,\epsilon})$	Image of $P_{Q,\epsilon}$
GD	Generational distance
IGD	Inverted generational distance
$int(A)$	Interior of A
KKT	Karush Kuhn Tucker
MaOP	Many objective optimization problem
MOEA	Multi-objective evolutionary algorithm
MOEA/D	Multi-objective Evolutionary Algorithm based on Decomposition
MOP	Multi-objective optimization problem
NSGA-II	Nondominated Sorting Genetic Algorithm II
$\partial(A)$	Boundary of A
P_Q	Pareto set
$P_{Q,\epsilon}$	Set of approximate solutions of a MOP
SMS-EMOA	S Metric Selection Evolutionary Multi-objective Algorithm

Chapter 1
Introduction

In a variety of applications, one is faced with the problem to optimize several conflicting and incommensurable objectives concurrently. Problems of that kind are termed *multi-objective optimization problems* (MOPs). As an example, consider the design of a certain product, where two objectives are in almost all cases (i) to maximize the "performance" of the given product while at the same time (ii) to minimize its cost. Hereby, the performance could be measured by one single objective if one particular performance aspect is desired leading to a MOP with two objectives (also called bi-objective optimization problem, BOP). However, depending on the complexity and importance of the product, one may consider even more performance-related measures at the same time leading, consequently, to a problem with more objectives. For instance, for the design of a car, one can think next to its cost of safety, environmental compatibility, engine power, interior capacity, energy consumption, useful life, perceived value and many others more that could all serve as individual objectives. Similar considerations hold for the design of other products, and we already see from this example that problems with multiple or even many objectives arise rather naturally from applications (e.g.., Stewart et al. 2008; Cui et al. 2017; Peitz and Dellnitz 2018b; Aguilera-Rueda et al. 2020; Moghadam et al. 2020).

One important characteristic of MOPs is that their objectives are typically in conflict with each other which has an important impact on the optimization of such problems. Taking again the example of a hypothetical car design, we can expect that e.g., the cheapest car is not the one with the highest safety value(s), engine power, or useful life (and vice versa). Similarly, the car with the lowest energy consumption is certainly not the one with the largest interior capacity or with the highest engine power, etc. That is, if we are given the bi-objective optimization problem "(cost, safety)", we already see that it is not enough to either minimize the cost function or to maximize the safety function. Analogously, the three-objective optimization problem "(energy consumption, interior capacity, engine power)" cannot be reduced to one of the scalar optimization problems "energy consumption", "interior capacity",

O. Schütze and C. Hernández, *Archiving Strategies for Evolutionary Multi-objective Optimization Algorithms*, Studies in Computational Intelligence 938,
https://doi.org/10.1007/978-3-030-63773-6_1

or "engine power", and so on. Instead, it is rather intuitive that there has to be more. In fact, it is known for continuous MOPs that contain k conflicting objectives, one can expect an entire *set* of optimal solutions, and furthermore, that this set forms at least locally an object of dimension $k - 1$ (Hillermeier 2001). This is in contrast to "classical" scalar optimization problems (SOPs, optimization problems with *one* objective), where one expects *one* optimal solution (e.g.., one cheapest car, one car with the largest interior capacity, etc.). The set of optimal solutions of a MOP is termed the *Pareto set*, and its image under the map of the objective functions the *Pareto front*. The Pareto front hence represents the set of all optimal values one can obtain for a given MOP. The knowledge of this *entire* set—or a suitable finite size representation of it—is hence of great interest for the decision maker (DM) as it can serve as a good basis for the related decision making process, i.e, the process to identify the "best" solution according to the particular scenario.

Historically, the first methods for the numerical treatment of MOPs are coming from Mathematical Programming (MP). One of the first methods is probably the Weighted Sum method from Gass and Saaty which dates back to 1955 (Gass and Saaty 1955). Until now, the most widely used sub-class of MP techniques for MOPs (including the Weighted Sum method and many more) is given by scalarization methods that transform the given problem into an auxiliary SOP (Roy 1971; Ehrgott 2005; Miettinen 2012). By nature of SOPs, the solution of each such auxiliary problem only yields one (ideally optimal) solution and not the entire solution set. As a remedy, many of these methods have design parameters that allow to incorporate user's preferences, and that help to guide the search towards a preferred solution. The proper adjustment of the design parameters to find the "ideal" solution, however, is a delicate problem in many cases. An alternative is to utilize these design parameters to construct a set of auxiliary SOPs so that the union of the resulting solutions form an approximation of the Pareto front (Das and Dennis 1998; Fliege 2004a; Eichfelder 2008). Though this approach is much more robust compared to the detection of one single ("ideal") solution, also here it is in some cases challenging to identify the proper set of design parameters to obtain a good Pareto front approximation. Further, these methods may get trapped in local minima since they are—as all MP techniques—of local nature.

An alternative to MP techniques is given by set based stochastic search algorithms, above all evolutionary algorithms (EAs) that evolve entire sets of solutions (called populations) over time (Rechenberg 1973; Holland 1975; Schwefel 1977; Kacprzyk and Pedrycz 2015). Specialized EAs for the treatment of MOPs, called multi-objective evolutionary algorithms (MOEAs), have been developed for more than three decades by now (Deb 2001; Coello Coello et al. 2007). Probably the first population based multi-objective algorithm has been proposed in 1997 by Peschel and Riedel (Peschel and Riedel 1977). The development of VEGA (Vector Evaluated Genetic Algorithm, Schaffer 1984, 1985) done by David Schaffer in the mid-1980s helped to increase the popularity of MOEAs. These methods have since then caught the interest of many researchers and practitioners since algorithms of that kind are of global nature, very robust, and require only minimal assumptions on the model (e.g., no gradient or Hessian information are needed for most MOEA variants). Furthermore, they are capable of computing a finite size approximation of the entire Pareto set/front in one single run of the algorithm. Algorithms such as NSGA-II (Deb et al. 2002a), SPEA2 (Zitzler et al. 2002), MOEA/D (Zhang and Li 2007), SMS-EMOA (Beume et al. 2007) and NSGA-III (Deb and Jain 2014; Jain and Deb 2014) have been further developed over the years, and have been successfully applied to solve numerous scientific and industrial problems which are very well documented in the literature (see e.g., Deb 2020 for an overview). However, while latest since the new millennium MOEAs are capable of reliably computing "good" Pareto front approximations for most MOPs with two and three objectives, it is fair to say that *convergence* (in the mathematical sense) has played a rather minor role in their development so far. In fact, none of the above mentioned algorithms indeed converges to a limit solution. For instance, the Hypervolume (HV, Zitzler et al. 2003) based algorithm SMS-MOEA uses $\lambda = 1$ child (offspring) from μ parents (μ being the population size) to construct the new population. In Bringmann and Friedrich 2014, it has been shown that for any strategy where $\lambda < \mu$ children are chosen from μ parents, there is no guarantee for convergence. Other studies coming from MP (e.g., Powell 1973; Nocedal and Wright 2006) indicate a similar result for any chosen performance indicator since $\lambda < \mu$ strategies in EA are equivalent to what is called cyclic search in MP. Another example is NSGA-II, the most widely known and used MOEA. Next to missing convergence toward an "optimal" population one cannot even expect a monotonic behavior in the sequence of archives that it generates. That is, for every given MOP there is a certain time step from which it is rather unclear that further computational times will lead to improvements in the solution set. Instead, it is very common to observe a kind of cyclic behavior in the approximation quality with respect to any chosen performance indicator.

The missing convergence properties (again, in the mathematical sense) do *not* show that the algorithms are not performing well (actually, state-of-the-art MOEAs are performing very well), however, rather indicate that algorithms of that kind are not tapping their full potential. In general, convergence properties significantly increase the reputation and reliability of the used algorithms and help to speed up the computations, e.g., facilitate the decision when to stop the algorithm. In particular, in the context of multi-objective optimization, one is faced with the problem of how to "suitably" discretize an entire manifold of solutions which changes with the scenario as well as the available resources. It is hence advantageous to a priori define several "suitable" finite size approximations of the set of interest, and to aim for one of them for a given problem according to the current situation.

The aim of this book is to contribute to fill the above mentioned gap. To this end, we will present several archiving strategies that were designed to approximate the Pareto set/front of a multi-objective optimization problem. All archivers will be designed for different requirements on the approximations, and will be investigated with respect to their monotonic as well as their limit behavior. The latter will show that already a very small change in the design of an archiver can have a significant impact on the final approximation qualities.

Most of the archivers will utilize the concept of ϵ-dominance (Loridan 1984; White 1986). By using this concept, single solutions s that are at least nearly optimal (with respect to ϵ) can "cover" entire regions of the Pareto front while the maximal deterioration (measured in objective space) for each "covered" solution compared to s is known: it is simply given by the entries of $\epsilon \in \mathbb{R}^k$ (more precisely: the value of ϵ_i represents the maximal deterioration with respect to the i-th objective). That is, ϵ-dominance yields a natural way to discretize the Pareto fronts while it gives the opportunity to adjust the desired approximation quality a priori. In turn, the magnitudes of the final archives depend both on ϵ and on the shape and size of the Pareto front which is in contrast to strategies that consider a fixed archive/population size (roughly speaking, larger Pareto fronts and smaller entries of ϵ result in larger archive sizes). We will analyze the theoretical upper bounds for each proposed archiver and compare them with the archive sizes obtained by test runs.

We stress that both the convergence analysis and the consideration of the upper bounds is done for continuous problems, but that all of the results also hold for discrete optimization problems.

Further, it is important to note that all of the discussed archiving strategies can be coupled with any stochastic algorithm such as a MOEA or any other set based multi-objective solver such as cell-to-cell mapping techniques (Hernández et al. 2013b; Xiong et al. 2014; Fernández et al. 2016; Sun et al. 2019) or subdivision techniques (Schütze et al. 2003; Dellnitz et al. 2005; Jahn 2006; Schütze et al. 2009; Gebken et al. 2019a; Peitz and Dellnitz 2018a; Gebken et al. 2019b). One straightforward but yet effective way is to use the considered strategies as "external archives" to the chosen base algorithm. That is, the base algorithm is run "as normal", but after each iteration the archive is fed with the newly generated candidate solutions. The external archive can be then used to decide when to stop the algorithm, and be used (either alone or in combination with the final solution of the base algorithm) to present the obtained results. This approach has the advantages that

- the design of the base algorithm remains unchanged. This is important since in many cases the interleaving of the different components of the algorithm is very sophisticated, and already small changes may lead to unwanted losses in the overall performance;
- the new hybrid (base algorithm plus external archive) takes over the convergence properties of the chosen external archive. The convergence analysis we provide for each archiver is based on minimal assumptions. For instance, all of the above mentioned MOEAs satisfy the assumptions that have been made for the consideration of the different limit archives in this book.

Certainly, also more sophisticated hybrids that utilize the discussed archiving strategies can be considered, which is, however, not within the scope of this book. Instead, we refer to some initial studies that have been made in this direction (Schütze et al. 2011b, c; Hernández et al. 2020).

This book is organized as follows: In Chap. 2, we briefly introduce to the main concepts of multi-objective optimization, and give a short overview of the numerical methods that deal with the solution of such problems. In Chap. 3, we address the related work in the design of algorithms for multi-objective optimization problems.

In Chap. 4, we present the framework we are using to investigate the archiving strategies, and in particular discuss the assumption we have to make for the convergence analysis.

In Chaps. 5 to 7 we will investigate archiving strategies that aim to compute approximations of the Pareto set/front of given MOPs. More precisely, in Chap. 5 we will first consider the archiver $ArchiveUpdateP_Q$ that maintains all non-dominated points out of the considered candidate solutions. This archiver will always yield the best performance qualities, however, is only of limited use since the sequence of archives generated by this strategy is unbounded. Next, we will in Chap. 6 consider the archivers $ArchiveUpdateEps1$ and $ArchiveUpdateEps2$ that compute discretizations of the Pareto fronts that are entirely based on the concept of ϵ-dominance. These archivers allow for finite size approximations that yield certain approximation qualities. However, even the limit archives generated by these strategies can reveal "gaps" which is unwanted in certain scenarios. We finally address in Chap. 7 the archivers $ArchiveUpdateTight1$ and $ArchiveUpdateTight2$ that aim for such "gap free" approximations.

Next to the computation of optimal solutions we will in Chap. 8 address the problem to approximate the entire set of nearly optimal solutions. Such solution can serve as an extended basis for the decision making process, and might in particular be useful in the detection of alternative or back-up solutions. After a detailed analysis of the set of approximation solutions, called $P_{Q,\epsilon}$, we will present and analyze four archivers that maintain all nearly optimal solutions out of the considered candidate solutions ($ArchiveUpdateP_{Q,\epsilon}$), aim for a uniform approximation of the solution set in decision variable or objective space ($ArchiveUpdateP_{Q,\epsilon}D_x$, respectively $ArchiveUpdateP_{Q,\epsilon}D_y$), or attempt to make a suitable discretization in both spaces ($ArchiveUpdateP_{Q,\epsilon}D_y$). In Chap. 9, we will again consider approximate solutions, however, this time for scalar optimization problems. While no explicit archiving strategy is needed if "only" the best solution of a SOP is sought (since one expects to have always exactly one best-found solution), the consideration of approximate solutions offers more opportunities where archiving becomes helpful.

Finally, we will discuss the enhancement of some state-of-the-art MOEAs with external archives as discussed above in Chap. 10 which indicates the usefulness of the archiving strategies that are presented in this book.

For convenience of the reader, we will briefly present in Appendix A all the test functions we use throughout this book as demonstrators and summarize in Appendix B the considered archiving strategies.

Chapter 2
Multi-objective Optimization

In this section, we give a brief introduction to continuous multi-objective optimization including the most important notations and definitions. After this, we will give a short overview on existing methods for the numerical treatment of such problems. For more in-depth discussions the reader is referred e.g. to Miettinen (2012), Deb (2001), Coello Coello et al. (2007), Ehrgott (2005).

In this work, we will mainly consider continuous multi-objective optimization problems (MOPs) that can be expressed mathematically as

$$\min_{x \in Q} F(x). \tag{MOP}$$

Hereby, the map F is defined by the individual objective functions f_i, i.e.,

$$F : Q \to \mathbb{R}^k, \qquad F(x) = (f_1(x), \ldots, f_k(x))^T. \tag{2.1}$$

We will in the following assume that the individual objectives $f_i : Q \to \mathbb{R}$, $i = 1, \ldots, k$, are continuous. We stress, however, that the archiving strategies we propose in this book can also be applied to discrete problems (see Tantar et al. 2008 for one example). MOPs with two objectives (i.e., $k = 2$) are termed bi-objective optimization problems (BOPs).

The domain or feasible set $Q \subset \mathbb{R}^n$ of F can in general be expressed by equality and inequality constraints,

$$Q = \{x \in \mathbb{R}^n \mid g_i(x) \leq 0, \ i = 1, \ldots, l, \ \text{and } h_j(x) = 0, \ j = 1, \ldots, m\}. \tag{2.2}$$

If $Q = \mathbb{R}^n$, we call the MOP unconstrained. For many problems, Q is given by the n-dimensional box

O. Schütze and C. Hernández, *Archiving Strategies for Evolutionary Multi-objective Optimization Algorithms*, Studies in Computational Intelligence 938,
https://doi.org/10.1007/978-3-030-63773-6_2

$$Q = B_{l,u} := \{x \in \mathbb{R}^n \ : \ l_i \le x_i \le u_i, \ i = 1, \ldots, n\}, \tag{2.3}$$

where $l, u \in \mathbb{R}^n$ with $l_i \le u_i, i = 1, \ldots, n$, are the lower and upper bounds of the problem. A point $x \in Q$ is called feasible.

In order to define optimality in multi-objective optimization, we cannot proceed as for scalar optimization problems (SOPs) which are optimization problems with one objective. Assume the objective function of a SOP is given by f, and we are given two feasible points x_1 and x_2 with objective values $y_i = f(x_i), i = 1, 2$. Then, it is either $y_1 = y_2$ or one of the values is better than the other one (i.e., either $y_1 < y_2$ or $y_2 < y_1$). That is, if $y_1 = y_2$, then both points x_1 and x_2 can considered to be peforming equally (with respect to f), else one of them is better then the other one. The situation gets more complicated for vector valued maps. As example, consider a hypothetical BOP defined by $F : Q \to \mathbb{R}^2$, $F(x) = (f_1(x), f_2(x))^T$ and two feasible points x_1 and x_2 with the objective vectors $y_1 = F(x_1) = (5, 10)^T$ and $y_2 = F(x_2) = (11, 4)^T$. In this case, x_1 is better than x_2 according to objective f_1, but worse according to the second objective (and, in turn, x_2 is worse than x_1 according to f_1 and better according to f_2). Hence, we cannot say that either of the two points is better than the other one according to both objectives.

To define optimality for MOPs, the concept of *dominance* Pareto 1927 is typically used.

Definition 2.1 [dominance]

(a) Let $v, w \in \mathbb{R}^k$. Then the vector v is *less than* w (in short: $v <_p w$), if $v_i < w_i$ for all $i \in \{1, \ldots, k\}$. The relation $\le_p$ is defined analogously.
(b) A vector $y \in Q$ is called *strictly dominated* (or simply *dominated*) by a vector $x \in Q$ ($x \prec y$) with respect to (MOP) if

$$F(x) \le_p F(y) \qquad \text{and} \qquad F(x) \neq F(y),$$

else y is called *non-dominated* by x.

If a feasible point x dominates another feasible point y, then we can consider x to be "better" than y with respect to the given MOP. Using this, the definition of optimality (i.e., the definition of the "best" solutions) in multi-objective optimization is now straightforward.

Definition 2.2 [Pareto point, Pareto set, Pareto front]

(a) A point $x \in Q$ is called *(Pareto) optimal* or a *Pareto point* of (MOP) if there exists no $y \in Q$ that dominates x.
(b) The set of all Pareto optimal solutions is called the *Pareto set*, i.e.,

$$\begin{aligned} P_Q :=& \{x \in Q \ : \ x \text{ is a Pareto point of } (MOP)\} \\ =& \{x \in Q \ : \ \nexists y \in Q \ : \ y \prec x\}. \end{aligned} \tag{2.4}$$

(c) The image $F(P_Q)$ of P_Q is called the *Pareto front*.

A point is called a *weak Pareto point* or *weakly optimal* if there does not exist a different point with a strict decrease in all of the individual objectives. Such points play a special role in the convergence analysis for MOPs, and in particular, for the analysis of the archivers we present in the sequel.

Definition 2.3 [weak dominance] A vector $y \in Q$ is called *weakly dominated* by a vector $x \in Q$ ($x \preceq y$) with respect to (MOP) if $F(x) \leq_p F(y)$.

Definition 2.4 [weak Pareto point] A point $x \in Q$ is called a *weak Pareto point* or *weakly optimal* if there exists no $y \in Q$ such that $F(\mathbf{y}) <_p F(\mathbf{x})$.

Example 2.1 We consider a hypothetical BOP and six feasible points $x_1, \ldots, x_6$. Their objective values are

$$y_i = F(x_i), \qquad i = 1, \ldots, 6, \tag{2.5}$$

where

$$y_1 = (17), \quad y_2 = (14), \quad y_3 = (35), \quad y_4 = (42), \quad y_5 = (66), \quad y_6 = (71) \tag{2.6}$$

(compare to Fig. 2.1). We see for instance that

(a) x_2 dominates x_3 and x_5
(b) x_4 dominates x_5
(c) x_5 is dominated by x_2, x_3 and x_4
(d) x_2 dominates x_1 but only weakly
(e) The points x_2, x_4, and x_6 are not dominated by any of the other points. That is, if the domain would be $Q = \{x_1, \ldots, x_6\}$, then the Pareto set would be

$$P_Q = \{x_2, x_4, x_6\}$$

and x_1 would be a weak Pareto point.

Example 2.2 Consider the following unconstrained MOP Köppen and Yoshida 2007; Schütze et al. 2011a:

$$\begin{aligned} &f_1, \ldots, f_k : \mathbb{R}^n \to \mathbb{R} \\ &f_i(x) = \|x - a^i\|_2^2 = \sum_{j=1}^n (x_j - a^i_j)^2, \end{aligned} \tag{2.7}$$

where $\mathbf{a}^i \in \mathbb{R}^n, i = 1, \ldots, k$. Each objective f_i is convex, and its minimizer is given by a^i. For n and k any positive integer can be chosen, and the Pareto set is equal to the $(k-1)$-simplex that contains all the a^i's as vertices, i.e.,

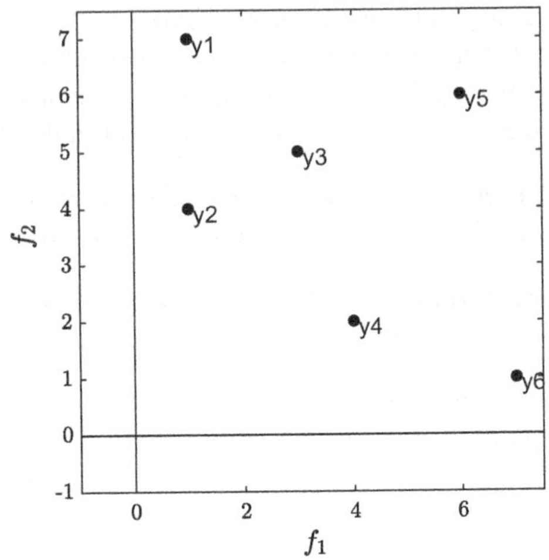

Fig. 2.1 Six objective vectors of a hypothetical BOP

$$S := S(a^1, \ldots, a^k) = \left\{ \sum_{i=1}^{k} \mu_i a^i \; : \; \mu_1, \ldots, \mu_k \geq 0 \text{ and } \sum_{i=1}^{k} \mu_i = 1 \right\}.$$

For the special case $n = 1$, $k = 2$, $a^1 = 0$, and $a^2 = 2$ we obtain the BOP that is also known as Schaffers problem Schaffer 1984:

$$\begin{aligned} F : \mathbb{R} &\to \mathbb{R}^2 \\ F(x) = &\ \left(x^2 (x-2)^2\right). \end{aligned} \tag{2.8}$$

Its Pareto set is equal to the interval $[0, 2]$, and the Pareto front is a curve connecting the points $[0, 4]^T$ and $[4, 0]^T$, see also Fig. 2.2.

Figure 2.3 shows the Pareto set and the Pareto front of the MOP that is of the form (2.7), where $n = 3$, $k = 3$, and where

$$a^1 = (-1, -1, -1)^T, \quad a^2 = (1, 1, 1)^T, \quad a^3 = (-1, 1, -1)^T.$$

The Pareto set forms a triangle with vertices a^1, a^2, and a^3. We call this problem CONV3 and will use it for the demonstrations of the archivers. See Appendix A for a brief description of all test problems we use throughout this book.

If all the objectives and constraint functions of the MOP are differentiable one can state a necessary condition for optimality which is analog for SOPs.

Theorem 2.1 *[Theorem of Kuhn and Tucker Kuhn and Tucker 1951] Let x^* be a Pareto point of (MOP), where Q is as in (2.2), and all objectives and constraint functions are differentiable in x^*. Then there exist vectors $\alpha^* \in \mathbb{R}^k$, $\lambda^* \in \mathbb{R}^m$, and $\mu^* \in \mathbb{R}^p$ such that the tuple $(x^*, \alpha^*, \lambda^*, \mu^*)$ satisfies*

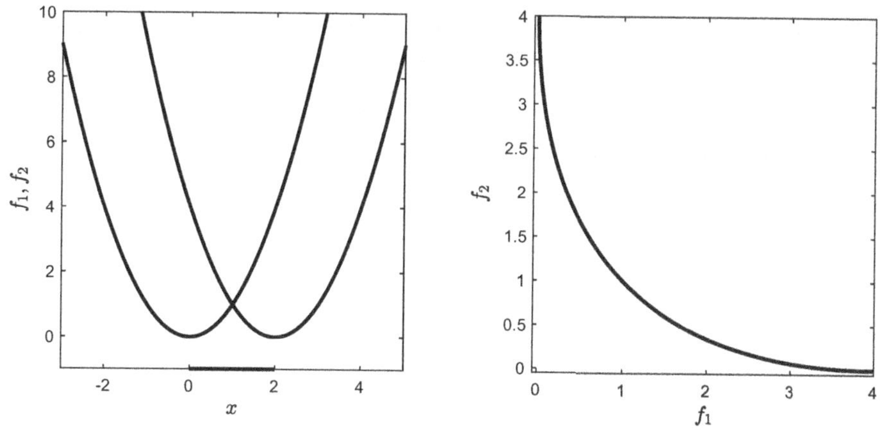

(a) The two objectives (b) Pareto front

Fig. 2.2 Illustration of Schaffers problem: Figure **a** shows the two objectives. The Pareto set is given by the interval [0, 2]. Figure **b** shows the Pareto front for this problem

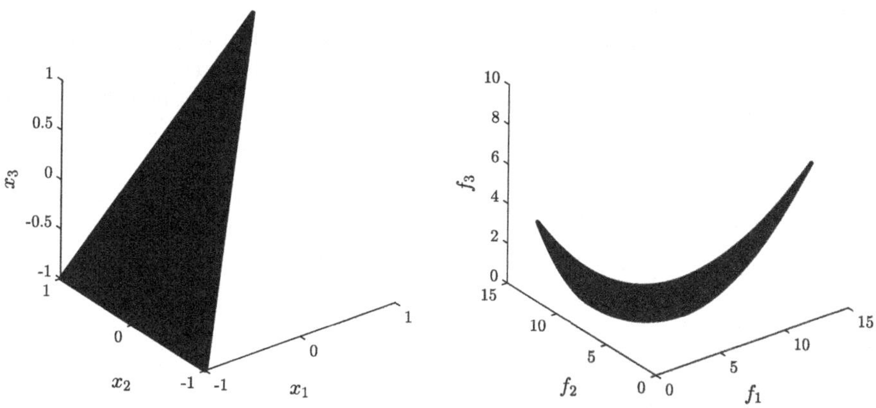

(a) The two objectives (b) Pareto front

Fig. 2.3 MOP of the form (2.7) that contains $n = 3$ decision variables and $k = 3$ objectives

$$\begin{aligned} \sum_{i=1}^{k} \alpha_i^* \nabla f_i(\mathbf{x}^*) + \sum_{i=1}^{l} \lambda_i^* \nabla g_i(\mathbf{x}) + \sum_{i=1}^{m} \mu_i^* \nabla h_i(\mathbf{x}^*) = 0 \\ \alpha_i^* \geq 0, \quad i = 1, \ldots, k \\ \sum_{i=1}^{k} \alpha_i^* = 1 \\ \lambda_i^* \geq 0, \quad i = 1, \ldots, l \\ \lambda_i^* g_i(x^*) = 0, \quad i = 1, \ldots, l. \end{aligned} \tag{2.9}$$

A point $x \in Q$ that satisfies. Equation (2.9) is called a *Karush–Kuhn–Tucker point*[1] or short KKT point. The equation in (2.9) are called the Karush–Kuhn–Tucker equations or KKT equations. These get significantly simplified in case the MOP is unconstrained.

Theorem 2.2 *[(Kuhn and Tucker, 1951)] Let x^* be a Pareto point of (MOP), where $Q = \mathbb{R}^n$, and all objectives are differentiable in x^*. Then there exists a vector $\alpha \in \mathbb{R}^k$ with $\alpha_i \geq 0, i = 1, \ldots, k$, and $\sum_{i=1}^k \alpha_i = 1$ such that*

$$\sum_{i=1}^{k} \alpha_i \nabla f_i(x^*) = 0. \tag{2.10}$$

We call a vector $\alpha \in \mathbb{R}^k$ a convex weight if $\alpha_i \geq 0, i = 1, \ldots, k$ and $\sum_{i=1}^k \alpha_i = 1$. Equation (2.10) hence claims that the zero vector can be expressed as a convex combination of the gradients $\nabla f_i(x^*)$ at each Pareto point x^*. Note that for $k = 1$ and $f = f_1$ as the unique objective the equation reads as $\nabla f(x^*) = 0$ which is a well-known fact from scalar optimization (e.g., Nocedal and Wright 2006).

Example 2.3 We consider again Schaffers problem, i.e., the unconstrained BOP with $f_1(x) = x^2$ and $f_2(x) = (x-2)^2$, and are interested in all KKT points. By non-negativity of the α_i's and due to $\alpha_1 + \alpha_2 = 1$ we see that each α_i can take any value between 0 and 1 – and no value outside this interval – and that the relation between the two values is given by

$$\alpha_2 = 1 - \alpha_1 \tag{2.11}$$

which holds for every BOP. Considering (2.10) we obtain

$$\begin{aligned} &\alpha_1 \nabla f_1(x) + \alpha_2 \nabla f_2(x) = 0 \\ \Leftrightarrow &\alpha_1 2x + \alpha_2 2(x-2) = 0 \end{aligned} \tag{2.12}$$

Using (2.11) and division by 2 leads to

$$\begin{aligned} &\Leftrightarrow \alpha_1 x + (1-\alpha_1)(x-2) = 0 \\ &\Leftrightarrow \alpha_1 x + x - \alpha_1 x - 2 + 2\alpha_1 = 0 \\ &\Leftrightarrow x = 2 - 2\alpha_1 \end{aligned} \tag{2.13}$$

Knowing that α_1 can take any value in [0, 1], we see that the set of KKT points for Schaffers problem is hence given by the interval [0, 2].

The above result can be used to see that the Pareto set typically forms an object of dimension $k-1$ which is in contrast to classical scalar optimization problems,

[1]Named after the seminal works of Karush Karush 1939 and Kuhn and Tucker (Kuhn and Tucker, 1951).

where one expects one single solution. For this, consider that we are given an equality constrained MOP (and no inequalities, as they typically do not affect the dimension of the solution set). For such problems define the following map:

$$\tilde{F} : \mathbb{R}^{n+k+m} \rightarrow \mathbb{R}^{n+m+1}$$

$$\tilde{F}(x, \alpha, \lambda) = \begin{pmatrix} \sum_{i=1}^{k} \alpha_i \nabla f_i(x) + \sum_{i=1}^{m} \mu_i \nabla h_i(x) \\ h(x) \\ \sum_{i=1}^{k} \alpha_i - 1 \end{pmatrix}, \tag{2.14}$$

where we have used the notation

$$h(\mathbf{x}) = \begin{pmatrix} h_1(x) \\ \vdots \\ h_m(x) \end{pmatrix}. \tag{2.15}$$

Let x^* be a Pareto point of the equality constrained MOP, then by Theorem 2.1 there exist vectors $\alpha^* \in \mathbb{R}^k$ and $\lambda^* \in \mathbb{R}^m$ with

$$\tilde{F}(x^*, \alpha^*, \lambda^*) = 0. \tag{2.16}$$

That is, the zero set of $\tilde{F}$ contains all the tuples (x, α, λ) that satisfy the KKT equations. By the Implicit Function Theorem (e.g., Loomis and Sternberg 1990) we can expect that $\tilde{F}^{-1}(\mathbf{0})$ forms at least locally a set of dimension $k - 1$ (which is the dimension of the domain of $\tilde{F}$ minus the dimension of its codomain). We refer to (Hillermeier, 2001) for a detailed discussion where it is shown that this is indeed the case under certain (mild) assumption on the MOP.

Multi-objective optimization is a very active research field. During the last decades, many different methods have been proposed for the numerical treatment of such problems. There exist, for instance, mathematical programming (MP) techniques. Methods of this kind are mainly point-wise iterative search techniques that generate a sequence of solutions that can converge locally toward *one* (weakly) optimal solution (e.g., Miettinen 2012; Ehrgott 2005 and references therein). The most prominent sub-class of multi-objective MP techniques is given by scalarization methods. Such methods replace the given multi-objective problem by a suitable SOP that can take into account certain preferences (e.g., Gass and Saaty 1955; Roy 1971; Gembicki and Haimes 1975; Pascoletti and Serafini 1984; Schandl et al. 2002). Hence, the application of a scalarization method can only yield one single solution. Via choosing a clever sequence of such scalarizations, however, one can obtain in certain cases a suitable approximation of the entire Pareto set/front (e.g., Das and Dennis 1998; Miettinen 2012; Fliege and Svaiter 2000; Klamroth et al. 2002; Fliege 2004a; Eichfelder 2008).

Another sub-class of the MP techniques is given by continuation-like methods that take advantage of the fact that the Pareto set forms—at least locally—a manifold. Methods of this kind start from a given initial solution and perform a search along the solution manifold (Hillermeier 2001; Recchioni 2003; Schütze et al. 2005; Ringkamp et al. 2012; Pereyra et al. 2013; Martin et al. 2013, 2014; Martín and Schütze 2018; Schütze et al. 2020). One potential drawback, however, of all the above mentioned techniques is that they are of local nature. That is, all of these methods may get trapped in locally optimal solutions of the given problem depending on the chosen initial starting point, the chosen SOP of the scalarization, and the chosen method to solve it numerically.

Next to these point-wise iterative search methods there exist subdivision (Dellnitz et al. 2005; Jahn 2006; Schütze et al. 2009) and cell-to-cell techniques (Hernández et al. 2013a; Xiong et al. 2014; Sun et al. 2019). These are set oriented MP techniques that use sets (n-dimensional boxes) to gradually approximate the entire solution sets in one run of the algorithm. These methods are of global nature and are in particular advantageous if a thorough investigation of the entire problem is desired. These methods, however, are restricted to small and moderate dimensions of the search space due to their exhaustive search mechanisms.

Another prominent class of methods that has caught the interest of many researchers and practitioners is given by specialized evolutionary algorithms, called multi-objective evolutionary algorithms (MOEAs). MOEAs use mechanisms inspired by biological evolution (such as reproduction, mutation, recombination, and selection) to evolve entire sets of solutions (populations) over time. Doing so, they yield a finite size approximation of the Pareto set/front after one run of the algorithms. See Fig. 2.4 for the general scheme of an evolutionary algorithm (which includes MOEAs). Currently, most state-of-the-art MOEAs can be classified into three main classes:

- **Dominance based methods**: MOEAs of this kind directly rely on the dominance relation. In general, these methods have two main components, a dominance ranking, and a diversity estimator. The dominance ranking sorts the solutions according to the partial order induced by Pareto dominance. This mechanism puts pressure towards the Pareto front. The diversity estimator is introduced to have a total order of the solutions, and hence, allows a comparison among solutions even if they are mutually non-dominated. The density estimators are mainly based on the idea that the decision-maker would like to obtain a good distribution of the solutions o along the Pareto front. The most popular dominance based MOEAs are NSGA-II (Deb et al. 2002a) and SPEA2 (Zitzler et al. 2002) for which for both numerous variants have been proposed over the years. Earlier MOEAs that have received a lot of attention include MOGA, NSGA, PAES, and VEGA (Schaffer 1985; Fonseca and Fleming 1993; Srinivas and Deb 1994; Fonseca and Fleming 1995; Knowles and Corne 2000a).
- **Decomposition based methods**: Such MOEAs rely on scalarization techniques to solve a given MOP. The idea behind these methods is that a set of well picked scalarization problems is capable of computing a suitable approximation of the

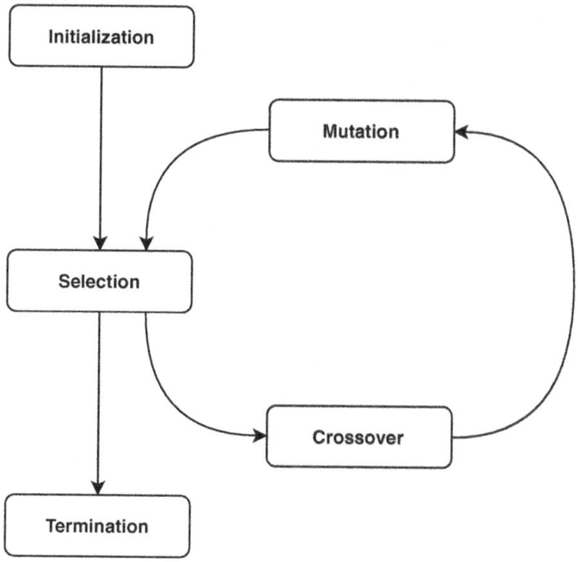

Fig. 2.4 General scheme of an evolutionary algorithm

Pareto front. During the search process, the evaluation of the scalar problem is used instead of the objective map F. These methods do not require density estimators since this is done via the selection of the scalar problems. The first method of this class is MOEA/D (Zhang and Li 2007), which is also the most prominent one. Another prominent MOEA designed for many objective problems is NSGA-III (Deb and Jain 2014; Jain and Deb 2014). Further decomposition based MOEAs are proposed in Martínez and Coello (2011), Zuiani and Vasile (2013), Moubayed et al. (2014), Zapotecas-Martínez et al. (2019).

- **Indicator based methods**: MOEAs of this kind rely on performance indicators to assign the contribution of an individual to the solution found by the algorithm. By using a performance indicator, these methods actually transform the MOP into a SOP (of higher dimension). In most cases, these methods do not need a density estimator, since the optimal distribution is implicitly given by the performance indicator. One of the methods that has received the most attention is SMS-EMOA (Beume et al. 2007) that uses the Hypervolume indicator to guide the search. Other hypervolume based MOEAs can be found in Zitzler et al. (2008), Wagner and Trautmann (2010), Sosa-Hernández et al. (2020). Methods that utilize the Δ_p indicator can be found in Rudolph et al. (2016), Schütze et al. (2016b).

Moreover, there exist several hybrid approaches, i.e., algorithms that hybridize MOEAs with local search coming in many cases from MP. Examples include hybridizations of the base MOEA together with scalarization methods (Knowles and Corne 2000b; Rigoni and Poles 2005; Zapotecas Martínez et al. 2011; Sindhya et al. 2012), local search based on multi-objective descent directions (Brown and Smith 2005; Bosman and De Jong 2005; Shukla 2007; Bosman 2011; Denysiuk et al. 2013; Schütze et al. 2016c; Uribe et al. 2020; Botello-Aceves et al. 2020), subdivision techniques (Schütze et al. 2003; Naranjani et al. 2017), continuation-like local search (Harada et al. 2006; Schütze et al. 2008a; Sindhya et al. 2011; Lara et al. 2009; Cuate et al. 2020), direct search methods and/or gradient-free local search (Zapotecas Martínez and Coello Coello 2008, 2012; Schütze et al. 2016a; Schütze et al. 2016c), and Estimation of Distribution Algorithms (Ahn et al. 2010; Shim et al. 2012; Fu et al. 2019).

Finally, for all of these set based multi-objective optimization algorithms the natural question that arises is the quality of the obtained candidate solution set. For this, several performance indicators have been proposed over the last decades such as the Hypervolume indicator (HV, Zitzler et al. 2003), the Generational Distance (GD, Van Veldhuizen 1999), the Inverted Generational Distance (IGD, Coello and Cortés 2005), R2 (Hansen and Jaszkiewicz 1998), DOA (Dilettoso et al. 2017), and the averaged Hausdorff distance Δ_p (Schütze et al. 2012a; Bogoya et al. 2019).

For further reading and overviews on MOEAs, the reader is referred e.g. to Zhou et al. (2011), Deb (2015).

Chapter 3
Archiving in Evolutionary Multi-objective Optimization: A Short Overview

As discussed in the previous chapter, one can expect that the solution set of a continuous MOP with k objectives forms—at least locally—an object of dimension $k-1$. That is, already for $k=2$ objectives the Pareto set/front typically contains uncountably many solutions which represents a challenge for set based multi-objective solvers. The main task of most MOEAs is to evolve an entire finite set of elements—either only the population or the population plus an additional external archive—toward a suitable representation of the Pareto front. Unless only relatively few function calls are spent during the search it is not possible to store all or even only all promising (e.g., all non-dominated) candidate solutions that have been computed. Instead, one has to make a selection of which points to keep and which ones to discard during the search in order to obtain a "suitable" representation of the solution set. This process is called *archiving*. We will give in this chapter a brief overview of the different archiving strategies in evolutionary multi-objective optimization. For details, the reader is referred to the cited works.

Archiving strategies are used within every MOEA, but normally referred to as the process to select the elements out of the generated candidate solutions that build the new generation (or in short *selection*). There are three main classes of MOEAs: (i) dominance based, (ii) decomposition based, and (iii) indicator based MOEAs. For MOEAs within the classes (ii) and (iii) the selection is a rather straightforward task: for decomposition based MOEAs, the selection is simply done via considering the values of the scalarizing functions that are used for the decomposition (recall that one can expect one best value for each SOP). The population size is hence entirely determined by the number of chosen scalarizing functions. For indicator based MOEAs, one can always choose the subset P out of the candidate solutions as the next population that contains μ elements (μ being the population size) and that yields the best indicator value. By doing so, in both cases monotonicity in the sequence of archives is obtained. That is, no deterioration in the approximation quality will be observed during the run of the algorithm measured by the values of the scalarizing functions

O. Schütze and C. Hernández, *Archiving Strategies for Evolutionary Multi-objective Optimization Algorithms*, Studies in Computational Intelligence 938,
https://doi.org/10.1007/978-3-030-63773-6_3

and the indicator values, respectively (assuming that neither the scalarizing functions nor the reference point/set for the performance indicator are changed). We stress that even if the selection process is straightforward for these algorithms, the use of an external archive may still be beneficial as we will see later in Chap. 10. The situation is more complicated for dominance based MOEAs due to the above described "curse of dimensionality". Early MOEAs such as MOGA (Fonseca and Fleming 1993), NSGA (Srinivas and Deb 1994), and NPGA (Horn et al. 1994) followed the advice of Goldberg (Goldberg 1989) and have coupled non-dominated sorting together with niching techniques. None of the algorithms possesses any convergence properties due to missing elite preservation. A later generation of MOEAs, including SPEA (Zitzler and Thiele 1999), SPEA-II (Zitzler et al. 2002), PAES (Knowles and Corne 2000a), NSGA-II (Deb et al. 2002a), and most of their variants incorporated elite preservation in the selection strategy leading to a much better performance which helped to significantly increase the popularity of evolutionary multi-objective optimization. However, still none of these methods converges in the mathematical sense which indicates that they still do not tap their full potential. In Deb (2001), a steady-state MOEA is presented that explicitly aims for uniformly spread solutions along the Pareto front. Though the numerical results that are presented are very promising, no convergence analysis is provided.

Though it is fair to say that convergence (in the mathematical sense) has played so far a rather minor role in the design of multi-objective evolutionary algorithms, there exist quite a few works that investigate archiving/selection strategies with respect to the limit behavior of the sequence of archives/populations they generate. We will in the following shortly summarize these works.

Rudolph and Hanne were probably the first who considered entire MOEA frameworks, and who have discussed and analyzed conditions under which the resulting sequences of archives/populations converges in the probabilistic sense.

Rudolph (Rudolph 1998a, b, 2001) and Rudolph and Agapie (Rudolph and Agapie 2000) suggested several algorithms with fixed size population sizes and have investigated under which conditions the entries of the populations converge toward Pareto optimal solutions. That is, they have considered the distance of the population toward the Pareto front. In turn, the distance of the Pareto front toward the population has not been considered (in other words: no statements could be made on the distribution of the images of the archive entries along the Pareto front). For the convergence analysis, finite Markov chains were used. Hence, the results hold for discrete (or discretized) problems, while an extension to continuous problems is at least not straightforward.

Hanne has proposed several MOEAs with Gaussian mutation distribution for continuous MOPs that are defined on compact domains (Hanne 1999, 2001, 2007a, b), and has provided convergence proofs of these algorithms in the probabilistic sense. In order to approximate the Pareto fronts he has used the concept of ϵ-dominance (which we will introduce in the next chapter, and which will be also the basis for most of our archiving strategies). Roughly speaking, ϵ-dominance allows to "cover" or represent an entire region of the Pareto front with only one solution that is optimal or at least nearly optimal. Further, the approximation quality of this "covered" region (measured in objective space) is simply given by the value of ϵ. Hence, via using this concept it is in principle possible to obtain a finite size approximation of the entire solution manifold with a certain approximation quality that is measured in objective space, and that can be adjusted a priori and according to the given scenario or preferences of the decision maker. The algorithms of Hanne provably converge toward such ϵ approximations in the limit. On the other hand, no statements on the distribution of the archive elements along the Pareto fronts are made, nor on the magnitudes of the limit archives.

Knowles and Corne analyzed archiving strategies with an a priori bounded archive size (Knowles and Corne 2003; Corne and Knowles 2003; Knowles and Corne 2004). They have shown that ϵ approximations as considered e.g., by Hanne or the ones in this book cannot be obtained in general for such bounded archives. Further, they have proposed and discussed several practical archiving strategies that are based on adaptive grid selections in order to obtain evenly spread solutions along the Pareto front.

Two related works focus on bounded archivers for Hypervolume approximations of the Pareto front, which allows to obtain a monotonic behavior in the sequence of obtained archives (Knowles et al. 2003; López-Ibáñez et al. 2011). In López-Ibáñez et al. (2011), several existing archivers are reviewed and compared with respect to their performance.

Laumanns et al. considered a very general class of stochastic search algorithms where the processes to generate new candidate solutions and to dynamically archive the incoming data are separated (Laumanns et al. 2001, 2002). Doing so, a (multi-objective) stochastic search algorithm is simply a generator and an archiver applied in a loop which helps to separately consider the archiving strategy under certain (mild) assumptions on the generator. This framework is also used throughout this book. Further, the two archivers that are proposed by Laumanns et al., and that aim for certain ϵ approximations of the Pareto front, are essentially identical to $ArchiveUpdateEps1$ and $ArchiveUpdateEps2$ as discussed in the sequel. In Laumanns et al. (2002), the sequences of archives generated by these strategies are discussed with respect to the obtained approximation qualities as well as their magnitudes. While the analysis in that work is restricted to finitely many iteration steps, we will consider here their respective limit behaviors and their limit archive sizes.

In a later study, Laumanns and Zenklusen considered a priori bounded archives (Laumanns and Zenklusen 2011). The authors investigate to which extent random search algorithms are capable of obtaining suitable Pareto front with the use of such archivers. Two archivers are proposed that aim for ϵ-approximations, the first one uses a hierarchical grid for a fixed value of ϵ and the second one is using an adaptation scheme to adjust the value of ϵ during the run of the algorithm.

In Luong and Bosman (2012), an adaptive archiving strategy is developed from a rigid-grid discretization mechanism that allows to perform a discretization in the objective space of a given MOP. This method allows the decision maker to decide the target magnitude of the final archive while all the details for the resulting update strategy are handled automatically. The competitiveness of the method is demonstrated on several experimental results.

Zapotecas Martìnez and Coello Coello propose a new archiving strategy based on the Convex Hull of Individual Minima (CHIM) with the aim to increase the diversity of the obtained solutions in objective space (Zapotecas Martìnez and Coello 2012). The performance of the resulting method is compared to the ones of NSGA-II and SPEA2 on some benchmark problems.

De Jong addresses the problem to obtain monotonicity in coevolutionary algorithms (de Jong 2007). To this end, he views opponents or tests as individual objectives leading to what is known as Pareto-coevolution. The presented algorithm, the Incremental Pareto-Coevolution Archive, guarantees the desired monotonicity for this solution concept, but, in turn, cannot be bounded by any threshold. In a next step of this work, strategies are investigated to limit the archive while offering efficiency in practice.

Some recent works that investigate and propagate the use of external archiving strategies in combination with MOEAs (i.e., in addition to the evolution of the populations) can be found in Cai et al. (2015), Wang et al. (2018), Tanabe and Ishibuchi (2018), Bezerra et al. (2019), Hernández et al. (2020), Patil (2020). Notably, some researchers propagate the use of unbounded archives (e.g., Brockhoff et al. 2015; Wang et al. 2016; Pang et al. 2020; Ishibuchi et al. 2020b, c).

One important problem related to archiving which we will not address here is to maintain the set of non-dominated solutions dynamically out of the incoming candidate solutions generated during the run of the algorithm. For this purpose, several data structures have been proposed and adapted so far including the Quad Tree (Habenicht 1983; Sun and Steuer 1996a, b; Mostaghim and Teich 2005), Elite Archives (Everson et al. 2002; Fieldsend et al. 2003), Dominance Decision Tree (Schütze 2003), and ND-trees (Jaszkiewicz and Lust 2018). For overviews of the different data structures and performance comparisons we refer to Altwaijry and El Bachir Menai (2012), Fieldsend (2020).

Chapter 4
The Framework

In this chapter, we discuss the framework we are using for the analysis of the archiving strategies we will discuss in the sequel. To this end, we first present the class of stochastic search algorithms we are considering in this book. Important in this context will be the assumption we have to make on the process to generate new candidate solutions in order to do the convergence analysis of the sequence of archives generated by the archivers. Further, we will shortly recall the Hausdorff distance d_H and the averaged Hausdorff distance Δ_p. The Hausdorff distance will be used for the convergence analysis, while we will use the averaged distance Δ_p for the evaluation of the numerical results.

The very general class of search algorithms we consider here is called *Generic Stochastic Search Algorithm* which have first been considered in (Laumanns et al., 2002). Algorithm 1 shows the pseudo code of a generic stochastic search algorithm (GSSA). This algorihm is simply an application of

- a process to generate a new set P_{j+1} of candidate solutions out of a given set P_j (called $Generate()$), and
- a strategy to update the current archive A_j by the new set of candidate solutions P_{j+1} leading to the new archive A_{j+1} (called $ArchiveUpdate()$)

in a loop. Since we are considering here also the limit behaviors of the sequences of archives A_i generated by such processes, we consequently consider the infinite loop or generation and archiving. Apparently, in practice, one has to stop the search after finitely many steps which we will not address here.

Figure 4.1 shows the scheme of a generic stochastic search algorithm. Note that for instance, all evolutionary algorithms fit in this framework (refer e.g., to Fig. 2.4), since one can regard the combination of Mutation and Crossover together as one generation step, while Selection can be viewed as the archiving strategy. Alternatively, if an external archive is used together with an evolutionary algorithm, one may regard the Mutation, Crossover and Selection together as one generation step while the archiving step is performed by the external archiving strategy.

O. Schütze and C. Hernández, *Archiving Strategies for Evolutionary Multi-objective Optimization Algorithms*, Studies in Computational Intelligence 938,
https://doi.org/10.1007/978-3-030-63773-6_4

Algorithm 1 Generic Stochastic Search Algorithm

1: $P_0 \subset Q$ drawn at random
2: $A_0 = ArchiveUpdate(P_0, \emptyset)$
3: **for** $j = 0, 1, 2, \ldots$ **do**
4: $\quad P_{j+1} = Generate(P_j)$
5: $\quad A_{j+1} = ArchiveUpdate(P_{j+1}, A_j)$
6: **end for**

The focus in this book will be on the design and analysis of the different archiving strategies while leaving out the generation process as much as possible. For this, however, we will have to make some assumption on $Generate()$ in order to do the convergence analysis for the limit behaviors of the sequences of the generated archives. To this end, we will have to assume that each neighborhood of each point x in the domain Q will be visited after finitely many steps by $Generate()$. That is, we will have to assume that the following condition is fulfilled:

$$\forall x \in Q \text{ and } \forall \delta > 0 : \quad P\left(\exists l \in \mathbb{N} \ : \ P_l \cap B_\delta(x) \cap Q \neq \emptyset\right) = 1, \qquad (4.1)$$

where $P(A)$ denotes the probability for the event A. Since the Pareto set for a given MOP is a priori not known we can hardly assume less: assume condition (4.1) is not true, then there exists with probability one a neighborhood N of a point $x \in Q$ that will *never* be "visited" by the search algorithm during the run of its algorithm. Hence, if at least a part of the set of interest (such as the Pareto set) of a given problem is contained in N, no convergence of the search process can be expected regardless of the choice of the archiver. In the case of (multi-objective) evolutionary algorithms this condition is ensured by the use of mutation operators such as Polynomial Mutation (Deb and Agrawal 1999; Deb and Deb 2014), where the support of the probability density functions is identical to the domain (at least for box-constrained problems).

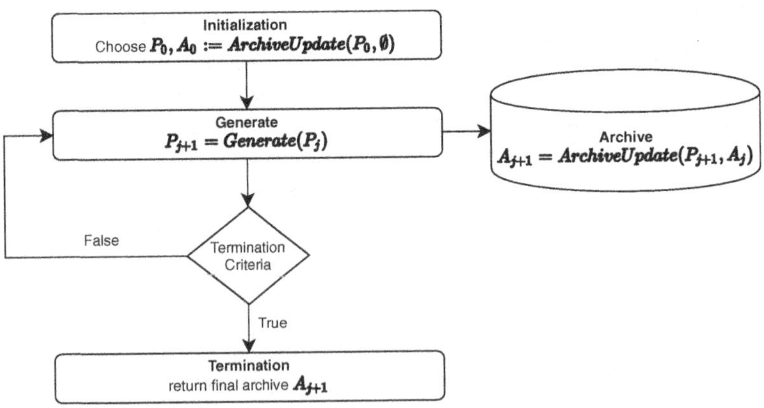

Fig. 4.1 Scheme of a generic stochastic search algorithm

Remark 1 The focus in this book will be on continuous optimization problems, in particular for the convergence analysis. We stress, however, that all our considerations also hold for discrete of discretized problems. For such problems, the condition (4.1) is given by

$$\forall x \in Q : \quad P(\exists l \in \mathbb{N} \, : \, x \in P_l) = 1. \tag{4.2}$$

This condition is satisfied in case *Generate()* is a homogeneous finite Markov chain with irreducible transition matrix (e.g., Rudolph 1994, 1996, 1997, 1998a; Eiben and Rudolph 1999; Rudolph and Agapie 2000).

For the convergence analysis of the sequence of archives obtained by the different archiving strategies in this book we will use the Hausdorff distance. This metric is widely used for theoretical investigations in many different fields, in particular when the object of interest is an entire set and not only a singleton (see, e.g., Jian 1992; Huttenlocher et al. 1993; Dellnitz and Hohmann 1997; Dellnitz and Junge 1999; Dellnitz et al. 2002b, a; Kirchberg et al. 2002; Chavent and Lechevallier 2002; Teytaud 2007; Kettner and Deng 2012; Aguirre et al. 2014).

Definition 1 Let $u \in \mathbb{R}^n$ and $A, B \subset \mathbb{R}^n$. The semi-distance $\text{dist}(\cdot, \cdot)$ and the *Hausdorff distance* $d_H(\cdot, \cdot)$ are defined as follows:

(a) $\text{dist}(u, A) := \inf_{v \in A} \|u - v\|$

(b) $\text{dist}(B, A) := \sup_{u \in B} \text{dist}(u, A)$

(c) $d_H(A, B) := \max\{\text{dist}(A, B), \text{dist}(B, A)\}$

The Hausdorff distance defines a metric in the mathematical sense on the set of compact subsets of $\mathbb{R}^n$. Here, we will mainly measure the distances between the sets of interest (mainly, the Pareto set, the Pareto front, and the set of approximate solutions and its image) and the computed archive A (respectively its image $F(A)$). The archives we consider here are given by finite sets of points and hence always compact. All of the considered sets of interest are also compact under certain mild assumptions on the model (but typically not finite nor countable), and hence, the use of the Hausdorff distance is a natural choice.

Let A_i, $i \in \mathbb{N}$, be a sequence of archives and assume that the limit archive

$$A^* = \lim_{i \in \mathbb{N}} A_i \tag{4.3}$$

exists. If the Hausdorff distance between the image of A^* and the Pareto front is zero, i.e., if

$$d_H((F(P_Q), F(A^*)) = 0 \tag{4.4}$$

it means that $F(P_Q)$ is identical to $F(A^*)$ in case the Pareto front is finite. Else (which is typically the case), it holds that the closure of $F(A^*)$ is identical to the closure of $F(P_Q)$. $F(A^*)$ can hence be seen as a perfect approximation of the Pareto front. Further, if the value $d_H((F(P_Q), F(A^*))$ is "small", we can expect that the approximation $F(A^*)$ of the Pareto front is still "good". d_H prefers, roughly speaking, evenly spread solutions around the Pareto front. Figure 4.2 shows some archives for four selected Pareto fronts. In all cases, the archives are approximations of the "best" archives for the given problem, i.e., these archives numerically solve the optimization problem

$$\min_{\substack{A \subset \mathbb{R}^n \\ |A| = \mu}} d_H(F(P_Q)), F(A)), \tag{4.5}$$

for the value $\mu = 20$ for the archive size.

One important characteristic of the Hausdorff distance is that a relatively large value for the distance does not necessarily mean a "bad" approximation quality. Consider, for example the scenario depicted in Fig. 4.3. The figure shows a hypothetical Pareto front of a bi-objective problem together with the approximation $F(A) = \{y_1, \ldots, y_{18}\}$ resulting from the archive $A = \{x_1, \ldots, x_{18}\}$. The Hausdorff distance between the Pareto front and $F(A)$ is relatively high owed to the point y_{18} which is far from the front. However, since the value of the first objective of x_{17} is much less than the respective value of x_{18} while the values are nearly the same for the second objective, i.e.,

$$f_1(x_{17}) \ll f_1(x_{18}) \quad \text{and} \quad f_2(x_{17}) \approx f_1(x_{18}) \tag{4.6}$$

it is very unlikely that any decision maker will ever be tempted to choose x_{18}. Further, via removing x_{18} one obtains the new archive $\bar{A} = \{x_1, \ldots, x_{17}\}$ those image $F(\bar{A})$ represents a "good" approximation of the entire Pareto front. The computations of such outliers are not unusual in particular when using evolutionary strategies, in particular when using dominance based selection strategies (e.g., Ikeda et al. 2001; Purshouse and Fleming 2007; Batista et al. 2011; Ishibuchi et al. 2020a).

One extreme example where outliers are frequently generated is the test function DTLZ1 which contains weakly optimal solutions that are not Pareto optimal (also called *dominance resistant solutions* Ikeda et al. 2001). Figure 4.4 shows some exemplary numerical results of NSGA-II on this test function. As can be seen, while the algorithm is capable of detecting the entire Pareto front, the final archives contain outliers even after a longer run of the algorithm.

To overcome this issue, we will in the following use the *averaged* Hausdorff distance Δ_p for the numerical evaluations of our archivers. In the following, we

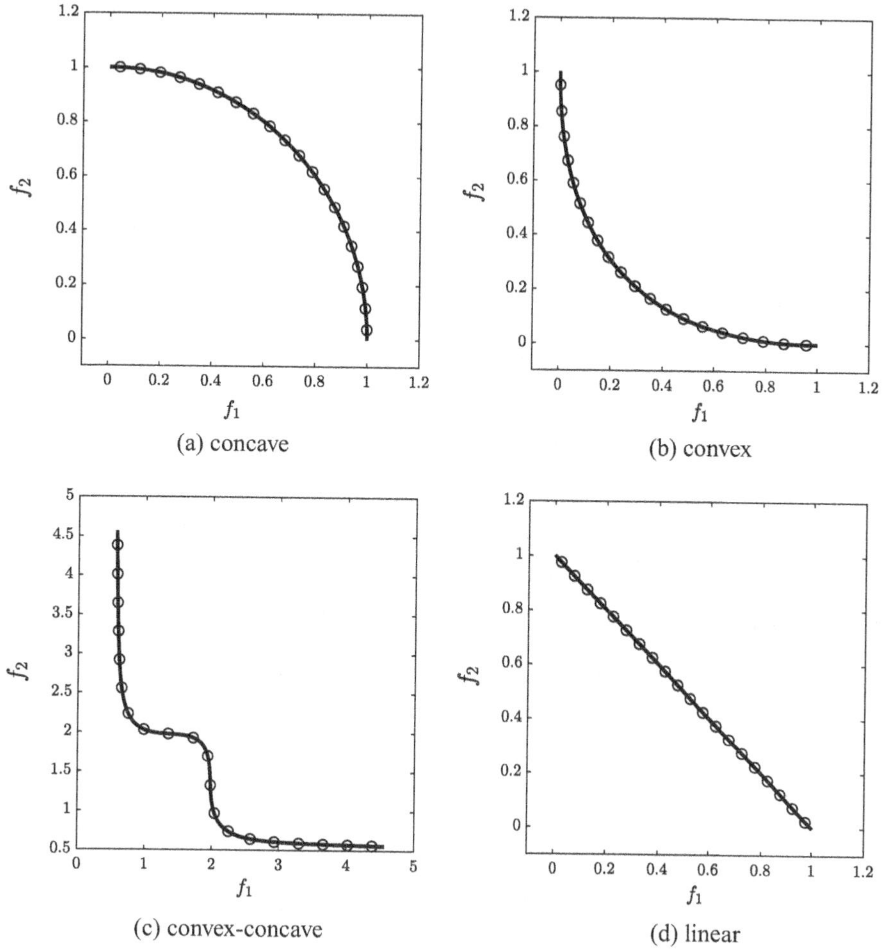

Fig. 4.2 Four Pareto fronts with different shapes together with the approximations of the best Δ_2 approximations for archive size $\mu = 20$

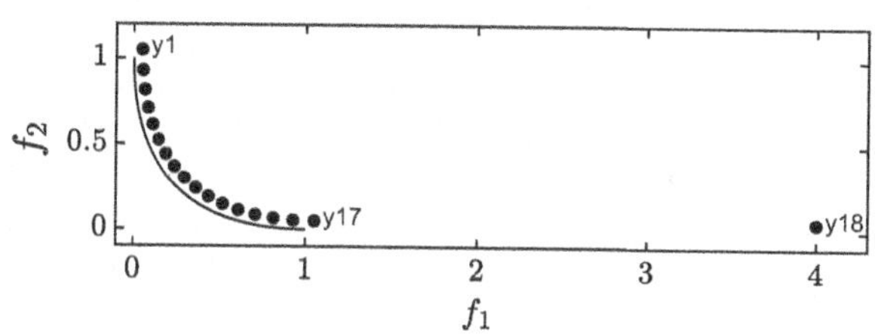

Fig. 4.3 Hypothetical example of a Pareto front approximation that contains one outlier

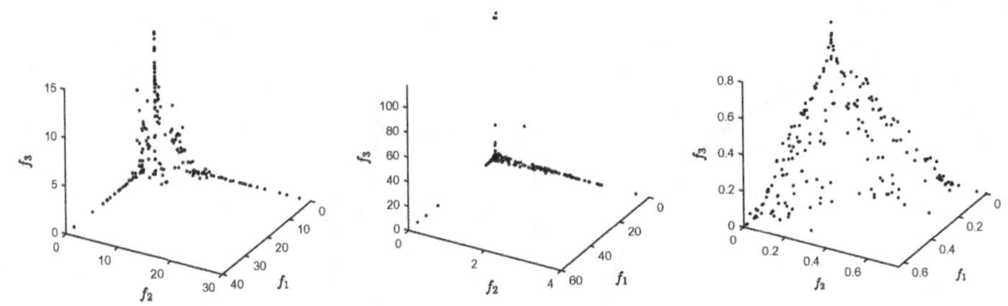

(a) 200 generations (b) 300 generations (c) 400 generations

Fig. 4.4 Numerical results of NSGA-II on the test problem DTLZ1 after several numbers of generations. The archive size has been set to 200 in all cases. The Pareto front of this problem is the triangle with vertices $(1/2, 0, 0)^T$, $(0, 1/2, 0)^T$ and $(0, 0, 1/2)^T$

define the indicator Δ_p for finite sets. For more discussion and the extension to continuous sets we refer to Schütze et al. (2012b), Rudolph et al. (2016), Bogoya et al. (2018), Bogoya et al. (2019).

Definition 2 Let $A, B \subset \mathbb{R}^n$ be finite sets. The value

$$\Delta_p(A, B) = \max(GD_p(A, B), IGD_p(A, B)), \tag{4.7}$$

where

$$\begin{aligned} GD_p(A, B) &= \left(\frac{1}{|A|} \sum_{a \in A} d(a, B)^p \right)^{1/p} \\ IGD_p(A, B) &= \left(\frac{1}{|B|} \sum_{b \in B} d(b, A)^p \right)^{1/p} \\ d(a, B) &= \inf_{b \in B} (\max_{i=1,2,..n} |a_i - b_i|), \end{aligned} \tag{4.8}$$

and $p \in \mathbb{N}$, is called the *averaged Hausdorff distance* between A and B.

The indicator Δ_p can be viewed as a composition of slight variations of the Generational Distance (GD, see Van Veldhuizen 1999) and the Inverted Generational Distance (IGD, see Coello and Cortés 2005). It is $\Delta_\infty = d_H$, but for finite values of p the indicator Δ_p averages the distances considered in d_H. Hence, as opposed to d_H, Δ_p does in particular not punish single (or few) outliers in a candidate set.

Chapter 5
Computing the Entire Pareto Front

5.1 The Set of Interest

In the following three chapters we will address archivers that aim for Pareto front approximations of a given MOP. In the first step, we are interested in maintaining *all* non-dominated solutions that were computed during the run of an algorithm. That is, if we are given a set of candidate solutions

$$C = \{x_1, \dots, x_m\} \tag{5.1}$$

those elements have e.g.. been computed by $Generate()$, the task of the archiver is to keep all non-dominated solutions within this set (and nothing else), i.e.,

$$P_C = \{x \in C \;:\; \nexists y \in C : y \prec x\}. \tag{5.2}$$

Note that if C where the (discrete) domain of the given MOP, then P_C would be the respective Pareto set. The archiver we consider in this chapter, $ArchiveUpdateP_Q$, is a straightforward realization of this idea that achieves the given goal. More precisely, Theorem 5.1 shows that an application of Algorithm 1, where $ArchiveUpdateP_Q$ is used to maintain the archives and the set C is generated by $Generate()$ leads to P_C as final archive.

The next question that arises—at least theoretically—is what happens with the sequence of archives A_i in the limit, i.e., when the set $C = \{x_1, \dots, x_{m_i}\}$ is considered for $m_i \to \infty$, and under the assumption (4.1) on the generation process. The 'ideal' result would certainly be to obtain the entire Pareto set, i.e.,

$$P_Q = \{x \in Q \;:\; \nexists y \in Q : y \prec x\}. \tag{5.3}$$

O. Schütze and C. Hernández, *Archiving Strategies for Evolutionary Multi-objective Optimization Algorithms*, Studies in Computational Intelligence 938,
https://doi.org/10.1007/978-3-030-63773-6_5

As we will see, this will not be possible in general if Q is a continuous set. This is due to the facts that the archiver uses the dominance relation which is defined in objective space and that the individual objectives f_i may be multi-modal. However, we will show in Theorem 5.2 that the distance $dist$ of the Pareto front comes arbitrarily close to the image of the archives A_i under iteration, i.e., that

$$\lim_{i \to \infty} dist(F(A_i), F(P_Q)) = 0 \quad \text{with probability one.} \tag{5.4}$$

If we additionally assume that there exists no weak Pareto point that is not already Pareto optimal, then the images of the archives converge toward the Pareto front in the Hausdorff sense, i.e.,

$$\lim_{i \to \infty} d_H(F(A_i), F(P_Q)) = 0 \quad \text{with probability one,} \tag{5.5}$$

which is the result of Theorem 5.3.

5.2 *ArchiveUpdateP*$_Q$

As discussed above, the task of the first archiver is to maintain all non-dominated solutions out of the considered candidate solutions computed by $Generate()$. The archiver $ArchiveUpdateP_Q$ is a straightforward realization of this task, where the dominance relation is used for the selection. The algorithm has two inputs: a set of candidate solutions (or population) P and the current archive A_0 which has to be updated. The only output is the updated archive A. In a first step, the current archive A_0 is copied into A. Further, for all elements $p \in P$ it is first checked if there already exists an element $a \in A$ that dominates p. If this is the case, p will be discarded from the archive, and else added to A. If p was added to the archive, all elements $a \in A$ that are dominated by p are discarded from A in order to only keep non-dominated solutions. After considering all elements $p \in P$ in this manner, the updated archive A is returned. See Algorithm 2 for a pseudo-code of $ArchiveUpdateP_Q$.

The following theorem shows that the archiver indeed fulfills its task.

Theorem 5.1 *Let (MOP) be given,* $l \in \mathbb{N}$, $P_0, \ldots, P_l \subset \mathbb{R}^n$ *be finite sets, and* $A_i, 0 = 1, \ldots, l$, *be the archives obtained by* $ArchiveUpdateP_Q$ *as in Algorithm 1, and* $C_l = \bigcup_{i=0}^{l} P_i$. *Then*

$$A_l = P_{C_l} = \{x \in C_l \ : \ \nexists y \in C_l : y \prec x\}. \tag{5.6}$$

Proof Let $C_l = \{x_1, \ldots, x_m\}$, where the points x_i are considered by the archiver in this order. Let $x \in C_l$, i.e., there exists an index $j \in \{1, \ldots, m\}$ such that $x = x_j$.

Algorithm 2 $A := ArchiveUpdateP_Q(P, A_0)$

Require: population P, archive A_0
Ensure: updated archive A
1: $A := A_0$
2: **for all** $p \in P$ **do**
3:    **if** $\nexists a \in A : a \prec p$ **then**
4:       $A := A \cup \{p\}$
5:    **end if**
6:    **for all** $a \in A$ **do**
7:       **if** $p \prec a$ **then**
8:          $A := A \backslash \{a\}$
9:       **end if**
10:    **end for**
11: **end for**
12: **return** A

Case A: let $x \in P_{C_l}$. Then x will be added to the archiver in the j-th archiving step (lines 3 and 4 of Algorithm 2), and will not be discarded further on (lines 7 and 8).

Case B: let $x \notin P_{C_l}$. That is, there exists an element $x_i \in P_{C_l}$ which dominates x. If $i < j$, then x_i will be added to the archive in the i-th step (lines 3 and 4) and not discarded further on. Hence, x will not be added to the archive since at least x_i dominates this point (line 3). If $i > j$, then x will be either not added to the archive or latest discarded in step i since $x_i \prec x$ (lines 7 and 8).
Hence, after m archiving steps the final archive is identical to P_{C_l} as claimed. □

If we would consider discrete or discretized problems we would be done already. When considering continuous MOPs, however, the domain Q is not even countable which changes the analysis. To see this, consider the BOP that is shown in Fig. 5.1. In this example, the objective f_2 is identical to objective f_1 except for a constant shift (i.e., $f_2 = f_1 + C$ for a constant $C \in \mathbb{R}$). If the problem is unconstrained or the domain is chosen large enough the Pareto set of this problem is given by $P_Q = \{x_1, x_2\}$, and both points have the same objective values (i.e., $F(x_1) = F(x_2)$). Hence, the Pareto front is given by $F(P_Q) = F(x_1) = F(x_2)$. If $Generate()$ would compute the two points x_1 and x_2, both points would be added to the archive, and all further points would be discarded from the archive leading to the perfect solution. The probability for this event, however, is zero. Assumption (4.1) only guarantees that each neighborhood is visited after finitely many steps with probability one. That is, during the run of the algorithm $Generate()$ will find points that are near to x_1 as well as other points that are near to x_2, however, the probability is one that in each step one of them is the best solution (measured by the dominance relation). Hence, in a typical run of Algorithm 1 together with $ArchiveUpdateP_Q$, the archive A_i in step i will consist of only one solution a_i, and a_i will be close either to x_1 or x_2. Further, the proximity to either of these points will switch over iteration. Hence, even in the limit, it is practically impossible to obtain the entire solution set P_Q which is due to the usage of the dominance relation which is defined in objective space. Note,

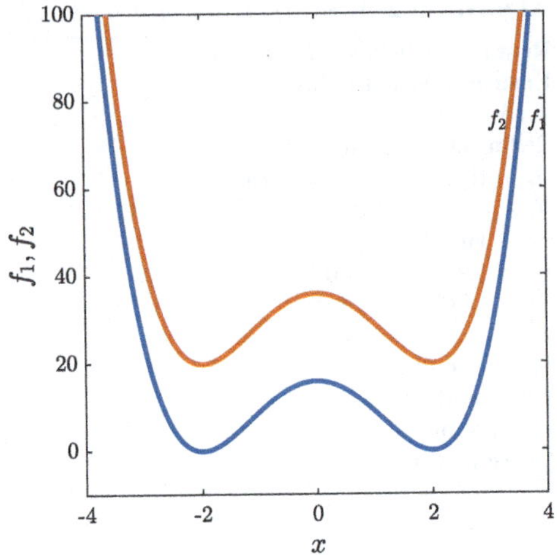

Fig. 5.1 Example of a BOP where $P_Q = \{x_1, x_2\}$. Algorithm 1 together with $ArchiveUpdateP_Q$ does not guarantee to approximate both solutions for this problem since $F(x_1) = F(x_2)$

however, that the *image* of the sequence of archives will converge for this example to the Pareto front.

In the following, we will more thoroughly investigate the limit behavior of the sequence of archives. The first result shows that the distance $dist$ of the Pareto front converges under certain assumptions to the image of the archives A_i during the run of the algorithm.

Theorem 5.2 *Let (MOP) be given, where F is continuous and Q compact. Then, under assumption (4.1), an application of Algorithm 1, where $ArchiveUpdateP_Q$ (Algorithm 2) is used to update the archive, leads to a sequence of archives A_i, $i \in \mathbb{N}$, with*

$$\lim_{i\to\infty} \operatorname{dist}(F(P_Q), F(A_i)) = 0 \quad \textit{with probability one.} \tag{5.7}$$

Proof Let $x \in P_Q$. By (4.1) it follows that there exists for every $i \in \mathbb{N}$ with probability one an index j_i and a point x_i such that

$$x_i \in P_{j_i} \cap B_{\frac{1}{i}}(x) \cap Q, \tag{5.8}$$

where $B_{\frac{1}{i}}(x)$ denotes the open ball around x with radius $1/i$:

$$B_{\frac{1}{i}}(x) = \left\{ y \in \mathbb{R}^n \ : \ \|y - x\|_2 < \frac{1}{i} \right\}. \tag{5.9}$$

Since x_i is considered by $ArchiveUpdateP_Q$ in the j_i-th archive step, there exists an element $d_i \in A_{j_i}$ such that $F(d_i) \leq_p F(x_i)$. Further, by construction of the archiver there exists for all archive steps $N > j_i$ an element d_i^N such that

$$F(d_i^N) \leq_p F(x_i^N). \tag{5.10}$$

Since $\lim_{i\to\infty} x_i = x$ and F is continuous it follows that $\lim_{i\to\infty} F(x_i) = F(x)$. Further, since $x \in P_Q$ it holds

$$\lim_{i\to\infty} F(d_i) = F(x). \tag{5.11}$$

Hence, we obtain

$$\lim_{i\to\infty} \text{dist}(F(x), F(A_i)) = 0 \quad \text{with probability one.} \tag{5.12}$$

Since $x \in P_Q$ has been chosen arbitrarily it thus holds

$$\lim_{i\to\infty} \text{dist}(F(P_Q), F(A_i)) = 0 \quad \text{with probability one,} \tag{5.13}$$

and we are done. □

The above result shows that the Pareto front $F(P_Q)$ comes arbitrarily close to the image of the archives $F(A_i)$ during the run of the algorithm. The opposite, however, does not have to be true since weak Pareto points can exist that are not Pareto optimal. To see this, consider the BOP shown in Fig. 5.2.

The points x_1 and x_2 are both weak Pareto points, but only x_2 is Pareto optimal (the Pareto set is given by the interval $[x_2, x_3]$ in case the domain Q is chosen large enough). x_1 is only dominated by x_2. In case x_1 is added to the archive, the assumption (4.1) does not assure that x_2 is selected during the search process, and hence, it cannot be guaranteed that x_1 will be discarded from the archive.

Under the additional assumption that there exist no weak Pareto points that are not Pareto optimal one can show convergence of the images of the archives toward the Pareto front in the Hausdorff sense.

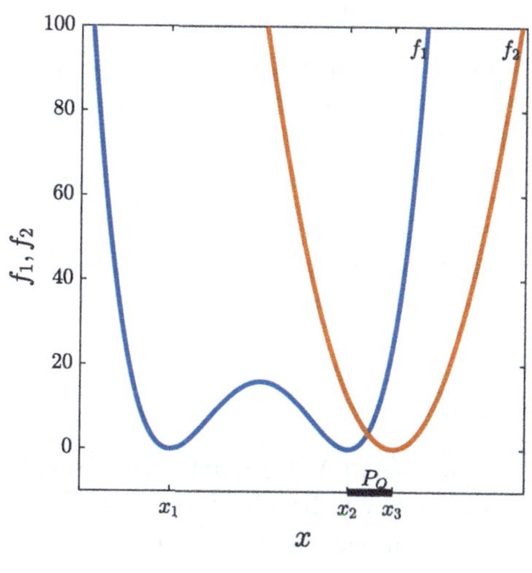

Fig. 5.2 The Pareto set of the BOP given by these two objectives is $P_Q = [x_2, x_3]$. x_1 is a weakly optimal solution with $f_1(x_1) = f_1(x_2)$ that is not Pareto optimal

Theorem 5.3 *In addition to the assumptions made in Theorem 5.2 assume that there is no weak Pareto point in $Q\backslash P_Q$. Then an application of Algorithm 1, where $ArchiveUpdateP_Q$ (Algorithm 2) is used to update the archive, leads to a sequence of archives A_i, $i \in \mathbb{N}$, with*

$$\lim_{i\to\infty} d_H(F(P_Q), F(A_i)) = 0 \quad \textit{with probability one.} \tag{5.14}$$

Proof Using Theorem 5.2 it remains to show that

$$\lim_{i\to\infty} \text{dist}(F(A_i), F(P_Q)) = 0 \quad \text{with probability one.} \tag{5.15}$$

For this let $x \in Q\backslash P_Q$. Since x is by assumption no weak Pareto point there exists a point $p \in P_Q$ such that $F(p) <_p F(x)$. Since F is continuous and P_Q is closed there exists an integer m such that for the neigborhood $B_{\frac{1}{m}}(p)$ of p it holds

$$F(y) <_p F(x) \quad \forall y \in B_{\frac{1}{m}}(p). \tag{5.16}$$

By (4.1) there exists with probability one an index j_p and an element d with

$$d \in P_{j_p} \cap B_{\frac{1}{m}}(p) \cap Q. \tag{5.17}$$

That is, x, if added to the archive, will be latest be discarded in step j_p, and by construction of the archiver never be added to the archive in subsequent archiving steps. That is, all elements $x \in Q\backslash P_Q$ will either not be added to the archive or eventually be discarded from it. Since

$$\text{dist}(F(A_i), F(P_Q)) = \max_{y\in F(A_i)} dist(y, F(P_Q)). \tag{5.18}$$

the claim hence follows. □

Finally, we address the magnitudes of the archives generated by $ArchiveUpdate P_Q$. Since P_Q typically forms at least locally a manifold we can hence by assumption (4.1) expect that

$$\lim_{i\to\infty} A_i = \infty. \tag{5.19}$$

As an example, consider a MOP with one decision variable where the Pareto set is given by the interval $[a, b]$. Then, for any threshold N the points

$$x_i = a + i\frac{b-a}{N}, \qquad i = 0, \ldots, N \tag{5.20}$$

form a set of $N + 1$ points that are evenly spread along the Pareto set (and hence, all points x_i will be added to the archive). The archive size of $ArchiveUpdateP_Q$ is hence unbounded in the limit.

In the next chapters, we will investigate archiving strategies that generate bounded archives even in the limit and that also yield certain approximation qualities.

5.3 Numerical Examples

In this section, we will illustrate the effect of $ArchiveUpdateP_Q$ on some selected test problems. More discussion together with a comparison to the other archivers that deal with the approximation of the Pareto set/front of a MOP will be done in Chapter 7. The details of all selected test problems can be found in Appendix A. We have chosen the test problems for the archivers presented in Chaps. 5 to 7 mainly by the shapes of their Pareto fronts.

In order to show the effect of the archivers independently of the chosen search algorithm, we have decided to feed the archiver with $N \in \mathbb{N}$ candidate solutions that have been chosen uniformly at random from the domains of each of the test problems. The numbers of candidate solutions that are required to obtain satisfactory approximations of the Pareto fronts are thus very high compared to the use of state-of-the-art searchers such as evolutionary algorithms. A discussion of the use of the archivers within particular MOEAs will be done in Chap. 10.

We start the illustration of the effect of $ArchiveUpdateP_Q$ by considering three bi-objective optimization problems. Figure 5.3 shows some typical results of the algorithm—$ArchiveUpdateP_Q$ fed with N candidate solutions that have been drawn uniformly drawn at random from the domain—on test problem CONV2 those Pareto front is convex. In all subfigures, the image of the final archive A_f is shown for different values of N (from $1,000$ to $100,000$). As it can be seen, the approximation quality increases with increasing values of N. For $N = 5,000$ candidate solutions the Pareto front approximation can be already considered good enough, at least from the practical point of view. It can also be seen that the magnitudes of the final archives increase. While the large archive sizes certainly come with a much higher computational effort (see also the discussion in Chap. 7), they do not, from a certain point on, yield any additional useful information for the decision maker.

Figures 5.4 and 5.5 show analog results for the bi-objective problems DENT and SSW those Pareto fronts are non-convex. Also for these two problems, we observe that the approximation quality increases leading to (nearly) perfect approximations of the Pareto fronts. On the other hand, we also see that the magnitudes of the final archives increase significantly. From the above discussion, we know that the magnitudes will exceed any given threshold if N is just taken large enough.

Figures 5.6 and 5.7 show the results for the two three-objective optimization problems CONV3 (convex Pareto front) and DTLZ7 (nonconvex and disconnected Pareto front). Since we are now dealing with the approximation of 2-dimensional

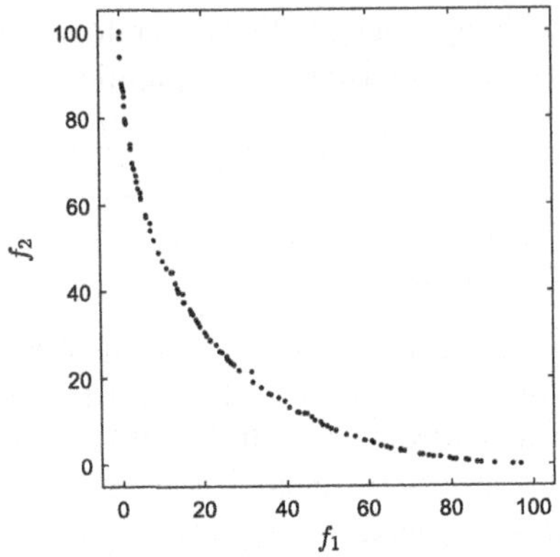

(a) $N = 1,000$, $|A_f| = 110$

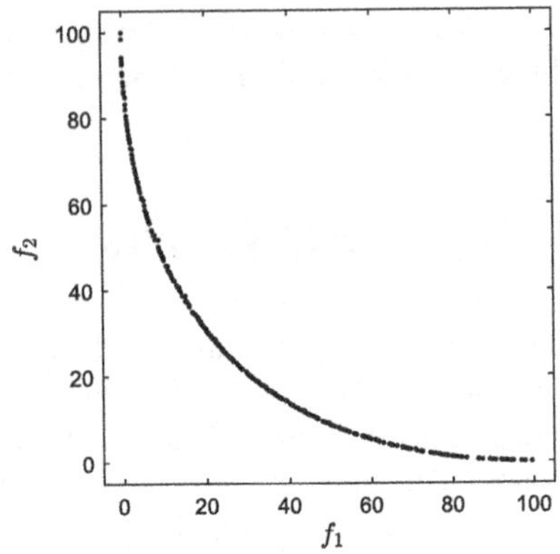

(b) $N = 5,000$, $|A_f| = 315$

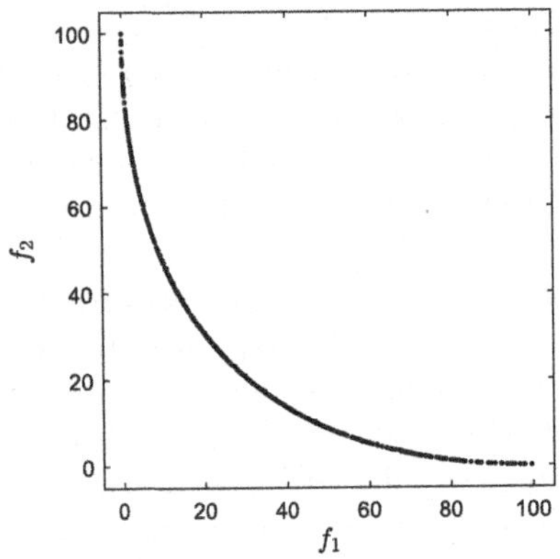

(c) $N = 10,000$, $|A_f| = 481$

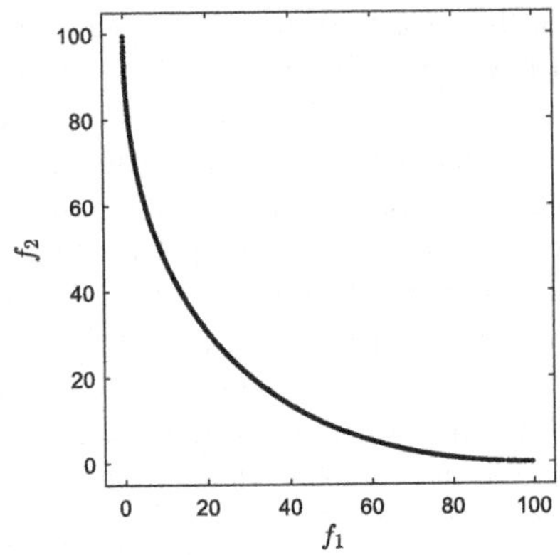

(d) $N = 25,000$, $|A_f| = 938$

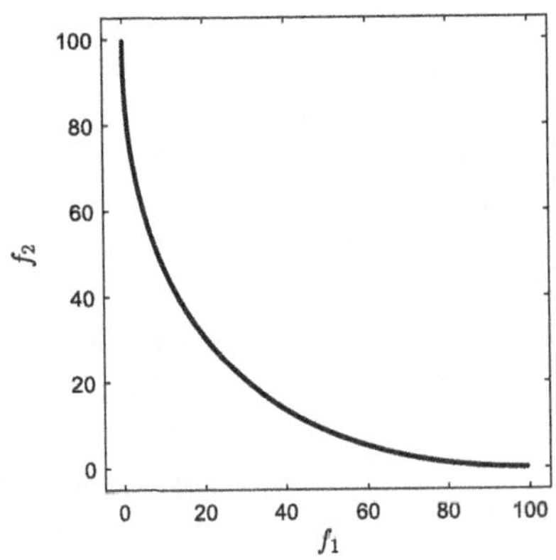

(e) $N = 50,000$, $|A_f| = 1453$

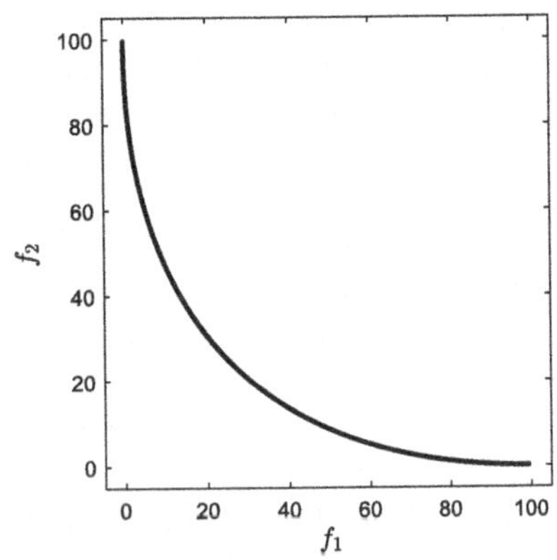

(f) $N = 100,000$, $|A_f| = 2349$

Fig. 5.3 Numerical results of $ArchiveUpdateP_Q$ for CONV2 for different numbers N of candidate solutions. A_f denotes in each case the final archive obtained by $ArchiveUpdateP_Q$

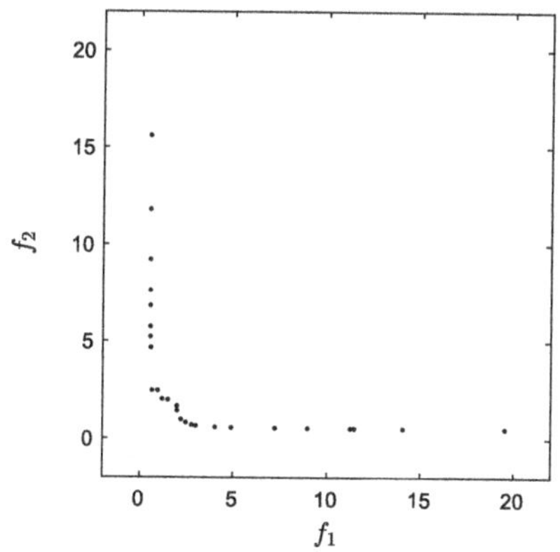

(a) $N = 1,000$, $|A_f| = 26$

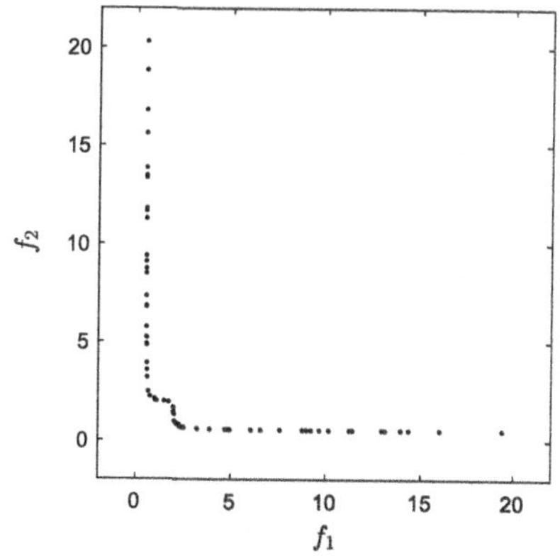

(b) $N = 5,000$, $|A_f| = 65$

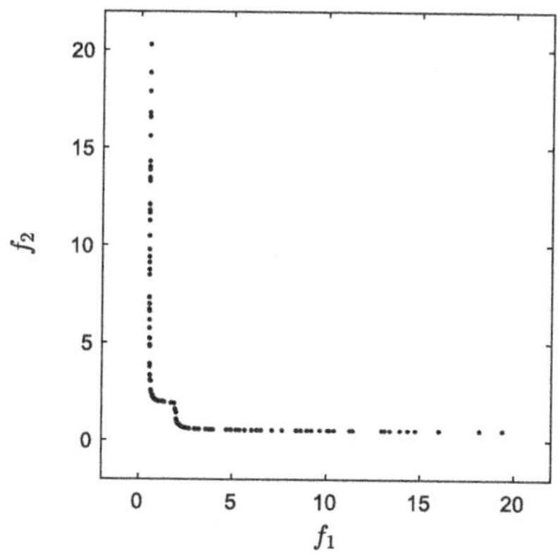

(c) $N = 10,000$, $|A_f| = 111$

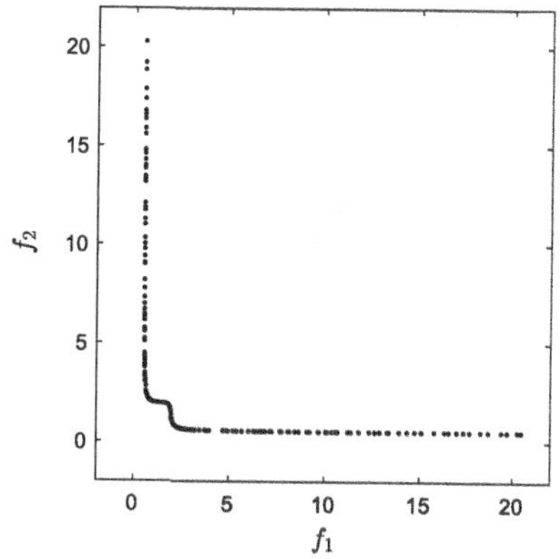

(d) $N = 25,000$, $|A_f| = 192$

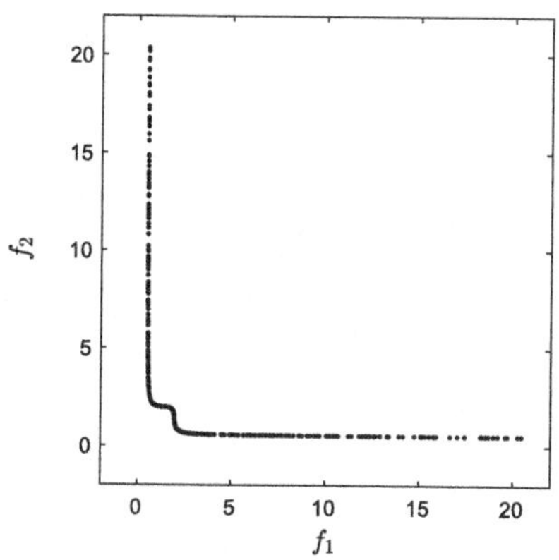

(e) $N = 50,000$, $|A_f| = 329$

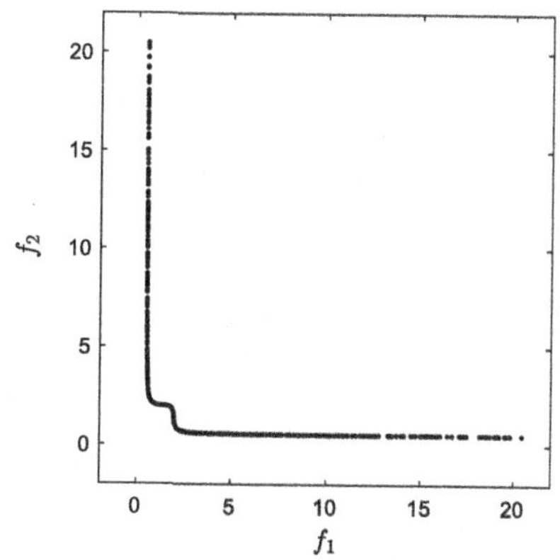

(f) $N = 100,000$, $|A_f| = 519$

Fig. 5.4 Numerical results of $ArchiveUpdateP_Q$ for DENT for different numbers N of candidate solutions. A_f denotes in each case the final archive obtained by $ArchiveUpdateP_Q$

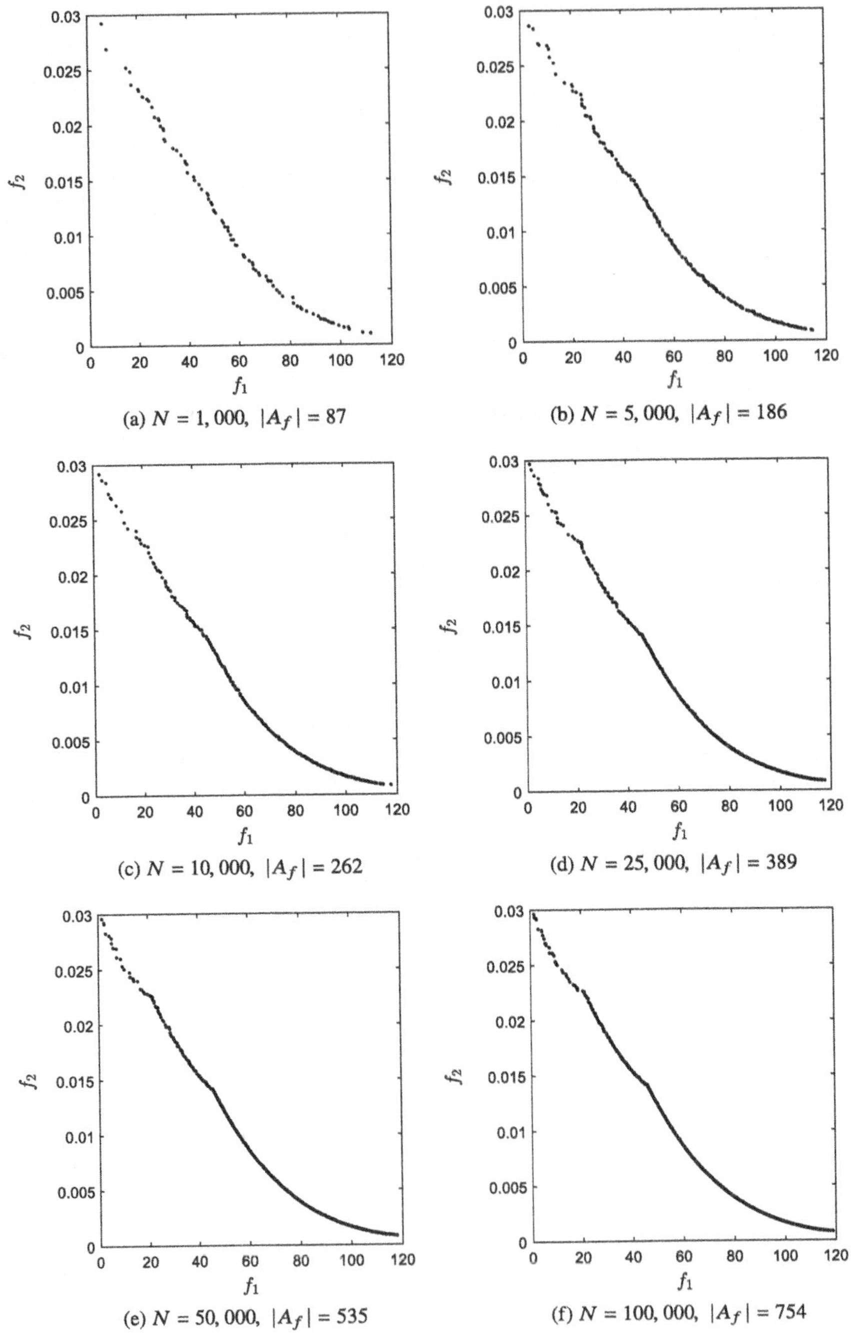

(a) $N = 1,000,\ |A_f| = 87$

(b) $N = 5,000,\ |A_f| = 186$

(c) $N = 10,000,\ |A_f| = 262$

(d) $N = 25,000,\ |A_f| = 389$

(e) $N = 50,000,\ |A_f| = 535$

(f) $N = 100,000,\ |A_f| = 754$

Fig. 5.5 Numerical results of $ArchiveUpdateP_Q$ for SSW of for different numbers N of candidate solutions. A_f denotes in each case the final archive obtained by $ArchiveUpdateP_Q$

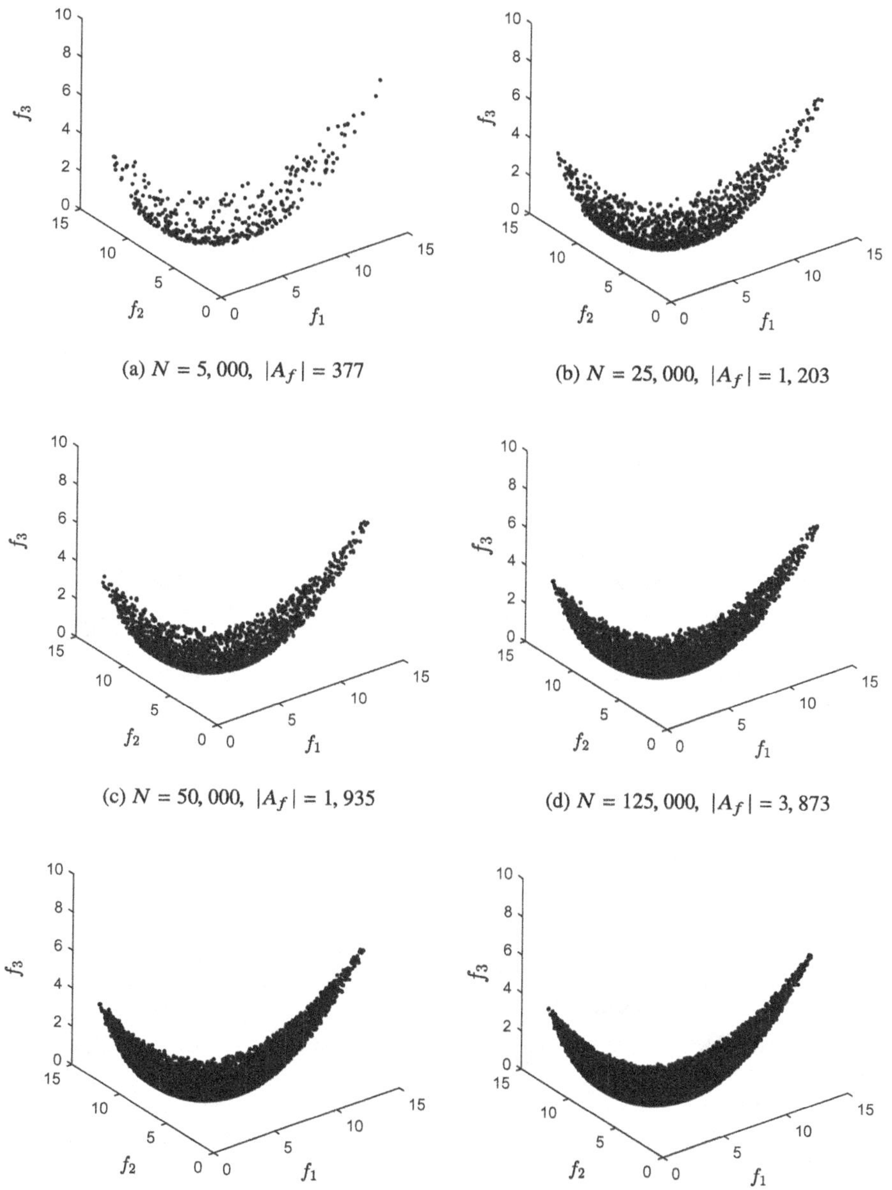

(a) $N = 5,000,\ |A_f| = 377$

(b) $N = 25,000,\ |A_f| = 1,203$

(c) $N = 50,000,\ |A_f| = 1,935$

(d) $N = 125,000,\ |A_f| = 3,873$

(e) $N = 250,000,\ |A_f| = 6,407$

(f) $N = 500,000,\ |A_f| = 10,786$

Fig. 5.6 Numerical results of $ArchiveUpdateP_Q$ for CONV3 of for different numbers N of candidate solutions. A_f denotes in each case the final archive obtained by $ArchiveUpdateP_Q$

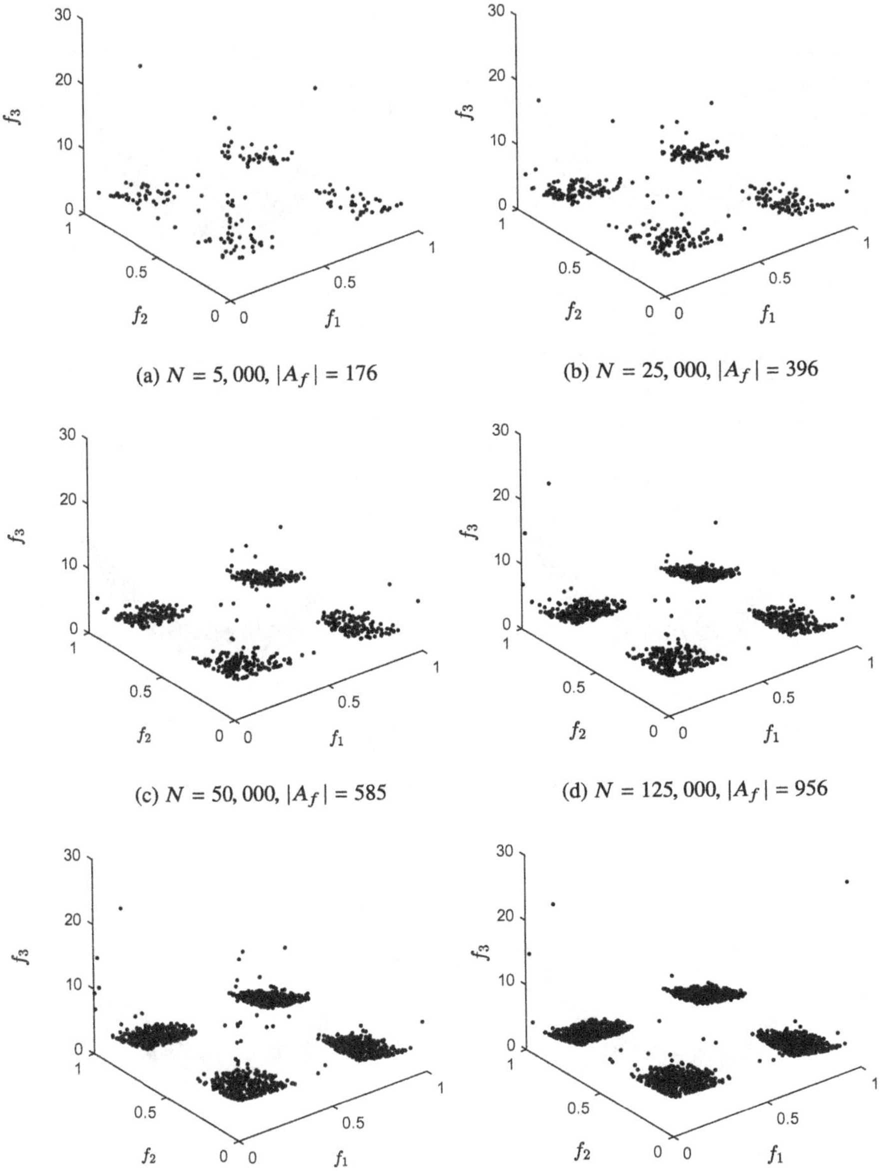

(a) $N = 5,000$, $|A_f| = 176$

(b) $N = 25,000$, $|A_f| = 396$

(c) $N = 50,000$, $|A_f| = 585$

(d) $N = 125,000$, $|A_f| = 956$

(e) $N = 250,000$, $|A_f| = 1473$

(f) $N = 500,000$, $|A_f| = 2197$

Fig. 5.7 Numerical results of $ArchiveUpdateP_Q$ for DTLZ7 for different numbers N of candidate solutions. A_f denotes in each case the final archive obtained by $ArchiveUpdateP_Q$

objects, we have increased the number of candidate solutions to $N_{max} = 500,000$. Similar observations can be made as for the bi-objective problems, with the difference that the magnitudes of the archive sizes grow even faster (as anticipated). Apparently, this problem will become more serious with increasing number of objectives in the given MOP.

The numerical results indicate that $ArchiveUpdateP_Q$ is capable to approximate the Pareto fronts, however, also show that it is imperative to utilize suitable discretization strategies which we will address in the sequel.

Chapter 6
Computing ϵ-(approximate) Pareto Fronts

6.1 The Sets of Interest

As we have seen in the last chapter, if *all* non-dominated solutions that have been found during the search are added to the archiver, we can expect the resulting sequences of archives to go beyond any given threshold—if the search process is just executed long enough. In order to obtain bounded archives that also allow to maintain a certain approximation quality (at least in the limit) we have to find suitable discretization strategies. In this chapter we will present two archivers that aim for *ϵ-approximate Pareto fronts* and *ϵ-Pareto fronts*, respectively. The basis for these discretizations will be the concept of ϵ-dominance which goes back to works of Loridan (1984) and White (1986).

Definition 6.1 (*ϵ-dominance*) Let $\epsilon = (\epsilon_1, \dots, \epsilon_k)^T \in \mathbb{R}^k_+$ and $x, y \in \mathbb{R}^n$. x is said to ϵ-dominate y (in short: $x \prec_\epsilon y$) with respect to (MOP) if

$$F(x) - \epsilon \leq_p F(y) \quad \text{and} \quad F(x) - \epsilon \neq F(y). \tag{6.1}$$

Figure 6.1 shows the region of dominance for the classical dominance relation as well as for ϵ-dominance. Hereby, let $y = F(x)$ for a feasible point x of a hypothetical BOP. Every point z such that $F(z)$ is located in the shaded region of the left figure is dominted by x. On the right, we see the analog region for ϵ-dominance. As we can see, the region of dominance is "enlarged" by the values of ϵ. ϵ-dominance can hence be seen as a weaker concept of dominance.

Using this new relation we can now define the two objects of interest that were first introduced in Laumanns et al. (2002) (and which were originally called ϵ-approximate Pareto sets and ϵ-Pareto sets, respectively).

O. Schütze and C. Hernández, *Archiving Strategies for Evolutionary Multi-objective Optimization Algorithms*, Studies in Computational Intelligence 938,
https://doi.org/10.1007/978-3-030-63773-6_6

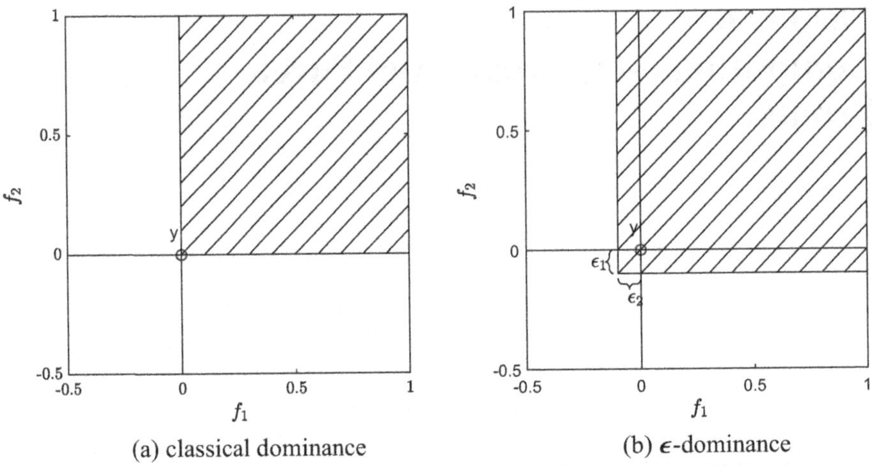

Fig. 6.1 Regions of dominance for the different dominance relations: Points those images lie in the shaded region are (ϵ-)dominated by points x with $F(x) = y$

Definition 6.2 (*ϵ-approximate Pareto front*) Let $\epsilon \in \mathbb{R}^k_+$. A set $F_\epsilon \subset \mathbb{R}^n$ is called an *ϵ-approximate Pareto set* of (MOP) if every point $x \in \mathbb{R}^n$ is ϵ-dominated by at least one $f \in F_\epsilon$, i.e.,

$$\forall x \in \mathbb{R}^n \; : \; \exists f \in F_\epsilon : \quad f \prec_\epsilon x. \tag{6.2}$$

Figure 6.2 shows two examples of ϵ-approximate Pareto fronts. Note that unions of the regions of ϵ-dominance "cover" the entire Pareto fronts. Hence, ϵ can be seen as the approximation quality of an ϵ-approximate Pareto front. In other words, if we allow a deterioration[1] of ϵ_i for each objective $f_i, i = 1, \ldots, k$, then an ϵ-approximate Pareto front can in a way be seen as a sufficiently complete approximation of the entire Pareto front.

Note that elements of an ϵ-approximate Pareto front are not required to be Pareto optimal. This may be unwanted and insatisfactory in certain cases, in particular if the values of ϵ are relatively large. If optimality of the elements is desired one can aim for an ϵ-Pareto front that we define in the following.

Definition 6.3 (*ϵ-Pareto front*) Let $\epsilon \in \mathbb{R}^k_+$. A set $F^*_\epsilon \subset \mathbb{R}^n$ is called an *ϵ-Pareto set* if F^*_ϵ is an ϵ-approximate Pareto set and if every point $f \in F^*_\epsilon$ is a Pareto point of (MOP).

Figure 6.3 shows the same Pareto fronts as in Fig. 6.2, but where the approximations form ϵ-Pareto fronts.

[1]For example, if one objective f_i is the speed (in km/h) of a vehicle that has to be designed and e_i is chosen to be 0.5, it means that we do not mind if the actual speed of the resulting vehicle is 180.0 km/h or 180.5 km/h.

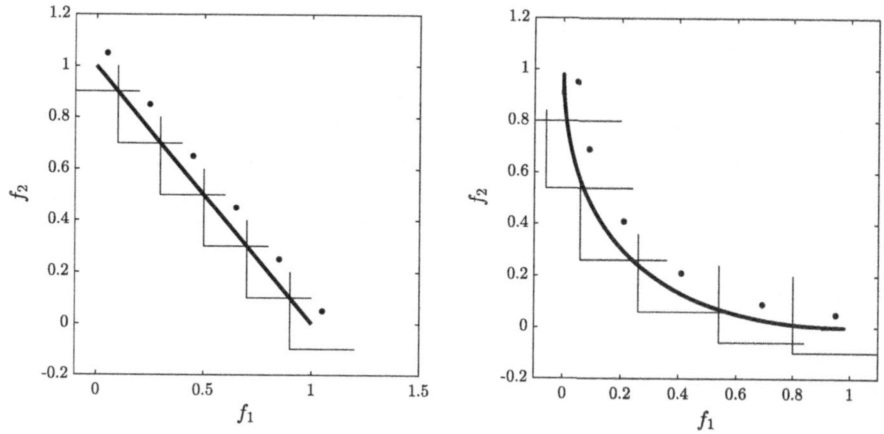

Fig. 6.2 Two examples of ϵ-approximate Pareto fronts. Note that in both examples none of the archive elements are optimal

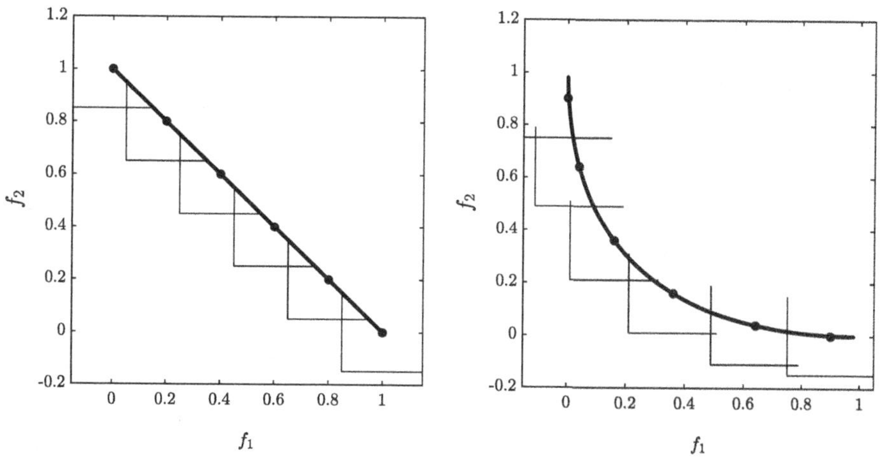

Fig. 6.3 Two examples of ϵ-Pareto fronts. All archive elements are optimal

We will proceed in the following section as in the previous chapter for every proposed archiving strategy: first, we will discuss the monotonic behavior of the archiver (i.e., the approximation quality of the final archive after finitely many iteration steps). Next, we will consider the limit behavior of the sequence of archives as well as the upper bounds on their magnitudes. Finally, we will show some numerical results that indicate the effect of each archiver.

6.2 *ArchiveUpdateEps*1

The first archiver we consider here, *ArchiveUpdateEps*1, aims for an ϵ-approximate Pareto front of a given MOP. The two inputs of this archiver are the set of candidate solutions P and the current archive A_0; the output is the updated archive A, see Algorithm 2 for a pseudo code of *ArchiveUpdateEps*1. For a given vector $\epsilon \in \mathbb{R}^k_+$ and a "safety factor" $\Theta \in (0, 1)$ (which will be explained below in more detail) the algorithm works as follows: first A_0 is copied into A. Then, for all elements $p \in P$ it is checked there already exists an element $a \in A$ that $\Theta\epsilon$-dominates p (line 3). If this is the case, p will not be accepted as new element for the archive A. Else, all elements that are dominated by p are removed from the archive (lines 7 and 8), and in turn p will be added to A (line 11).

Algorithm 3 $A :=$ *ArchiveUpdateEps*1 (P, A_0)

Require: P: current population, A_0: current archive, $\epsilon \in \mathbb{R}^k_+$, $\Theta \in (0, 1)$
Ensure: A: updated archive
1: $A := A_0$
2: **for all** $p \in P$ **do**
3:     **if** $\exists a \in A : a \prec_{\Theta\epsilon} p$ **then**
4:         CONTINUE ▷ *do not execute lines* 6 – 11
5:     **end if**
6:     **for all** $a \in A$ **do**
7:         **if** $p \prec a$ **then**
8:             $A := A\backslash\{a\}$
9:         **end if**
10:     **end for**
11:     $A := A \cup \{p\}$
12: **end for**
13: **return** A

We first investigate the monotonic behavior of *ArchiveUpdateEps*1. An analog statement as for *ArchiveUpdate*P_Q cannot be expected as we have used $\Theta\epsilon$-dominance for the disretization (and hence, not all non-dominated solutions found during the search process will be stored any more). However, it can be shown that for all elements p that were considered in the algorithm there exists an element a in the final archive that $\Theta\epsilon$-dominates p.

Lemma 6.1 *Let (MOP) be given,* $A_0, P \subset \mathbb{R}^n$ *be finite sets,* $\epsilon \in \mathbb{R}^k_+$*, and* $A :=$ *ArchiveUpdateEps1* (P, A_0)*. Then it holds:*

$$\forall x \in P \cup A_0 : \quad \exists a \in A : a \prec_{\Theta\epsilon} x. \tag{6.3}$$

Proof Loosely speaking, the claim holds since elements a are only discarded from the archive if in turn another element p with $p \prec a$ is inserted to the archive (this "replacement" is realized in lines 7, 8 and 11 in Algorithm 2).

More precisely, let $P = \{p_1, p_2, \ldots, p_l\}$, $l \in \mathbb{N}$, where we assume that all points p_i are considered in this order (i.e., in the for-loop in line 2 of Algorithm 3: first p_1, then p_2, and so on). There are two cases we have to distinguish. Case A: $x \in A_0$. Define $p'_0 := x$ and

$$p'_i := \begin{cases} p_i & \text{if } p_i \text{ "replaces" } p'_{i-1} \\ p'_{i-1} & \text{else} \end{cases}, \quad i = 1, \ldots, l.$$

It holds that $p'_l \in A$ and either $p'_l = x$ or $p'_l \prec x$ (due to the transitivity of $\prec$). In both cases it is $p'_l \prec_{\Theta\epsilon} x$.

Case B: $x \in P$. Let $x = p_j$, $j \in \{1, \ldots, l\}$. After the j-th iteration of the outer for-loop in Algorithm 3 there exists an element $a_j \in A$ with $a_j \prec_{\Theta\epsilon} p_j$ (line 3 resp. line 11 of Algorithm 3). Define $p'_j := a_j$ and p'_i, $i = j+1, \ldots, l$, as above. It follows that $p'_l \in A$ and $p'_l \prec_{\Theta\epsilon} x$ as claimed. □

Lemma 6.2 *Let (MOP) be given, $l \in \mathbb{N}$, $P_0, \ldots, P_l \subset \mathbb{R}^n$ be finite sets, and A_i, $i = 1, \ldots, l$, be the archives obtained by ArchiveUpdateEps1 as in Algorithm 3, and $C_l = \bigcup_{i=0}^{l} P_i$. Then*

$$\forall x \in C_l : \quad \exists a \in A_l : a \prec_{\Theta\epsilon} x. \tag{6.4}$$

Proof Follows immediately by Lemma 6.1. □

We next address the limit behavior of the sequence of archives. From Theorem 6.1 it follows that we can expect to obtain an ϵ-approximate Pareto front after finitely many iteration steps for any given $\epsilon \in \mathbb{R}^k_+$.

Theorem 6.1 *Let (MOP) be given, where F is continuous and Q compact, and $\epsilon \in \mathbb{R}^k_+$. Then, under assumption (4.1), an application of Algorithm 1, where ArchiveUpdateEps1 (Algorithm 3) is used to update the archive, leads to a sequence of archives A_i, $i \in \mathbb{N}$, such that there exists with probability one an $l_0 \in \mathbb{N}$ such that A_l is an ϵ-approximate Pareto front for all $l \geq l_0$.*

Proof Let

$$m_i = \min_{x \in Q} f_i(x) \quad \text{and} \quad M_i = \max_{x \in Q} f_i(x), \qquad 1 \leq i \leq k.$$

Define

$$Q' = [m_1 - \Theta\epsilon_1, M_1] \times \cdots \times [m_k - \Theta\epsilon_k, M_k]$$

and

$$D(A, \epsilon) := \{y \in Q' \mid \exists a \in A : F(a) - \Theta\epsilon \leq_p y\}.$$

A point p is only inserted into the given archive A if there exists no $a \in A$ which $\Theta\epsilon$-dominates p. Assume A and p are given, and that p is inserted into A. Denote by A' the resulting archive. By Lemma 6.1 it follows that

$$D(A, \epsilon) \subset D(A', \epsilon)$$

whether or not p dominates points $a \in A$. Define

$$B_p := \{y \in \mathbb{R}^k \mid y \leq_p F(p) \text{ and } F(p) - \Theta\epsilon \leq_p y\}.$$

It is $int(B_p) \cap D(A, \epsilon) = \emptyset$, where $int(B_p)$ denotes the interior of B_p (assume there exists an $y \in int(B_p) \cap D(A, \epsilon)$, i.e., there exists an $a \in A$ with $F(a) - \Theta\epsilon \leq_p y <_p F(p)$, and thus $a \prec_{\Theta\epsilon} p$, a contradiction). Since $B_p = B(F(p) - \frac{\Theta\epsilon}{2}, \frac{\Theta\epsilon}{2})$, i.e., B_p is a k-dimensional box with center $c = F(p) - \frac{\Theta\epsilon}{2}$ and radius $r = \frac{\Theta\epsilon}{2}$, we have

$$Vol(D(A', \epsilon)) \geq Vol(D(A, \epsilon)) + \underbrace{Vol(B_p)}_{=\prod_{i=1}^{k} \Theta\epsilon_i > 0}, \tag{6.5}$$

where $Vol(D)$ denotes the k-dimensional volume of a set $D \subset \mathbb{R}^k$. Since $0 \leq Vol(D(A, \epsilon)) \leq Vol(Q') \leq \infty$, it follows only finitely many insertions can be done.

Using this observation we can now show the existence of a number l_0 such that the A_{l_0} is an ϵ-approximate Pareto set. For this, we assume that there exists no such number l_0. That is, for every $l \in \mathbb{N}$ there exists a point $x = x(l) \in Q$ which is not ϵ-dominated by any of the elements of the archive A_l.

Let $l \in \mathbb{N}$. Since there exists no $a \in A_l$ which ϵ-dominates $x(l)$ and since $\Theta\epsilon <_p \epsilon$ there exists also no $a \in A$ which $\Theta\epsilon$-dominates $x(l)$. Moreover, since $\Theta < 1$ and F is continuous there exists a neighborhood U_l of $x(l)$ such that

$$\nexists a \in A_l : a \prec_{\Theta\epsilon} u \quad \forall u \in U_l. \tag{6.6}$$

By (4.1) it follows that there exists with probability one a number $i_l \in \mathbb{N}$ and a point $x_{i_l} \in P_{i_l} \cap Q$. By (6.6) and the construction of $ArchiveUpdateEps1$ (lines 3 resp. 11 of Algorithm 3) it follows that either this point has to be inserted, or that another point $\tilde{x}_l$ has been inserted in one of the iteration steps $l + 1, \ldots, i_l$. In any case, at least one insertion must have occurred from iteration step $l + 1$ to i_l.

Proceeding in this manner, a sequence of infinitely many insertions can be constructed, which contradicts the observation made above. Thus, there must exist with probability one an l_0 such that A_{l_0} forms an ϵ-approximate Pareto set as claimed. Further, all subsequent archives are ϵ-approximate Pareto sets due to Lemma 6.1, and the proof is complete. □

In the following me make some remarks on the assumptions we have made to obtain the results, as well as on the approximation qualities of the limit archive.

Remark 6.1 (*purpose of* Θ) In line 3 of Algorithm 3 a test for $\Theta\epsilon$-dominance is made, where $\Theta \in (0, 1)$, instead of using ϵ-dominance. Θ is required to guarantee the convergence of the sequence of archives. As counter example, consider the unconstrained BOP given by $F : \mathbb{R} \to \mathbb{R}^2$, where

$$F(x) = (x^2, x^2)^T. \tag{6.7}$$

Consider $\epsilon = (1, 1)^T$, and let $\Theta = 1$ (i.e., line 3 is checking for ϵ-dominance). Since both objectives are identical, and $x^* = 0$ is the only minimizer of each objective, the Pareto set is also given by x^*, and the Pareto front by the point $(0, 0)^T$.

Next, consider the archive A which consists only of the point $a = 1$. A does not form an ϵ-approximate Pareto front of the given BOP, and the probability to complete this set is zero since the since x^* is the only point which is not ϵ-dominated by a, and the probability to select this point is zero.

A direct consideration why this cannot happen if Θ is chosen in $(0, 1)$ can be found in Schütze et al. (2008d).

Apart from these theoretical considerations that are needed to guarantee convergence, it is fine to use $\Theta = 1$ which is indeed done for all the computations that are presented in the sequel.

Remark 6.2 (*compact domain*) We have considered the domain Q of F to be compact. The following two small examples show that this was needed to guarantee the existence of a finite approximation. First, consider the following unconstrained problem given by

$$F(x) = (x, -x)^T. \tag{6.8}$$

The Pareto set P of this problem is the entire set of real numbers $\mathbb{R}$, and the Pareto front is given by the line

$$F(P) = \left\{ \begin{pmatrix} x \\ -x \end{pmatrix} \; : \; x \in \mathbb{R} \right\}. \tag{6.9}$$

Let $\epsilon = (\epsilon_1, \epsilon_2)^T$, where $\epsilon_1, \epsilon_2 > 0$. Then every point $x \in \mathbb{R}$ ϵ-dominates exactly all points

$$y \in (x - \epsilon_1, x + \epsilon_2) \tag{6.10}$$

(see also Fig. 6.4). Hence, there exists no finite size ϵ-approximate Pareto front for this problem.

Next, consider the BOP

$$\begin{aligned} F &: \mathbb{R}^+ \to \mathbb{R}^2 \\ F(x) &= \left(-x, -\frac{1}{x}\right)^T. \end{aligned} \tag{6.11}$$

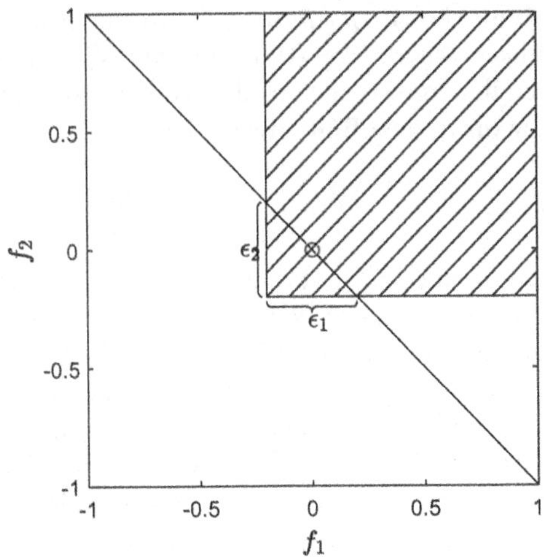

Fig. 6.4 Region of ϵ-dominance for a point x (here $x = 0$) for MOP (6.8)

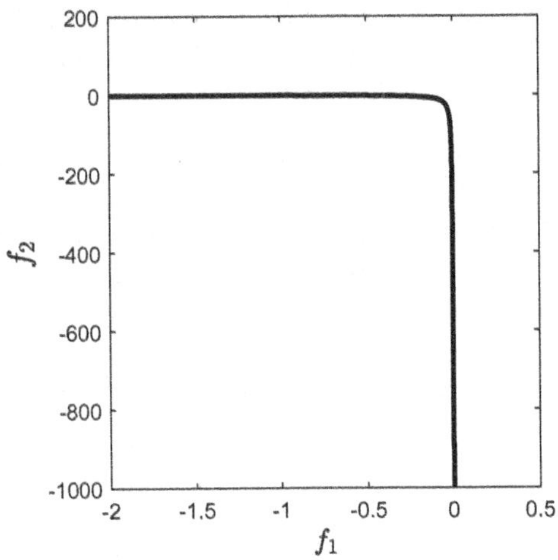

Fig. 6.5 The Pareto front of BOP (6.11) is unbounded to the left and below

The Pareto set for this problem is given by $P_Q = \mathbb{R}^+$, and the Pareto front is as indicated in Fig. 6.5. Since the Pareto front is neither bounded to the left nor below, it is as for the previous example not possible to select a finite size ϵ-approximate Pareto front for this problem. If we e.g., choose the bounded domain $Q = (0, b)$ for any value $b > 0$ we are still in a similar situation as the Pareto front is still unbounded below.

The situation changes if we choose the compact domain $Q = [a, b]$, where $a < b$ and $a, b > 0$. In this case, the Pareto set P is identical to Q. Since both individual objectives are Lipschitz continuous on Q we can choose the Lipschitz constants

$$\begin{aligned} L_1 &= \max_{x\in[a,b]} |f_1'(x)| = 1 \\ L_2 &= \max_{x\in[a,b]} |f_2'(x)| = \frac{1}{a^2}. \end{aligned} \tag{6.12}$$

Due to the Lipschitz continuity of both objectives we have

$$|f_i(x) - f_i(y)| \leq \epsilon_i \quad \text{for} \quad |x - y| \leq \frac{\epsilon_i}{L_i}, \quad i = 1, 2. \tag{6.13}$$

Let

$$N = \lceil \frac{b - a}{\min(\frac{\epsilon_1}{L_1}, \frac{\epsilon_2}{L_2})} \rceil, \tag{6.14}$$

then the set $\{x_0, \ldots, x_N\}$, where

$$x_j = a + \frac{j}{N}(b - a), \qquad j = 0, \ldots, N, \tag{6.15}$$

forms an ϵ-approximate Pareto front—even an ϵ-Pareto front since since all elements x_j are Pareto optimal.

It is hence required for the convergence analysis to assume compactness of the domain of F. In practise, however, boundedness will be sufficient.

Remark 6.3 (*Approximation in objective space*) Similar as for the previous archiver the obtained result yields a certain approximation quality of the limit archive measuring the coverage of the Pareto *front*. This is due to the use of the ϵ-dominance which is as the classical dominance relation defined in objective space. Figure 5.1 and the related discussion shows that also here we would have to postulate injectivity of F in order to be able to obtain certain approximation qualities in decision variable space. We will further address the issue of a distribution of the archive elements in decision variable space in Chap. 6.

Remark 6.4 (*Approximation quality of the limit archive*) A new candidate solution p is only added to the new archive via $ArchiveUpdateEps1$ if there does not exist an element a in the current archive A_0 that $\Theta\epsilon$-dominates p (lines 3 and 4 of Algorithm 3). On the one hand, this "“exclusion strategy” warrants that the sequence A_i reaches after finitely many steps a steady state archive that builds a an ϵ-approximate Pareto front.

On the other hand, this strategy is also the reason why the archive elements do not converge toward Pareto optimal solutions, and hence, that the A_i's do not approach an ϵ-Pareto front. We will discuss this issue with the next archiver.

Another effect of the exclusion strategy is that—at least theoretically—the distance $dist(F(A), F(P_Q))$ can get large (that is, images $F(a)$ of archive entried $a \in A$ can be far away from the Pareto front, even for steady state archives). To see this, consider the scenario in Fig. 6.6. Shown are a hypothetical Pareto front and the images of three elements a_1, a_2 and a_3. Assume that they have been inserted into

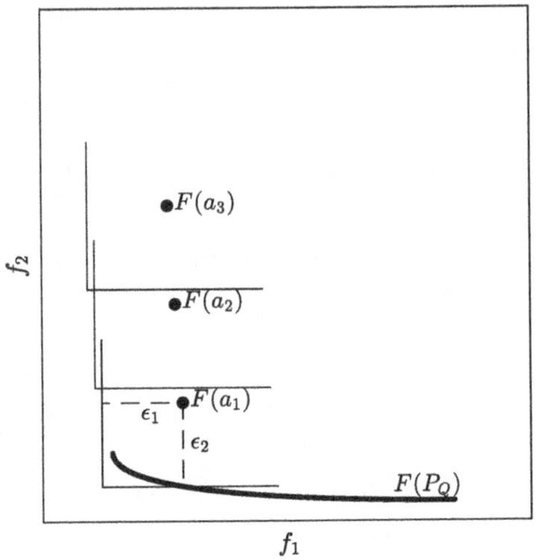

Fig. 6.6 The distance of $F(A)$ to the Pareto front can get large in certain cases. If a_1, a_2 and a_3 are inserted into the archive in this order, the distance $dist(F(A), F(P_Q))$ is given by $dist(F(a_1), F(P_Q)$ which is approximately $3\epsilon_2$. If, on the other hand, a_3 would be inserted first, both a_1 and a_2 would discarded from the archive and $dist(F(A), F(P_Q))$ would be given by ϵ_2

the archive in this order (and for simplicity that and without loss of generality that $\Theta = 1$). Due to the construction of $ArchiveUpdateEps1$, none of these elements will be discarded from subsequent archives due to the exclusion strategy. Similarly one can construct examples where $dist(F(A), F(P_Q))$ can get very large. We stress, however, that we have never observed such extreme cases in our computations.

Figure 6.6 also illustrates that the order of the consideration of the candidate solution has an impact on the archive. If, for instance, a_3 would have been considered first, neither a_1 nor a_2 would have been added to the archiver in later steps.

Next, we study the bounds on the magnitude of the archives generated by $ArchiveUpdateEps1$ with respect to the discretization value $\epsilon \in \mathbb{R}^k_+$. For this, we assume that $|P_0| = 1$, and hence, that also $|A_0| = 1$. The lower bound for the archive size even in the limit is 1. To see this, consider the MOP where all objectives are the same, i.e., $f_1 = f_2 = \cdots = f_k$, and where this objective is a strongly convex function that takes its unique minimizer inside the domain Q.

The following result is on the upper bound of the limit archive.

Theorem 6.2 *Let* $|A_0| = 1$, *and*

$$m_i = \min_{x \in Q} f_i(x) \quad \text{and} \quad M_i = \max_{x \in Q} f_i(x), \qquad 1 \leq i \leq k. \tag{6.16}$$

Then, when using $ArchiveUpdateEps1$, *the archive size maintained in Algorithm 1 for all* $l \in \mathbb{N}$ *is bounded above as follows*

$$|A_l| \leq \sum_{\substack{i_1,\ldots,i_{k-1}=1 \\ i_1 > \ldots > k_{k-1}}}^{k} \prod_{j=1}^{k-1} \left\lceil \frac{M_{i_j} - m_{i_j}}{\Theta \epsilon_{i_j}} \right\rceil, \tag{6.17}$$

Proof Consider a sequence $p_1, p_2, \ldots$ of points which are all accepted by *ArchiveUpdateEps*1 in this order (i.e., starting with $A_0 = \{p_1\}$). Consider the i-th step and let $A_i = \{a_1, \ldots, a_l\}$ with $l \leq i$. Define $B_j := B(F(a_j) - \frac{\Theta\epsilon}{2}, \frac{\Theta\epsilon}{2}),\ j = 1, \ldots, l$ (compare to proof of Theorem 6.1). Using inductive arguments we see that (a) all elements in A_i are mutually nondominating, and that (b) the interiors of all the boxes $B_j,\ j = 1, \ldots, l$, are mutually non-intersecting. Since the points a_j are the upper right corners of the boxes B_j and since the interiors of these boxes are mutually non-intersecting the minimal distance between two points a_{j_1} and a_{j_2}, $j_1 \neq j_2$, is determined by $\Theta\epsilon$ (see Fig. 6.7). Hence, we are able to bound the number of entries in the archives if we can bound the number of such boxes which can be placed in the image space such that their lower left corners are mutually nondominating and the boxes are mutually non-intersecting.

Let us first consider a bi-objective problem (i.e., $k = 2$), since in this case the proof is geometrically descriptive and already captures the basic idea. Since all points a_j are mutually nondominating, the images of these points are all located on a (virtual) continuously differentiable curve

$$\begin{aligned} c &: [m_1, M_1] \to \mathbb{R}^2 \\ u &\mapsto (u, f(u)) \end{aligned} \tag{6.18}$$

where $f : [m_1, M_1] \to [m_2, M_2]$ is a strictly monotonically decreasing (but not necessarily surjective) function. The length of this curve can be bounded as follows:

$$\begin{aligned} L(c) &= \int_{m_1}^{M_1} \|c'(u)\| du = \int_{m_1}^{M_1} \sqrt{|1|^2 + |f'(u)|^2} du \\ &\leq \int_{m_1}^{M_1} 1 du + \int_{m_1}^{M_1} |f'(u)| du = \int_{m_1}^{M_1} 1 du - \int_{m_1}^{M_1} f'(u) du \\ &\leq (M_1 - m_1) + (M_2 - m_2). \end{aligned} \tag{6.19}$$

Thus, for $k = 2$ we see that

$$|A_i| \leq \left\lceil \frac{(M_1 - m_1)}{\Theta\epsilon_1} \right\rceil + \left\lceil \frac{(M_2 - m_2)}{\Theta\epsilon_2} \right\rceil, \quad i \in \mathbb{N}, \tag{6.20}$$

as claimed above.

Now we turn our attention to the general case, i.e., let $k \geq 2$ be given. Define

$$\begin{aligned} K &:= [m_1, M_1] \times \cdots \times [m_{k-1}, M_{k-1}], \\ K_{(i)} &:= [m_1, M_1] \times \cdots \times [m_{i-1}, M_{i-1}] \times [m_{i+1}, M_{i+1}] \times \cdots \times [m_{k-1}, M_{k-1}], \\ u_{(i)} &:= (u_1, \ldots, u_{i-1}, u_{i+1}, \ldots, u_{k-1}),\ i = 1, \ldots, k-1. \end{aligned} \tag{6.21}$$

In analogy to the bi-objective case, the images of the elements of the archives are located in the graph of a map Φ which is characterized as follows:

$$\begin{aligned} \Phi &: K \to \mathbb{R}^k \\ \Phi(u_i, \ldots, u_{k-1}) &= (u_1, \ldots, u_{k-1}, f(u_1, \ldots, u_{k-1})), \end{aligned} \tag{6.22}$$

where $f : K \to [m_k, M_k]$ is a sufficiently smooth function satisfying the monotonicity conditions $\frac{\partial f}{\partial u_i} u < 0$, $\forall u \in K$ and $\forall i = 1, \ldots, k-1$. Then, the $(k-1)$-dimensional volume of Φ with parameter range K can be bounded as follows:

$$\begin{aligned} Vol(\Phi) &= \int_K \sqrt{||\nabla f||^2 + 1} du \\ &= \int_K \sqrt{\left(\frac{\partial f}{\partial u_1}\right)^2 + \cdots + \left(\frac{\partial f}{\partial u_{k-1}}\right)^2 + 1} du \\ &\leq \int_K \left|\frac{\partial f}{\partial u_1}\right| du + \cdots + \int_K \left|\frac{\partial f}{\partial u_{k-1}}\right| du + \int_K 1 du \\ &= \sum_{i=1}^{k-1} \left(\int_{K_{(i)}} \left(\int_{m_i}^{M_i} \left|\frac{\partial f}{\partial u_i}\right| du_i \right) du_{(i)} \right) + \int_K 1 du \\ &= \sum_{i=1}^{k-1} \left(\int_{K_{(i)}} \left(- \int_{m_i}^{M_i} \frac{\partial f}{\partial u_i} du_i \right) du_{(i)} \right) + \int_K 1 du \\ &\leq \sum_{\substack{i_1, \ldots, i_{k-1} = 1 \\ i_1 > \ldots > k_{k-1}}}^{k} \prod_{j=1}^{k-1} (M_{i_j} - m_{i_j}) \end{aligned} \tag{6.23}$$

This bound of the volume leads directly to the bound of the cardinality of the archives as stated above which completes the proof. □

One interesting aspect is the growth of the limit archives with respect to the discretization vector ϵ. Choosing m_ϵ as the smallest entry of ϵ, i.e., $m_\epsilon = \min_{i=1,\ldots,k} \epsilon_i$, we obtain

$$|A_l| \in O\left(\left(\frac{1}{m_\epsilon}\right)^{k-1}\right) \quad \text{for} \quad m_\epsilon \to 0 \text{ and } l \to \infty. \tag{6.24}$$

This is a satisfying result since we can expect that the dimension of the Pareto set is $(k-1)$.

The scenario in Fig. 6.7 shows how to construct an example an archive for the bi-objective case those magnitude reaches the maximal bound. This value, however, is typically by far not reached, mainly since the volume of the Pareto front is normally much smaller than estimated above. Further, flat regions of the Pareto front can be covered by just a few archive elements.

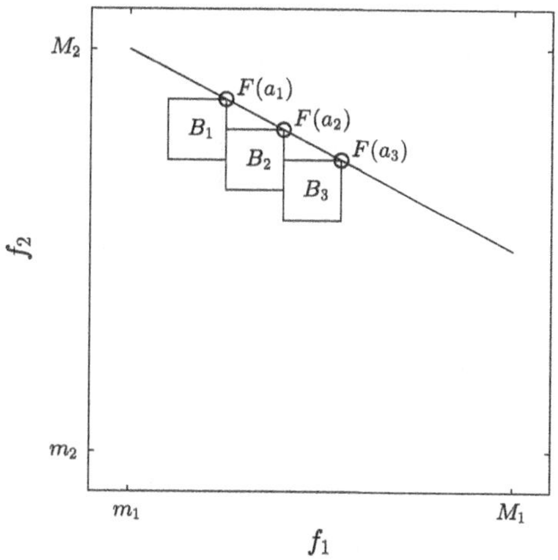

Fig. 6.7 Approximating the upper bound of the archives for *ArchiveUpdateEps*1

The next example shows the potential advantage of the use of the archiver in the context of knapsack problems, where the number of the Pareto optimal solutions can get huge.

Example 6.1 Consider bi-objective {0, 1}-knapsack problems of the following form (e.g., Gomes da Silva et al. 2004a, b):

$$f_1, f_2 : \{0, 1\}^n \to \mathbb{R}$$
$$f_i(x) = \sum_{j=1}^{n} c_j^i x_j, \; i = 1, 2 \tag{6.25}$$

such that

$$\sum_{j=1}^{n} w_j x_j \leq W.$$

Hereby, each scalar $c_j^i \in [0, 1]$, $j = 1, \ldots, n$, $i = 1, 2$, stands for the (normalized) value of item j and value i, w_j stands for the weight of item j, and W denotes the overall capacity of the knapsack. It is known that the Pareto sets and fronts for such problems are typically very large. For instance, for the special case of equal weighted items where the sums of the value coefficients are constant (i.e., $c_j^1 + c_j^2 = 1$ and $w_j = c$, $j = 1, \ldots, n$), the cardinality of the Pareto set P is from Gomes da Silva et al. (2004a) known to be

$$|P| = \binom{n}{l} = \frac{n!}{(n-l)!l!}, \tag{6.26}$$

where $l := \lfloor \frac{W}{c} \rfloor \leq n$.

For instance, for the following (arbitrarily chosen) values for n and l we obtain the following magnitudes of the Pareto sets/fronts:

$$\begin{aligned}
(n,l) &= (100,20) \qquad |P_{(100,20)}| = 5.35 \cdot 10^{20} \\
(n,l) &= (100,40) \qquad |P_{(100,40)}| = 1.37 \cdot 10^{28} \\
(n,l) &= (500,30) \qquad |P_{(500,30)}| = 1.45 \cdot 10^{48} \\
(n,l) &= (500,50) \qquad |P_{(500,50)}| = 2.31 \cdot 10^{69}
\end{aligned} \tag{6.27}$$

That is, it is practically intractable to compute all optimal solutions for such instances. Instead, it is desired to compute a finite size approximation of the solution sets that yield a certain approximation quality in the limit as for $ArchiveUpdateEps1$.

To compute these upper bounds let us assume for simplicity that all entries of ϵ are the same, i.e., that $\epsilon = (\epsilon_0, \ldots, \epsilon_0)^T$ for a $\epsilon_0 > 0$. It is known (Gomes da Silva et al. 2004a) that for all optimal solutions it holds $|x| := x^T x = l$. Further, since $c_j^i \in [0, 1]$ we can deduce that $f_i(a) \in [0, l]$, $\forall a \in A$, and hence, that $m_i = 0$ and $M_i = l$. Using these values, and taking $\Theta = 1$, we can now compute the upper bounds for the archives: it is

$$|A| \le \left(\frac{l}{\epsilon_0} + \frac{l}{\epsilon_0}\right) = \frac{2l}{\epsilon_0}. \tag{6.28}$$

Choosing for instance the value $\bar{\epsilon} = 0.1$, which corresponds to 10% of the value of each item, we obtain for the upper bounds for the archives of the above chosen instances

$$\begin{aligned}
(n,l) &= (100,20) \qquad |A_{(100,20)}| \le 400 \\
(n,l) &= (100,40) \qquad |A_{(100,40)}| \le 800 \\
(n,l) &= (500,30) \qquad |A_{(500,30)}| \le 600 \\
(n,l) &= (500,50) \qquad |A_{(500,50)}| \le 1000.
\end{aligned} \tag{6.29}$$

As discussed above, these bounds are pessimistic and will typically not be reached in computations. Nevertheless, it becomes apparent that the approximation of the entire Pareto front becomes a feasible task, at least when consdiering the required number of elements that have to be stored when accepting the discretization error of ϵ_0 in objective space.

6.3 *ArchiveUpdateEps2*

The previous archiver, $ArchiveUpdateEps1$, is aiming for ϵ-approximate Pareto fronts, and the "exclusion strategy" (line 3 of Algorithm 4) prevents that the archive entries converge toward Pareto optimal solutions. While this is acceptable in case all entries of ϵ are small, it may become an unwanted effect if the entries of the discretization vector are getting larger which in particular happens for many objective problems (MaOPs), i.e., MOPs with a larger number of objectives (e.g., $k > 3$). Recall that for a continuous MOP with k objectives we can expect the Pareto set/front to be an object of dimension $k - 1$. The points needed to obtain an ϵ-(approximate) Pareto front depends hence next to the values of ϵ also on the number of objectives. Hence, if k is "high", the values of ϵ have to be chosen "large" so that such approximations can be obtained with a reasonable number of elements.

The archiver $ArchiveUpdateEps2$ is a modification of $ArchiveUpdateEps1$ which aims for ϵ-Pareto fronts. As we will see, this archiver allows for the approximation of an ϵ-Pareto front in the limit, but in turn requires to store more archive entries compared to $ArchiveUpdateEps1$.

Algorithm 4 shows the pseudo code of $ArchiveUpdateEps2$. The key difference to $ArchiveUpdateEps2$ is that elements p that are already $\Theta\epsilon$-dominated by an element of the current archive A are not necessarily discarded. Instead, p is added to the archive if (i) no element from A exist that $\Theta\epsilon$-dominates p (lines 3 and 4 of 4), or (ii) it dominates an existing element a of A. In that case, a is replaced in the archive by p.

We first consider the monotonic behavior of $ArchiveUpdateEps2$ which is identical to the previous archiver.

Algorithm 4 $A := ArchiveUpdateEps2\ (P, A_0)$

Require: P: current population, A_0: current archive, $\epsilon \in \mathbb{R}^k_+$, $\Theta \in (0, 1)$
Ensure: A: updated archive
1: $A := A_0$
2: **for all** $p \in P$ **do**
3:     **if** $\nexists a \in A : a \prec_{\Theta\epsilon} p$ **then**
4:         $A := A \cup \{p\}$
5:     **end if**
6:     **for all** $a \in A$ **do**
7:         **if** $p \prec a$ **then**
8:             $A := A \cup \{p\} \backslash \{a\}$
9:         **end if**
10:     **end for**
11: **end for**
12: **return** A

Lemma 6.3 *Let (MOP) be given,* $A_0, P \subset \mathbb{R}^n$ *be finite sets,* $\epsilon \in \mathbb{R}^k_+$, *and* $A := ArchiveUpdateEps2\ (P, A_0)$. *Then it holds:*

$$\forall x \in P \cup A_0 : \quad \exists a \in A : a \prec_{\Theta\epsilon} x. \tag{6.30}$$

Proof Analog to the proof of Lemma 6.1. □

Lemma 6.4 *Let (MOP) be given, $l \in \mathbb{N}$, $P_0, \ldots, P_l \subset \mathbb{R}^n$ be finite sets, and A_i, $i = 1, \ldots, l$, be the archives obtained by $ArchiveUpdateEps2$ as in Algorithm 4, and $C_l = \bigcup_{i=0}^{l} P_i$. Then*

$$\forall x \in C_l: \quad \exists a \in A_l : a \prec_{\Theta\epsilon} x. \tag{6.31}$$

Proof Follows immediately by Lemma 6.3. □

Next, we investigate the limit behavior of the archiver. The following results shows that as for the previous archive we can expect that the archives form an ϵ-approximate Pareto front after finitely many steps when using $ArchiveUpdateEps2$. Further, all archive elements converge toward Pareto optimal solutions, and hence, that the archives come arbitrarily close to an ϵ-Pareto front under iteration.

Theorem 6.3 *Let (MOP) be given and $Q \subset \mathbb{R}^n$ be compact, and let there be no weak Pareto points in $Q \backslash P_Q$. Further, let F be injective. Then, under assumption (4.1), an application of Algorithm 1, where $ArchiveUpdateEps2$ is used to update the archive, leads to a sequence of archives A_l, $l \in \mathbb{N}$, where the following holds:*

(a) There exists with probability one an $l_0 \in \mathbb{N}$ such that A_l is an ϵ-approximate Pareto front for all $l \geq l_0$.

(b)

$$\lim_{l \to \infty} dist(A_l, P_Q) = 0, \quad \textit{with probability one.}$$

Proof (a) Analog to the proof of Theorem 6.1a.

(b) A new element p is added to the archive A by $ArchiveUpdateEps2$ only in one of the following two cases: (1) if there exists no $a \in A$ which $\Theta\epsilon$-dominates p (denote by *type 1 insertion* for the remainder of this proof), and (2), if p dominates an element $a \in A$, which will in turn be discarded from the archive (denote by *type 2 insertion*). Since A is finite and P_Q is compact it follows that

$$dist(A, P_Q) = \max_{a \in A} \min_{p \in P_Q} ||a - p||. \tag{6.32}$$

We prove the claim in the following way: We show that for every $a \in A$ there exists a sequence a_i of dominating points with $a_i \to a^* \in P_Q$ which will be inserted in future instances of the archive. Since $dist(a^*, P_Q) = 0$ and since all the points a_i, $i \in \mathbb{N}$, do not have to be considered for the limit archive due to the insertion of type 2 it remains to show that there are only finitely many insertions of type 1 (which is the only manipulation of the archive by which the value of $dist(A, P_Q)$ can be increased). This will be done in the second part.

Let $l_0 \in \mathbb{N}$ and $a_0 \in A_{l_0}$. If $a_0 \in P_Q$ it follows that $a_0 \in A_{l+m}$, $\forall m \in \mathbb{N}$. In this case set $a_i := a_0$. Now assume that $a_0 \notin P_Q$. Define

$$\begin{aligned} M &: Q \to \mathbb{R} \\ M(x) &:= \max_{p \in P_Q} \min_{i=1,\ldots,k} (f_i(x) - f_i(p)). \end{aligned} \tag{6.33}$$

Under the assumptions made above it holds that

$$M(x) \geq 0 \; \forall x \in Q \quad \text{and} \quad M(x) = 0 \Leftrightarrow x \in P_Q.$$

Let $p_0 \in P_Q$ be the argument of the maximum of $M(a_0)$. Since $a_0 \notin P_Q$ and a_0 is not a weak Pareto point it follows that $M(a_0) > 0$ and $F(p_0) <_p F(a_0)$. Since F is continuous there exists a neighborhood U_{p_0} of p_0 such that

$$F(y) <_p F(p_0) + \frac{M(a_0)}{2} \cdot (1, \ldots, 1) \quad \forall y \in U_{p_0},$$

and thus, that $F(y) <_p F(a_0)$, $\forall y \in U_{p_0}$. By (4.1) it follows that *Generate* (see Algorithm 1) generates with probability one after finitely many steps a point $b \in U_{p_0} \cap Q$. Now there are two cases: (1) b is added to the archive (in this case set $a_1 := b$), and (2), a_0 has already been replaced by an element $\tilde{a} \in \mathbb{R}^n$ such that b and $\tilde{a}$ are mutually nondominating (in this case set $a_1 := \tilde{a}$). In both cases there exists a $j \in \{1, \ldots, k\}$ such that

$$f_j(a_1) < f_j(p_0) + \frac{M(a_0)}{2}.$$

Proceeding in an analogous way we obtain a sequence $\{a_i\}_{i \in \mathbb{N}}$ of dominating points. Since the sequence $\{F(a_i)\}_{i \in \mathbb{N}}$ is bounded below and F is injective it follows that $a_i \to a^* \in Q$ for $i \to \infty$.
Next we show that $a^* \in P_Q$. For this, assume that $a^* \notin P_Q$. Define p^* as the argument of the maximum of $M(a^*)$. Since $a^* \notin P_Q$ and a^* is not a weak Pareto point it follows that $F(p^*) <_p F(a^*)$ and $M(a^*) > 0$. Proceeding further as above we obtain a point a^{**} and an element $j \in \{1, \ldots, k\}$ such that

$$\begin{aligned} f_j(a^{**}) &< f_j(p^*) + \frac{M(a^*)}{2} \leq f_j(p^*) + \frac{f_j(a^*) - f_j(p^*)}{2} \\ &= \frac{f_j(p^*) + f_j(a^*)}{2} < f_j(a^*) \end{aligned}$$

This is a contradiction to the assumption of the convergence of the sequence, and thus it must be that $a^* \in P_Q$.

Further, it can be shown analogous to *ArchiveUpdateEps*1 (see proof of Theorem 6.1) that only finitely many insertions of type 1 can be done during the run of the algorithm. Thus, only finitely many times the value of $dist(A, P_Q)$ can be increased (an insertion of type 2 decreases the value), and the claim for the limit archive follows. □

We finally consider the upper bounds for archives that are generated by *ArchiveUpdateEps2*.

Theorem 6.4 *Let* $m_i = \min_{x \in Q} f_i(x)$ *and* $M_i = \max_{x \in Q} f_i(x)$, $1 \leq i \leq k$, *and* $|A_0| = 1$. *Then, when using ArchiveUpdateEps2, the archive size maintained in Algorithm 1 is bounded for all* $l \in \mathbb{N}$ *as*

$$|A_l| \leq \prod_{i=1}^{k} \left\lceil \frac{M_i - m_i}{\Theta \epsilon_i} \right\rceil. \tag{6.34}$$

Proof We can consider the process of including solutions into the archive over time as a process for constructing a directed graph G. Starting with an empty graph, we add a new node for each new solution p that is added to the archive A in line 4 or line 8 of the algorithm. If p is added in line 8 (meaning the condition in line 7 is true), we add arcs (p, a) from p to each solution a that is discarded in line 8 due to $p \prec a$. Let $V_t := \bigcup_{1 \leq j \leq t} A_j$ be the union of all archives up to iteration t and $V_t' \subseteq V_t$ the subset of those archive members that have been added in line 4. Thus, the node set of G_t after iteration t is V_t, and G_t itself is a forest whose roots are the current archive members A_t and whose leafs are the elements of V_t'. Obviously, the number of roots must be smaller than the number of leafs, so $|A_t| \leq |V_t'|$.

To bound $|V_t'|$, the number of elements that ever entered the archive in line 4, we again consider the boxes $B_v := B(F(v) - \frac{\Theta\epsilon}{2}, \frac{\Theta\epsilon}{2})$ for all $v \in V_t'$. Due to line 3, a solution p generated in iteration $t' \leq t$ cannot be accepted in line 4 if $F(p)$ lies inside the box B_v of any previously accepted element of $v \in V_t'$, otherwise $a \prec_{\Theta\epsilon} p$ for some current archive member $a \in A_t$ as there exists $a \in A_t$ with $F(a) \leq F(v)$ and $v \prec_{\Theta\epsilon} p$. If p was accepted in line 4, then $F(p)$ cannot lie inside the box B_v of any subsequently accepted element of $v \in V_t'$ neither, as this would entail $p \prec v$. Hence, the interiors of the boxes B_v must be mutually non-intersecting. The maximum number of non-intersecting boxes with side length $\Theta\epsilon$ and centers c with $m_i \leq c_i \leq M_i$ is $\prod_{i=1}^{k} \lceil (M_i - m_i)/(\Theta\epsilon_i) \rceil$, thus the claimed bound on the archive size follows. □

We consider again the growth of the limit archives with respect to ϵ. Using $m_\epsilon = \min_{i=1,\dots,k} \epsilon_i$ we obtain

$$|A_l| \in O\left(\left(\frac{1}{m_\epsilon}\right)^k\right) \quad \text{for} \quad m_\epsilon \to 0 \text{ and } l \to \infty. \tag{6.35}$$

i.e., onc order of magnitude more than for *ArchiveUpdateEps*1, which seems to be the price we have to pay to compute an ϵ-Pareto front in the limit. In Schütze et al. (2008d), an example can be found for $n = 2$ where the bound on the archive size is tight.

6.4 Numerical Results

As in the previous section, we illustrate the impact of the archiving strategies via feeding them with N candidate solutions that are chosen uniformly at random from the domain of each problem. The different entries of ϵ are chosen such that an effect of the discretization parameters can be seen. We stress, however, that in real-world problems meaningful values for ϵ normally arise naturally from the given application.

We are first interested in the effect of $ArchiveUpdateEps1$. Figure 6.8 shows some typical results of the algorithm on CONV2 for $\epsilon = (\frac{10}{3}, \frac{10}{3})$. Recall that next to the choice of the candidate solutions the result of the algorithm also depends on the order of the considered points, i.e., different runs will likely lead to different results. The figures, however, already show a completely different behavior than for the previous archiver $ArchiveUpdateP_Q$. The algorithm is capable of detecting a finite size approximation of the Pareto front, and the solutions are more or less evenly spread along this set. In order to see an evolution in the sequence of final archives, we had to reduce the number N of considered candidate solutions siginficantly compared to the previous algorithm. Most notably, the magnitudes of the archives are all relatively low, and do not exceed any threshold as for $ArchiveUpdateP_Q$. The maximal archive size, as discussed above, depends of course on the choice of ϵ. Further, once the archive builds an ϵ-approximate Pareto font, it stays static, i.e., neither do replacements occur, nor are further elements are added to it.

Figure 6.9 shows analog results but this time for problem DENT and for $\epsilon = (\frac{1}{12}, \frac{1}{12})$. Also in this case, the resulting archive quickly forms an ϵ-approximate Pareto front, and the archive remains static after that (which is not shown, but which follows by the construction of $ArchiveUpdateEps1$). As it can be seen, some parts of the Pareto front are not "covered" by the archiver, even for larger numbers N of candiate solutions. Note that this happens where the Pareto front is "flat" which makes it possible that single solutions can ϵ-dominate entire (large) regions of the solution set. Though the archives form an ϵ-approximate Pareto front as anticipated, the existence of such gaps is not wanted in certain applications. We will address the computation of gap free Pareto front approximations in the next chapter.

Figure 6.10 shows some archives for the problems SSW, CONV3 and DTLZ7 obtained by $ArchiveUpdateEps1$. For all cases, $N = 100,000$ candidate solutions have been used to feed the archive, and in all cases the final archives (even for less candidate solutions) form ϵ-approximate Pareto fronts. All approximations can hence considered to be perfect in the defined sense and for the chosen values of ϵ. One issue with the archiver, however, is that it does not aim for Pareto optimal solutions which may become a drawback for larger entries of ϵ.

In order to see the difference between the archivers $ArchiveUpdateEps1$ and $ArchiveUpdateEps2$ we consider the next two figures. Figures 6.11 and 6.12 show comparisons of the two archivers on CONV2 and SSW for different values of ϵ. For $\epsilon = 0$, both archivers conicide with $ArchiveUpdateP_Q$, leading hence to the same result. The difference between the two archivers becomes apparent for increasing values of ϵ. While $ArchiveUpdateEps1$ just keeps ϵ-approximate solutions, the other archiver aims for Pareto opimal ones. This explains the different approximations. On the other hand, no significant differences in the archive sizes can be observed (which has been confirmed with other empirical results by the authors) as opposed to the consideration of the theoretical upper bounds on the archive sizes done in the previous section. Hence, while for small values of ϵ in practise both archivers perform equally, for larger values of ϵ the archiver $ArchiveUpdateEps2$ seems to be preferable.

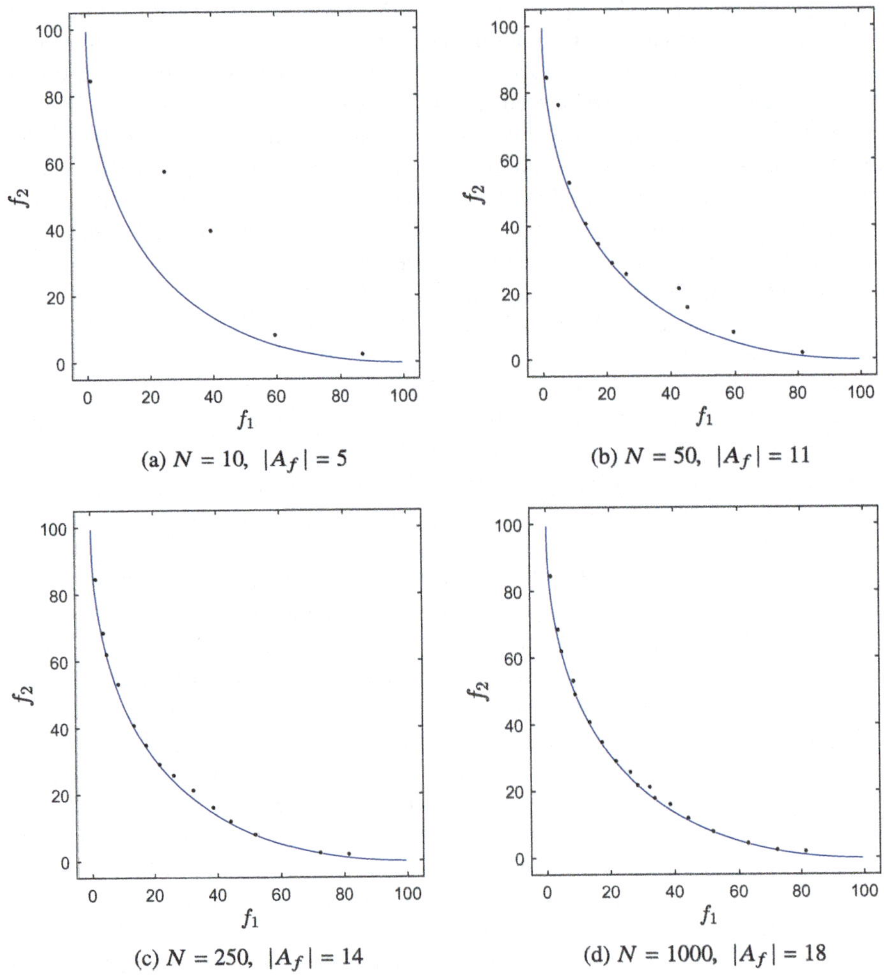

(a) $N = 10, \ |A_f| = 5$

(b) $N = 50, \ |A_f| = 11$

(c) $N = 250, \ |A_f| = 14$

(d) $N = 1000, \ |A_f| = 18$

Fig. 6.8 Numerical results of $ArchiveUpdateEps1$ for CONV2 and $\epsilon = (\frac{10}{3}, \frac{10}{3})$ for different numbers N of candidate solutions. A_f denotes in each case the final archive obtained by $ArchiveUpdateEps1$. Note that the archives for $N_1 = 1,000$ and $N_2 = 100,000$ are identical. This is the case since the archiver for N_1 already forms an ϵ-approximate Pareto font as can be seen bottom right

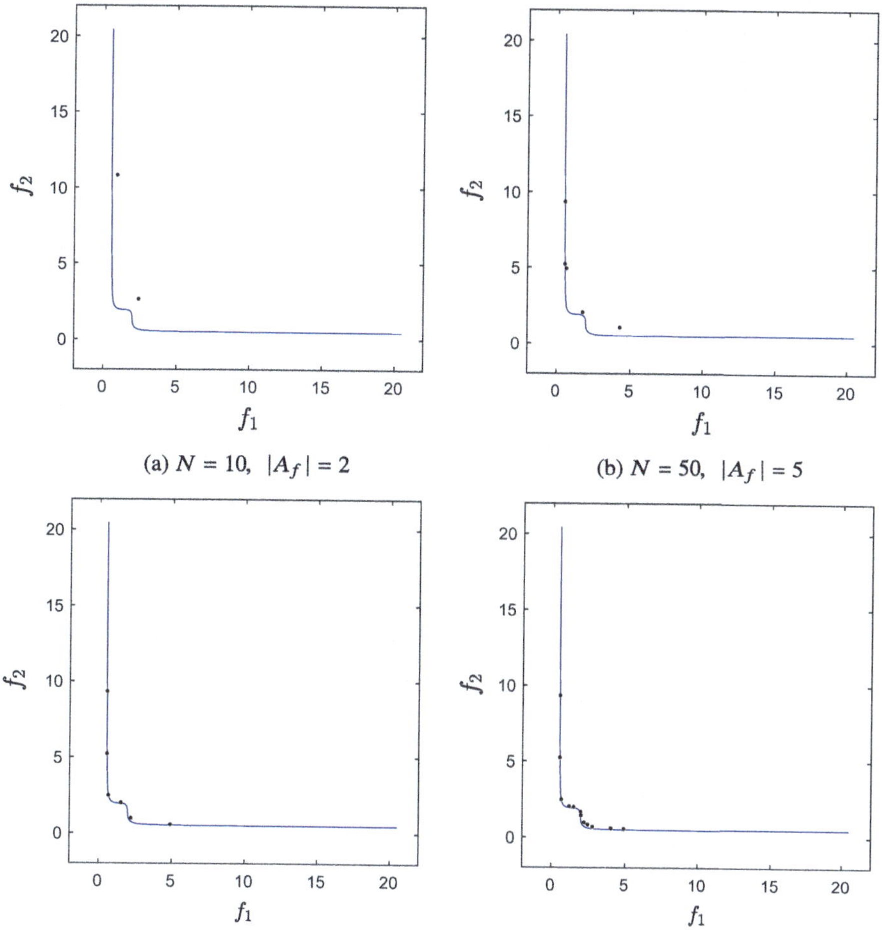

(a) $N = 10, \;\; |A_f| = 2$

(b) $N = 50, \;\; |A_f| = 5$

(c) $N = 250, \;\; |A_f| = 6$

(d) $N = 1,000, \;\; |A_f| = 12$

Fig. 6.9 Numerical results of *ArchiveUpdateEps*1 for DENT and $\epsilon = (\frac{1}{12}, \frac{1}{12})$ for different numbers N of candidate solutions. A_f denotes in each case the final archive obtained by *ArchiveUpdateEps*1

Finally, Figs. 6.13 and 6.14 demonstrate the effect of the choice of ϵ for *ArchiveUpdateEps*2 on the three-objective problems CONV3 and DTLZ7. While, as anticipated, larger values of ϵ lead to smaller final archvies, in all cases the shape of the Pareto front is still captured adequately.

Fig. 6.10 Numerical results of *ArchiveUpdateEps*1 for SSW, CONV3, and DTLZ7 for $N = 100,000$

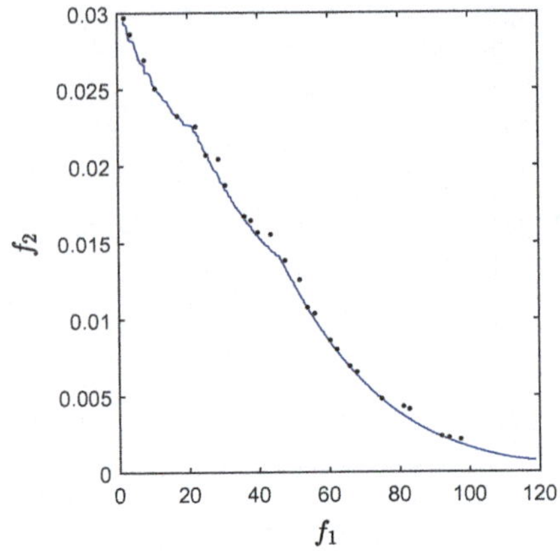

(a) SSW, $\epsilon = \frac{1}{3}(5, 0.005)$, $|A_f| = 27$

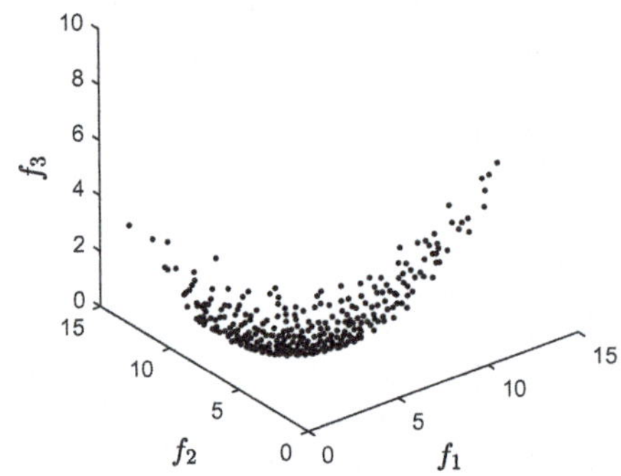

(b) CONV3, $\epsilon = \frac{1}{3}(0.5, 0.5, 0.5)$, $|A_f| = 303$

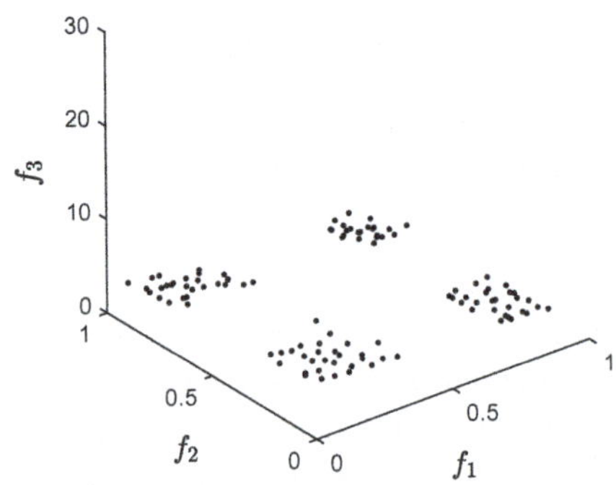

(c) DTLZ7, $\epsilon = \frac{1}{3}(0.15, 0.15, 0.15)$, $|A_f| = 110$

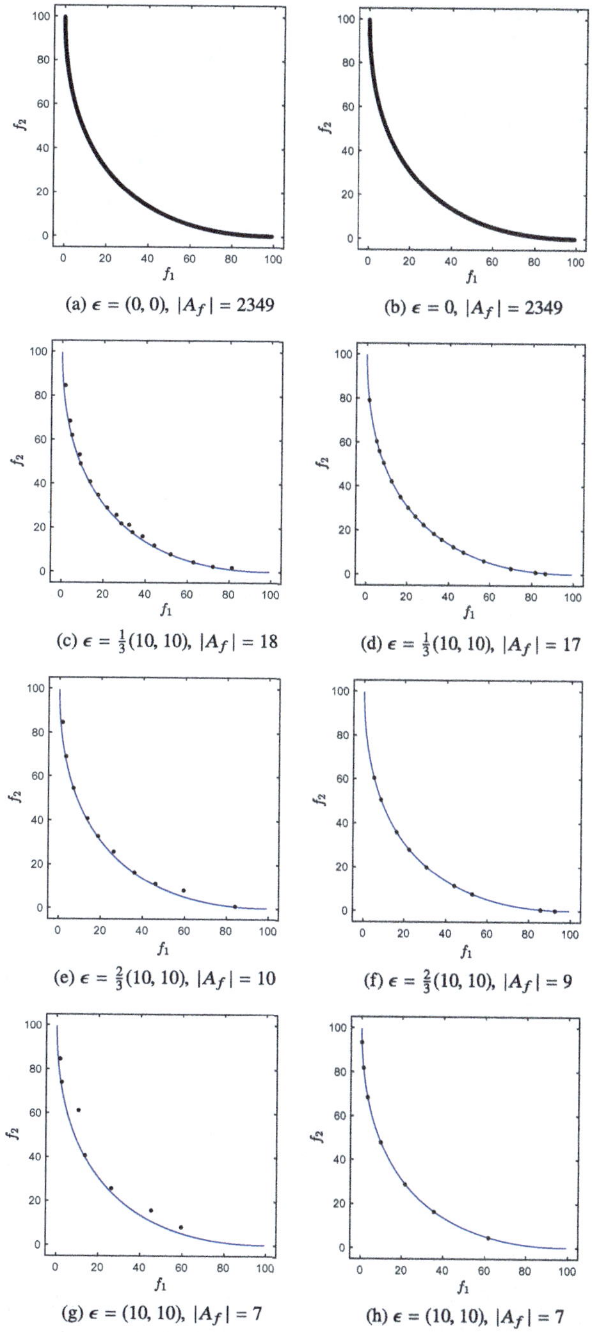

(a) $\epsilon = (0, 0)$, $|A_f| = 2349$

(b) $\epsilon = 0$, $|A_f| = 2349$

(c) $\epsilon = \frac{1}{3}(10, 10)$, $|A_f| = 18$

(d) $\epsilon = \frac{1}{3}(10, 10)$, $|A_f| = 17$

(e) $\epsilon = \frac{2}{3}(10, 10)$, $|A_f| = 10$

(f) $\epsilon = \frac{2}{3}(10, 10)$, $|A_f| = 9$

(g) $\epsilon = (10, 10)$, $|A_f| = 7$

(h) $\epsilon = (10, 10)$, $|A_f| = 7$

Fig. 6.11 Comparison of the results of *ArchiveUpdateEps*1 (left) and *ArchiveUpdateEps*2 (right) on CONV2 for different values of ϵ. For each problem we have chosen $N = 10,000$

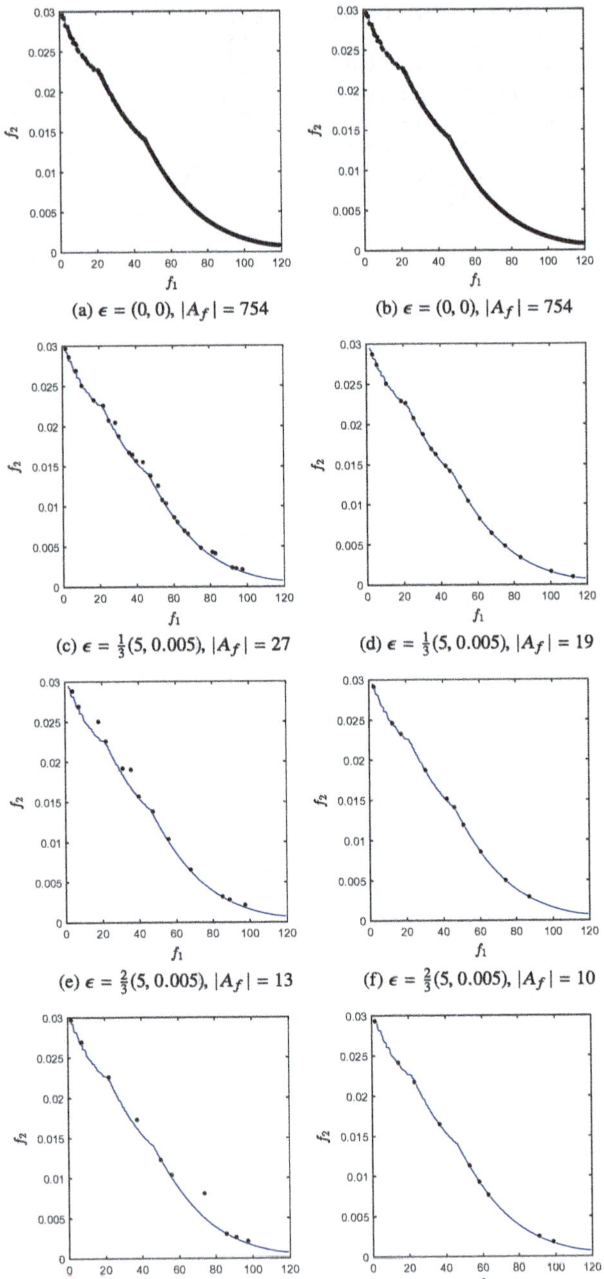

Fig. 6.12 Comparison of the results of $ArchiveUpdateEps1$ (left) and $ArchiveUpdateEps2$ (right) on SSW for different values of ϵ

(a) $\epsilon = (0, 0)$, $|A_f| = 754$

(b) $\epsilon = (0, 0)$, $|A_f| = 754$

(c) $\epsilon = \frac{1}{3}(5, 0.005)$, $|A_f| = 27$

(d) $\epsilon = \frac{1}{3}(5, 0.005)$, $|A_f| = 19$

(e) $\epsilon = \frac{2}{3}(5, 0.005)$, $|A_f| = 13$

(f) $\epsilon = \frac{2}{3}(5, 0.005)$, $|A_f| = 10$

(g) $\epsilon = (5, 0.005)$, $|A_f| = 10$

(h) $\epsilon = (5, 0.005)$, $|A_f| = 9$

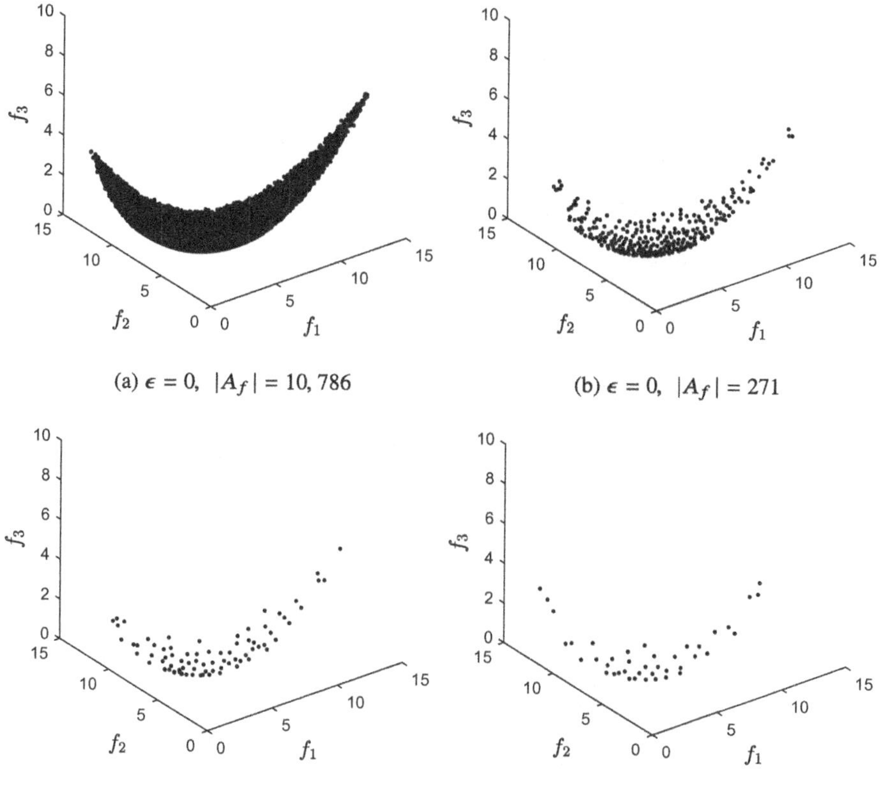

(a) $\epsilon = 0, \ |A_f| = 10,786$

(b) $\epsilon = 0, \ |A_f| = 271$

(c) $\epsilon = 0, \ |A_f| = 79$

(d) $\epsilon = 0, \ |A_f| = 41$

Fig. 6.13 Numerical results of $ArchiveUpdateEps2$ on CONV3 for different values of ϵ

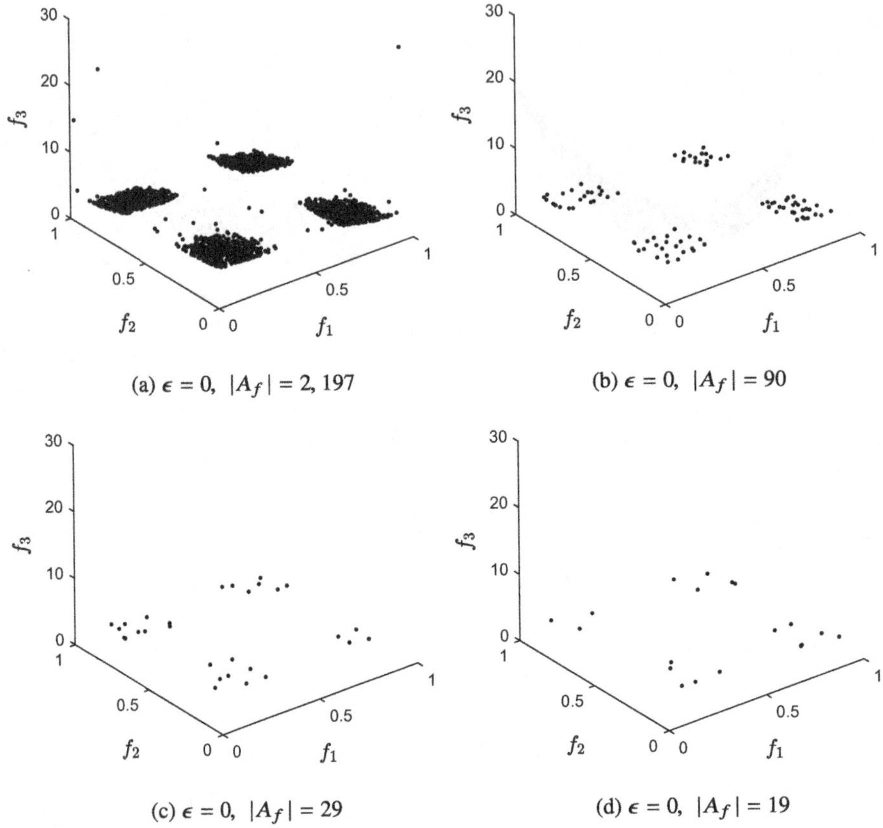

(a) $\epsilon = 0,\ |A_f| = 2,197$

(b) $\epsilon = 0,\ |A_f| = 90$

(c) $\epsilon = 0,\ |A_f| = 29$

(d) $\epsilon = 0,\ |A_f| = 19$

Fig. 6.14 Numerical results of $ArchiveUpdateEps2$ on DLTZ7 for different values of ϵ

Chapter 7
Computing Gap Free Pareto Fronts

So far, we have discussed the archiver that stores all non-dominated solutions out of the set of candidate solutions, and two archivers that are entirely based on the concept of ϵ-dominance. While the applicability of $ArchiveUpdateP_Q$ is restricted since it stores too many points during the run of the search process, the opposite can happen for the two ϵ-dominance based archivers. To see the latter, consider the extreme examples depicted in Fig. 7.1. Both sub-figures show Pareto fronts that contain flat parts and an "approximation" of this set that consists of three elements. Both approximations indeed form ϵ-Pareto fronts for relatively small entries of ϵ (compared to the length of the entire Pareto front). That is, given certain applications together with suitable vectors ϵ, these two Pareto fronts could indeed be reduced to the three elements without loss of (significant) information. We have also seen in the last chapter that such approximations could even be obtained in extreme cases by $ArchiveUpdateEps1$ or $ArchiveUpdateEps2$. The reason for this is that single solutions can ϵ-dominate even large portions of the Pareto front at its flat parts.

On the other hand, such approximations may be incomplete or at least appear to be. Note that the proper choice of ϵ might be a delicate task. If an entry is chosen to be too large (even sightly), over reductions such as in Fig. 7.1 can happen. Even if such a reduction is "correct", it still leaves uncertainty to the decision maker if the computation is complete, or if a significant part of the Pareto front is still missing. For instance, in the Evolutionary Multi-objective Optimization (EMO) community it is commonly agreed that a set of candidate solutions should be presented to the decision maker that is ideally evenly spread around the Pareto front ("spread and convergence", e.g.., Deb 2001; Deb et al. 2002a; Zitzler et al. 2002; Coello Coello et al. 2007; Zhang and Li 2007). Figure 7.2 shows the same Pareto fronts as in Fig. 7.1, however, this time with "gap free" approximations, which may be more suitable in certain situations. Scenarios and applications where gap free approximations are desired are reported in Hestermeyer and Oberschelp (2003), Pottharst et al. (2004), Fliege (2004b), Abbass (2004), Dellnitz et al. (2005), Fernández et al. (2014),

O. Schütze and C. Hernández, *Archiving Strategies for Evolutionary Multi-objective Optimization Algorithms*, Studies in Computational Intelligence 938,
https://doi.org/10.1007/978-3-030-63773-6_7

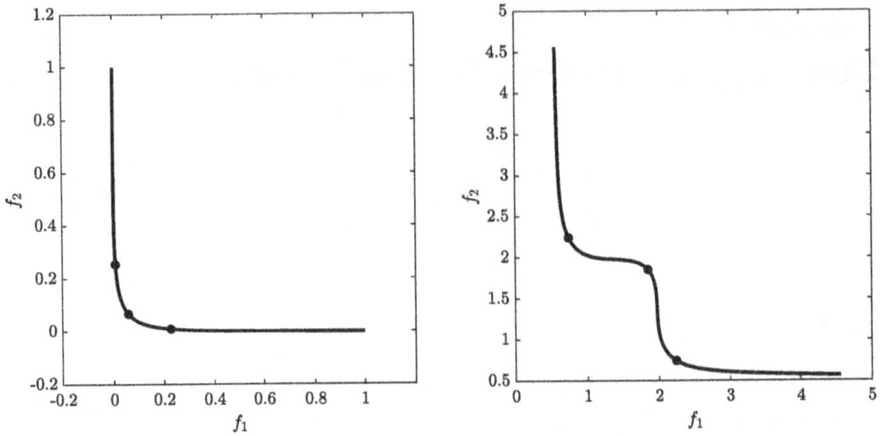

Fig. 7.1 Two Pareto fronts together with two three-element approximations. Both approximations may form ϵ-Pareto fronts (depending on the choice of ϵ), but do not cover some large portions of the Pareto front

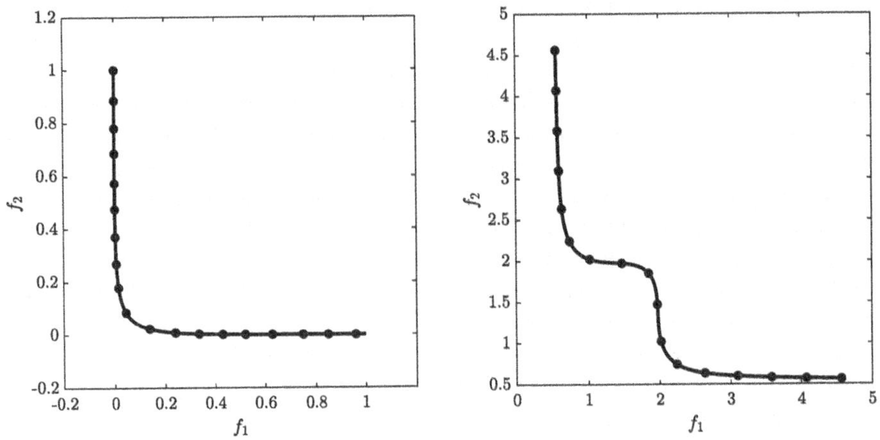

Fig. 7.2 Same Pareto fronts as in Fig. 7.1, but this time with "gap free" approximations which may be more suitable in certain situations

Valledor Pellicer et al. (2020) which includes one example from the steel-making industry and two examples of online adjustments of particular mechatronical systems. For the latter, finite size approximations A of the Pareto fronts, which are computed offline, are used as the basis for an online strategy. More precisely, during the operation of the system, points $a \in A$ are chosen according to the given situation (which changes time). In both applications, abrupt changes in the given mechatronical systems have to be avoided. Consequently, gaps in the approximations have to be avoided.

In this chapter, we will investigate two archiving strategies that aim for "gap free" approximations that will be also applicable to Pareto fronts that contain flat regions. We will further address the uniformity of the archived solutions, which has also been neglected so far.

7.1 The Sets of Interest

In this section, we will define two new sets of interest. Motivated by the above discussion, we will first define Δ_M-tight ϵ-(approximate) Pareto fronts.

Definition 7.1 (*Δ_M-tight ϵ-(approximate) Pareto front*) Let $\epsilon \in \mathbb{R}^k_+$.

(a) A set $A \subset \mathbb{R}^n$ is called a *Δ_M-tight ϵ-approximate Pareto front* of (MOP) if A is an ϵ-approximate Pareto front of (MOP) and if in addition it holds

$$dist(F(P_Q), F(A)) \leq \Delta_M. \tag{7.1}$$

(b) A set $A \subset \mathbb{R}^n$ is called a *Δ_M-tight ϵ-Pareto front* if A is an ϵ-Pareto front of (MOP) and if in addition it holds

$$d_H(F(P_Q), F(A)) \leq \Delta_M. \tag{7.2}$$

The difference of the above objects to an ϵ-(approximate) Pareto front is given by condition (7.1) (respectively (7.2) which has the same meaning). This condition is fulfilled for an archive A if for every point y of the Pareto front there exists an archive entry a of A such that $\|y - F(a)\|_\infty \leq \Delta_M$. Denote by $B^\infty_\Delta(x)$ the closed box with center x and radius Δ, i.e.,

$$B^\infty_\Delta(x) := \{y \in \mathbb{R}^k \ : \ \|x - y\| < \Delta\}. \tag{7.3}$$

Condition (7.1) is hence fulfilled if the Pareto front $F(P_Q)$ is entirely contained in the collection of boxes $B^\infty_\Delta(F(a))$ of all images $a \in A$ (see also Fig. 7.3):

$$F(P_Q) \subset C_{A,\Delta} := \bigcup_{a \in A} B^\infty_\Delta(F(a)). \tag{7.4}$$

That is, if condition (7.1) is satisfied the largest possible "gap" in the Pareto front approximation $F(A)$ is given by the value of Δ. Hence, the approximation can be considered to be "tight" or "gap free" if Δ is sufficiently small.

Note that if an archive A is an ϵ-(approximate) Pareto front, this is the case for every archiver B such that $A \subset B$. That is, the definition of the above objects does not include any information about the spread of the images of the archive entries along the Pareto front. Since uniform distributions are ideal in most cases we add in the following a uniformity level Δ_m to the objects defined in Definition 7.1.

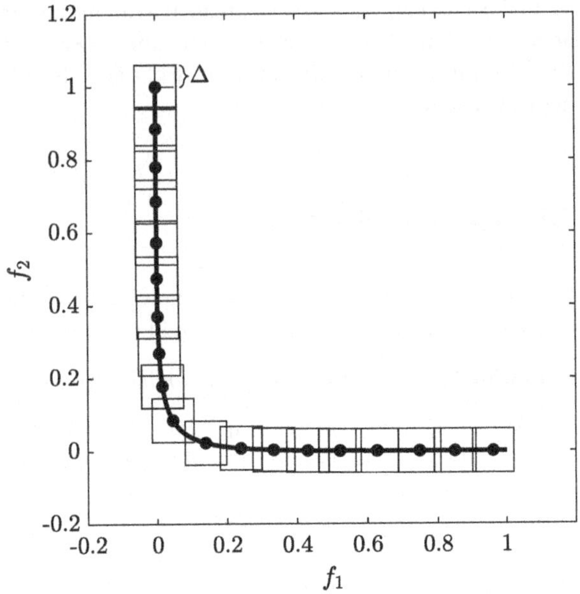

Fig. 7.3 Example where the Pareto front is contained in the union of all Δ-boxes with center $F(a)$ and $a \in A$

Definition 7.2 Let $\epsilon \in \mathbb{R}^k_+$.

(a) A set $A \subset \mathbb{R}^n$ with $|A| \geq 2$ is called a (Δ_M, Δ_m)*-tight* ϵ*-approximate Pareto front* if A is an Δ_M*-tight* ϵ*-approximate Pareto front* of (MOP) and if

$$dist(F(a), F(A \backslash \{a\})) \geq \Delta_m, \quad \forall a \in A. \tag{7.5}$$

(b) A set A with $|A| \geq 2$ is called a (Δ_M, Δ_m)*-tight* ϵ*-Pareto front* if A is an Δ_M*-tight* ϵ*-approximate Pareto front* of (MOP) and if

$$dist(F(a), F(A \backslash \{a\})) \geq \Delta_m, \quad \forall a \in A.$$

As example, consider a BOP those Pareto front is the line segment connecting the points $(0, 4)^T$ and $(4, 0)^T$ (see Fig. 7.4). Further, let the points $x_1, \ldots, x_6$ be given with $y_i = F(x_i), i = 1, \ldots, 6$, where

$$y_1 = \begin{pmatrix} 0.5 \\ 3.5 \end{pmatrix}, \quad y_2 = \begin{pmatrix} 1.5 \\ 2.5 \end{pmatrix}, \quad y_3 = \begin{pmatrix} 2.5 \\ 1.5 \end{pmatrix}, \quad y_4 = \begin{pmatrix} 3.5 \\ 0.5 \end{pmatrix}, \quad y_5 = \begin{pmatrix} 2 \\ 2 \end{pmatrix}, \quad y_6 = \begin{pmatrix} 2.1 \\ 2.1 \end{pmatrix}, \tag{7.6}$$

and let $\epsilon = (1, 1)^T$. First, consider the archive

$$A_1 = \{x_1, x_2, x_3, x_4\}. \tag{7.7}$$

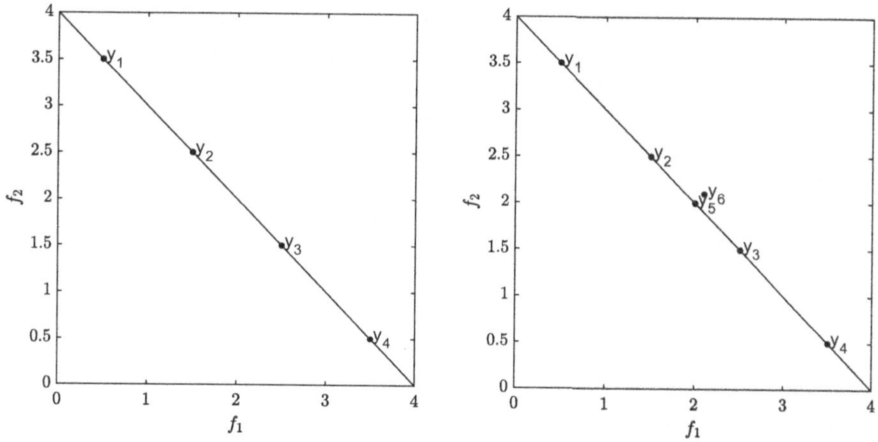

Fig. 7.4 Left: the archive A_1 forms a $(\frac{1}{2}, 1)$-tight ϵ-Pareto front. Right: if further elements are added to the archive the uniformity level Δ_m shrinks

All elements of A_1 are Pareto optimal and since ϵ is chosen large enough (for all distance calculations we use here the maximum norm), the set forms an ϵ-Pareto front. Further, Δ_M is given by $\frac{1}{2}$ and the uniformity level by Δ_m is given by 1. A_1 hence forms a $(\frac{1}{2}, 1)$-tight ϵ-Pareto front. Note that A_1 is a four-element approximation of a Pareto front that has length 4. The uniformity level of 1 is hence perfect. Indeed, A_1 is also the "best" four-element approximation of this Pareto front in the Hausdorff sense.

If we consider the archive

$$A_2 = A_1 \cup \{x_5\} \tag{7.8}$$

the uniformity level reduces to $\Delta_m = 0.5$ since for instance $\|y_2 - y_5\| = 0.5$. A_2 forms a $(\frac{1}{2}, \frac{1}{2})$-tight ϵ-Pareto front.

If we further add the point x_6, i.e., if we consider the archive

$$A_3 = A_2 \cup \{x_6\} \tag{7.9}$$

then the uniformity level shrinks to $\Delta_m = 0.1$. Since x_6 is not Pareto optimal, A_3 is a $(\frac{1}{2}, 0.1)$-tight ϵ-approximate Pareto front.

7.2 *ArchiveUpdateTight*1

The first archiving strategy we consider is *ArchiveUpdateTight*1, see Algorithm 5 for the pseudo code. For given values of $\epsilon \in \mathbb{R}^k_+$ and $\Delta \in \mathbb{R}_+$, an element $p \in Q$ is accepted either if (i) there does not exist an element a of the current archive A that

$\Theta\epsilon$-dominates p, or if (ii) there does not exist an element $a \in A$ which dominates p *and* the distance of $F(a)$ to $F(p)$ for all $a \in A$ is larger than the threshold $\tilde{\Delta}$ (line 3 of Algorithm 5). If one of the conditions (i) or (ii) holds, then the candidate p is added to the archive (line 11), and finally, all elements $a \in A\backslash\{p\}$ that are dominated by p are discarded (lines 7 to 10). Hence, also this archiver only stores elements that are mutually non-dominated. Similar as for the above archivers, the values of $0 < \Theta < 1$ and $0 < \tilde{\Delta} < \Delta$ are only required to guarantee the convergence of the sequence of archives in the limit (see Theorem 7.2), but in practice the values $\Theta = 1$ and $\tilde{\Delta} = \Delta$ can be chosen.

Algorithm 5 $A := ArchiveUpdateTight1_{\epsilon,\tilde{\Delta}}\ (P, A_0)$

Require: P: current population, A_0: current archive, $\epsilon \in \mathbb{R}^k_+$, $\Theta \in (0, 1)$, $0 < \tilde{\Delta} < \Delta$
Ensure: A: updated archive

```
1: A := A0
2: for all p ∈ P do
3:   if (∃a ∈ A : a ≺ p) or (∃a1 ∈ A : a1 ≺Θϵ p and ∃a2 ∈ A : d∞(F(a2), F(p)) ≤ Δ̃) then
4:     CONTINUE                          ▷ do not execute lines 6 – 11
5:   end if
6:   for all a ∈ A do
7:     if p ≺ a then
8:       A := A\{a}
9:     end if
10:  end for
11:  A := A ∪ {p}
12: end for
13: return A
```

Again, we first consider the monotonic behavior of the archiver which his analog to the archivers we have discussed in the previous chapter.

Lemma 7.1 *Let* $A_0, P \subset \mathbb{R}^n$ *be finite sets,* $\epsilon \in \mathbb{R}^k_+$, $0 < \Theta < 1$, $0 < \tilde{\Delta} < \Delta$, *and* $A :=$ *ArchiveUpdateTight1* (P, A_0). *Then it holds*

$$\forall x \in P \cup A_0 : \quad \exists a \in A : a \prec_{\Theta\epsilon} x. \tag{7.10}$$

Proof Let $P = \{p_1, p_2, \ldots, p_l\}$, $l \in \mathbb{N}$. Without loss of generality we assume that all points p_i are considered in this ordering (i.e., in the for-loop in line 2 of Algorithm 5). Denote by A_i the resulting archive after p_i has been considered, and thus, $A = A_n$.

Let $x \in P \cup A_0$. There are two cases we have to distinguish.

Case A: $x \in A_0$. In that case the statement follows since points a are only discarded from the archive if in turn another point p with $p \prec a$ is inserted (see lines 7, 8 and 11 of Algorithm 5), and by the transitivity of $\prec$.

Case B: $x \in P$, i.e., there exists $j \leq l$ such that $x = p_j$, and thus, x is considered in the jth step of the algorithm. In case x is added to A_j, the statement follows analogously to Case A since $A =$ *ArchiveUpdateTight1* $(\{p_{j+1}, \ldots, p_l\}, A_j)$, and $x \in A_j$. If x is not added to A_j there are two possibilities (see line 3 of Algorithm

5): (i) there exists an $a \in A_j$ such that $a \prec x$, or (ii) there exists an $a_1 \in A_j$ with $a_1 \prec_{\Theta\epsilon} x$. In both cases x gets $\Theta\epsilon$-dominated by a point $\tilde{a} \in A_j$. Thus, again analog to Case A, a point $a^* \in A$ exists which is equal to $\tilde{a}$ or dominating it, and thus $\Theta\epsilon$-dominating x. □

Theorem 7.1 *Let $l \in \mathbb{N}$, $\epsilon \in \mathbb{R}^k_+$, $0 < \Theta < 1$, and $0 < \tilde{\Delta} < \Delta$. Further, let A_i, P_i, $i = 0, \ldots, l$, be as defined in Algorithm 1, where $ArchiveUpdateTight1_{\epsilon,\tilde{\Delta}}$ is used to update the archive. Let $C_l = \bigcup_{i=0}^{l} P_i$, then*

$$\forall x \in C_l : \quad \exists a \in A_l : a \prec_{\Theta\epsilon} x. \tag{7.11}$$

Proof The proof is done via induction over l. For $l = 1$ we have

$$A_1 = ArchiveUpdateTight1_{\epsilon,\tilde{\Delta}}(P_0, A_0),$$

and the claim follows by Lemma 7.1. For the induction hypothesis, suppose the claim (7.11) is right for $l - 1 > 1$. Let $x \in \bigcup_{i=1}^{l} P_i$. If $x \in \bigcup_{i=1}^{l-1} P_i$ there exists by induction hypothesis an element $a \in A_{l-1}$ such that $a \prec_{\Theta\epsilon} x$. Thus, a is also a member of the subsequent archive A_l or is replaced by an element $p \in P_l$ with $p \prec a$. In both cases there exists an element in A_l which $\Theta\epsilon$-dominates x. In case $x \in P_l$ the claim follows again by Lemma 7.1 since

$$A_l = ArchiveUpdateTight1_{\epsilon,\tilde{\Delta}}(P_l, A_{l-1}),$$

and the proof is complete. □

We next consider the limit behavior of the sequence of archives generated by $ArchiveUpdateTight1$. The following results show that we can expect—under some assumptions—that from an iteration l_1 on all the Archivers A_l, where $l \geq l_1$, will form a $(\Delta, \Theta\epsilon_m)$-tight ϵ-approximate Pareto front.

Theorem 7.2 *Let (MOP) be given, where F is continuous, let $Q \subset \mathbb{R}^n$ be a compact set and $\epsilon \in \mathbb{R}^k_+$. Let $\epsilon_m := \min_{i=1,\ldots,k} \epsilon_i$, $\epsilon_M := \max_{i=1,\ldots,k} \epsilon_i$, further let $\Delta, \tilde{\Delta} \in \mathbb{R}_+$ be given such that $\epsilon_M < \tilde{\Delta} < \Delta$, let $0 < \Theta < 1$, and let assumption (4.1) be fulfilled. Then an application of Algorithm 1, where $ArchiveUpdateTight1_{\epsilon,\tilde{\Delta}}$ is used to update the archive, leads to a sequence of archives, such that there exists with probability one an $l_1 \in \mathbb{N}$ such that A_l is a $(\Delta, \Theta\epsilon_m)$-tight ϵ-approximate Pareto front for all $l \geq l_1$.*

Proof First, we turn our attention to the question of which elements are added to the archive. The crucial expression $\mathcal{E}$ (line 3 of Algorithm 5) reads as follows:

$$\underbrace{(\exists a \in A : a \prec p)}_{\mathcal{A}} \text{ or } (\underbrace{\exists a_1 \in A : \ a_1 \prec_{\Theta\epsilon} p}_{\mathcal{B}_1} \text{ and } \underbrace{\exists a_2 \in A : d_\infty(F(a_2), F(p)) \leq \tilde{\Delta})}_{\mathcal{B}_2}$$

Since $\neg\mathcal{E} = (\neg\mathcal{A}$ and $\neg\mathcal{B}_1)$ or $(\neg\mathcal{A}$ and $\neg\mathcal{B}_2)$ and since $\neg\mathcal{B}_1$ implies $\neg\mathcal{A}$ it follows that points $p \in \mathbb{R}^n$ are added to a given archive A if (and only if) one of the two following expressions is true

$$\begin{aligned} &(\mathcal{E}_1) \quad \nexists a \in A : a \prec_{\Theta\epsilon} p, \text{ or} \\ &(\mathcal{E}_2) \quad \nexists a \in A : a \prec p \text{ and } \forall a \in A : d_\infty(F(a), F(p)) > \tilde{\Delta}. \end{aligned} \tag{7.12}$$

Now we are in the position to prove the theorem. By $\mathcal{E}_1$ it follows that all points, which are added by $ArchiveUpdateEps1$ to the archive are also added by $ArchiveUpdateTight1$. Thus, by Theorem 6.1 it follows that there exists with probability one a $l_0 \in \mathbb{N}$ such that A_l is an ϵ-approximate Pareto front w.r.t. (MOP) for all $l \geq l_0$, since points $a \in A_l$ are only removed from the archive if in turn another point $\tilde{a}$ is added which dominates a (if $x \prec y$ and $y \prec_\epsilon z$ it follows that $x \prec_\epsilon z$).

It remains to show the "tightness" of the limit archive. The uniformity level ϵ_m follows directly by an inductive argument and using the "exclusion strategy" (7.12). This and the fact that $F(Q)$ is bounded is the reason that the size of the archive is bounded above for a given MOP by a number $n_0 = n_0(\epsilon, F(Q)))$.

As mentioned above (see Eq. (7.4) and related discussion) the claim is right for an archive A_l if

$$F(P_Q) \subset C_{A_l,\Delta}. \tag{7.13}$$

Assume that A_l is an ϵ-approximate Pareto front for all $l \geq l_0$ and let $l \geq l_0$. By construction of $ArchiveUpdateTight1$ it follows that if $F(P_Q) \subset C_{A_{l_1},\Delta}$ this inclusion holds for all $l \geq l_1$ since in this case no further point will be added to the archive (since the expressions $\mathcal{E}_1$ and $\mathcal{E}_2$ in (7.12) will be false for all further candidates). That is, it is sufficient to show the existence of such a number l_1. In the following we will do this by contradiction: first, we show that by using $ArchiveUpdateTight1$ and under the assumptions made above only finitely many replacements can be done during the run of the algorithm. Then—under the assumption that there exists no number l_1 with the above property—we construct a contradiction by showing that infinitely many replacements have to be done during the run of the algorithm with the given setting.

Let a finite archive A_0 be given. If a point $p \in \mathbb{R}^n$ replaces a point $a \in A_0$ (see lines 8 and 11 of Algorithm 5) it follows by construction of $ArchiveUpdateTight1$ (see also (7.12)) that

$$\exists i \in 1, \ldots, k: \quad f_i(p) < f_i(a) - \Theta\epsilon_i. \tag{7.14}$$

Since the relation $\prec$ is transitive, there exists for every $a \in A$ a "history" of replaced points $a_i \in A_{l_i}$ where Eq. (7.14) holds for a_i and a_{i-1}. Since $F(Q)$ is bounded there exist $l_i, u_i \in \mathbb{R}, i = 1, \ldots, k$, such that $F(Q) \subset [l_1, u_1] \times \ldots \times [l_k, u_k]$. After r replacements there exists at least one $a \in A_{l(r)}$ such that the length h of the history of a is at least $h \geq \lceil r/n_0 \rceil$, where n_0 is the maximal number of entries in the archive (see above). Denote by $a_0 \in A_0$ the root of the history. For a, a_0 it follows that

$$\exists i \in 1, \ldots, k: \quad f_i(a) < f_i(a_0) - s\Theta\epsilon_i,$$

where $s \geq \lceil h/k \rceil$. For $\tilde{s} > d_{max} := \Theta^{-1} \max_{i=1,\ldots,k} \frac{u_i - l_i}{\epsilon_i}$ (which is given for $\tilde{r} > n_0 k d_{max} + n_0 + 1$) we obtain a contradiction since in that case $f_i(a) < l_i$ and thus $F(a) \notin F(Q)$. Hence it follows that there can be done only finitely many such replacements during the run of an algorithm.

Assume that such an l_1 as claimed above does not exist, that is, that $F(P_Q) \not\subset C_{A_l,\Delta}$ for all $l \in \mathbb{N}$. Hence there exists a sequence of image points

$$y_i \in F(P_Q) \backslash C_{A_i,\Delta} \quad \forall i \in \mathbb{N}. \tag{7.15}$$

Since $F(Q)$ is compact there exists an accumulation point $y^* \in F(P_Q)$, that is, there exists a subsequence $\{i_j\}_{j \in \mathbb{N}}$ with

$$y_{i_j} \rightarrow y^* \text{ for } j \rightarrow \infty. \tag{7.16}$$

Since $y^* \in F(P_Q)$ there exists a neighborhood U_1 of y^* such that the following holds

$$\forall (y, \tilde{y}) \in F(Q) \times U_1: \; y \leq_p \tilde{y} \Rightarrow d_\infty(y, \tilde{y}) \leq \tilde{\Delta}. \tag{7.17}$$

Let $\tilde{U}_1 := U_1 \cap B^\infty_{(\Delta - \tilde{\Delta})/2}(y^*)$. By (4.1) it follows that there exists with probability one an $l_1 \in \mathbb{N}$ and an $\tilde{x}_1 \in P_{l_0 + l_1}$ with $\tilde{y}_1 = F(\tilde{x}_1) \in \tilde{U}_1$. By construction of *ArchiveUpdateTight1* there exists an element $a_1 \in A_{l_0+l_1}$ such that $d_\infty(F(a_1), \tilde{y}_1) < \tilde{\Delta}$ (due to (7.12) there are three possibilities: $\mathcal{E}_2$ is false and thus there already exists an $a_1 \in A_{l_0+l_1}$ which (a) dominates $\tilde{x}$—in this case the claim follows with (7.17)—or (b) where $d_\infty(F(a_1), \tilde{y}_1) \leq \tilde{\Delta}$, or $\mathcal{E}_2$ is true and thus (c) $a_1 = \tilde{x}_1$ is added to the archive). Thus we have

$$d_\infty(F(a_1), \tilde{y}) \leq d_\infty(F(a_1), \tilde{y}_1) + d_\infty(\tilde{y}_1, \tilde{y}) < \tilde{\Delta} + 2\frac{\Delta - \tilde{\Delta}}{2} = \Delta \quad \forall \tilde{y} \in \tilde{U}_1. \tag{7.18}$$

By (7.15) and (7.16) there exist $j_1, \tilde{l}_1 \in \mathbb{N}$ with

$$y_{i_{j_1}} \in \tilde{U}_1 \backslash C_{l_0 + l_1 + \tilde{l}_1, \Delta}.$$

Since by (7.18) it holds that $d_\infty(y_{i_{j_1}}, F(a_1)) < \Delta$ it follows that $a_1 \notin A_{l_0+l_1+\tilde{l}_1}$, which is only possible via a replacement in Algorithm 5 (lines 8 and 11).

In an analogous way a sequence $\{a_i\}_{i \in \mathbb{N}}$ of elements can be constructed which have to be replaced by other elements. Since this leads to a sequence of infinitely many replacements this is a contradiction to the assumption, and the proof is complete. □

Note that similar as for $ArchiveUpdateEps1$, the exclusion strategy (7.12) prevents convergence of the archive elements toward Pareto optimal solutions. We will address this issue in the following with the next archiving strategy.

Finally, we consider the archive bounds. As for the above examples, the lower bound is given by one if e.g. all objectives are identical and have one isolated minimum.

For the maximal archive size, it makes sense to assume that the value of Δ is larger than any entry of ϵ. Doing so, the archive size is entirely determined by ϵ, and the upper bound maintained by $ArchiveUpdateTight1$ is in fact the same one as for $ArchiveUpdateEps1$:

$$|A_l| \leq \sum_{i=1}^{k} \prod_{\substack{j=1 \\ j \neq i}}^{k} \left\lceil \frac{M_j - m_j}{\Theta \epsilon_j} \right\rceil, \tag{7.19}$$

where $m_i = \min_{x \in Q} f_i(x)$, $M_i = \max_{x \in Q} f_i(x)$, $1 \leq i \leq k$, and $|A_0| = 1$. That is, also for this archiver we obtain

$$|A_l| \in O\left(\left(\frac{1}{m_\epsilon}\right)^{k-1}\right) \quad \text{for} \quad m_\epsilon \to 0 \quad \text{and} \quad l \to \infty, \tag{7.20}$$

where $m_\epsilon = \min_{i=1,\ldots,k} \epsilon_i$ is the smallest entry of ϵ.

7.3 *ArchiveUpdateTight2*

As for $ArchiveUpdateEps1$, the "exclusion strategy" of $ArchiveUpdateTight1$ (line 3 of Algorithm 5 respectively condition (7.12)) prevents that the archive entries converge to Pareto optimal solutions. If convergence of the archive entries toward optimal solutions is desired, one can e.g. choose the archiver $ArchiveUpdateTight2$ which we will discuss in this section. This archiver allows its entries to converge toward optimal solutions, however, in turn, the uniformity of the obtained solutions cannot be guaranteed any more.

Algorithm 6 shows the pseudo code of $ArchiveUpdateTight2$. The difference of this archiver and $ArchiveUpdateTight1$ lies in the strategy to accept a candidate solution p. For an given archive A, $ArchiveUpdateTight2$ accepts this point if either (i) one of the terms ($\mathcal{E}_1$) or ($\mathcal{E}_2$) of (7.12)) is true (line 3 of Algorithm 6) or (ii) if there exists an element of A that is dominated by p (line 7 of Algorithm 6). Hence, the difference between $ArchiveUpdateTight1$ and $ArchiveUpdateTight2$ is given by case (ii) which is not considered by $ArchiveUpdateTight1$.

Algorithm 6 $A := ArchiveUpdateTight2_{\epsilon,\tilde{\Delta}}\ (P, A_0)$

Require: P: current population, A_0: current archive, $\epsilon \in \mathbb{R}^k_+$, $\Theta \in (0, 1)$, $0 < \tilde{\Delta} < \Delta$
Ensure: A: updated archive

```
A := A_0
for all p ∈ P do
    if ℰ_1 is true or ℰ_2 is true then                ▷ see (7.12)
        A := A ∪ {p}
    end if
    for all a ∈ A do
        if p ≺ a then
            A := A ∪ {p}\{a}
        end if
    end for
end for
return A
```

First, we consider the monotonic behavior which is as for the previous archiver.

Theorem 7.3 *Let* $l \in \mathbb{N}$, $\epsilon \in \mathbb{R}^k_+$, $0 < \Theta < 1$, *and* $0 < \tilde{\Delta} < \Delta$. *Further, let* $A_i, P_i, i = 0, \ldots, l$, *be as defined in Algorithm 1, where* $ArchiveUpdateTight2_{\epsilon,\tilde{\Delta}}$ *is used to update the archive. Let* $C_l = \bigcup_{i=0}^{l} P_i$, *then*

$$\forall x \in C_l : \quad \exists a \in A_l : a \prec_{\Theta\epsilon} x. \tag{7.21}$$

Proof Analog to proof of Theorem 7.1. □

Next, we investigate the limit behavior of the sequence of archives. We can expect to obtain a Δ-tight ϵ-approximate Pareto front after finitely many steps (i.e., the uniformity level compared to $ArchiveUpdateTight1$ is not given any more). In turn, the archive elements of A_l come arbitrarily close to the Pareto set P_Q under iteration. For the convergence, we have to assume injectivity of F. In our computations, however, we have observed that this property is not relevant.

Theorem 7.4 *Let (MOP) be given and* $Q \subset \mathbb{R}^n$ *be compact, and let there be no weak Pareto points in* $Q \backslash P_Q$. *Further, let* F *be injective and let assumption (4.1) be fulfilled. Then an application of Algorithm 1, where* $ArchiveUpdateTight2_{\epsilon,\tilde{\Delta}}$ *is used to update the archive, leads to a sequence of archives* $A_l, l \in \mathbb{N}$, *where the following holds:*

(a) There exists with probability one a $l_1 \in \mathbb{N}$ *such that* A_l *is a* Δ*-tight* ϵ*-approximate Pareto front w.r.t. (MOP) for all* $l \geq l_1$.

(b)

$$\lim_{l \to \infty} dist(A_l, P_Q) = 0, \quad \textit{with probability one.}$$

Proof All parts of the proof are analog to parts in proofs of Theorems 6.3 and 7.2.

Analogous to $ArchiveUpdateTight1$, if we assume that Δ is larger than any entry of ϵ, then the archive size is bounded above by

$$|A_l| \leq \prod_{i=1}^{k} \left\lceil \frac{M_i - m_i}{\Theta\epsilon_i} \right\rceil, \tag{7.22}$$

where m_i and M_i are as defined above and $|A_0| = 1$, which is the same bound as for $ArchiveUpdateEps2$. That is, we obtain

$$|A_l| \in O\left(\left(\frac{1}{m_\epsilon}\right)^k\right) \quad \text{for} \quad m_\epsilon \to 0 \text{ and } l \to \infty, \tag{7.23}$$

where $m_\epsilon = \min_{i=1,\ldots,k} \epsilon_i$ is the smallest entry of ϵ.

As for previous archivers, the lower bound is also given for this one by 1, however, also for this archiver only for pathological cases.

7.4 Numerical Results

We will first show some numerical results to demonstrate the behavior of the archivers $ArchiveUpdateTight1$ and $ArchiveUpdateTight2$. Since this is the last chapter that exclusively deals with Pareto front approximations, we will further on make a detailed comparison of the five archivers we have discussed so far.

Figures 7.5 and 7.6 show the evolution of the final archives obtained by $ArchiveUpdateTight1$ on the problems DENT and SSW. In both cases, gap free approximations are obtained after a certain time. In fact, both final approximations form $(\Delta, \Theta\epsilon_m)$-tight ϵ-approximate Pareto fronts for the values of Δ and ϵ (and hence ϵ_m) as given in the captions of the figures. While for DENT all final archive elements are (nearly) Pareto optimal, this does not hold for the final approximation of SSW.

Figure 7.7 shows a comparison of the archivers $ArchiveUpdateTight1$ and $ArchiveUpdateTight2$ on the problems CONV2, SSW, CONV3, and DTLZ7. As it can be seen, all archivers yield gap free approximations. In particular, on SSW it can be seen that the images of the archive elements obtained by $ArchiveUpdateTight2$ are closer to the Pareto front. We can also observe that not all points have yet converged since $N = 100{,}000$ randomly chosen points within the domain Q were not sufficient for this problem. We will address this problem in Chap. 10, where we discuss the usage of the archivers within MOEAs. The results on DTLZ7 show another characteristic of the new archivers, namely that the intent to obtain a gap free approximation may lead to some troubles for problems with disconnected Pareto fronts. As we can see, even for $N = 3{,}000{,}000$ uniformly chosen candidate solutions the final approximation does not give a sharp approximation of the set of interest, in particular at the borders of the components of the Pareto front. We will also reconsider this

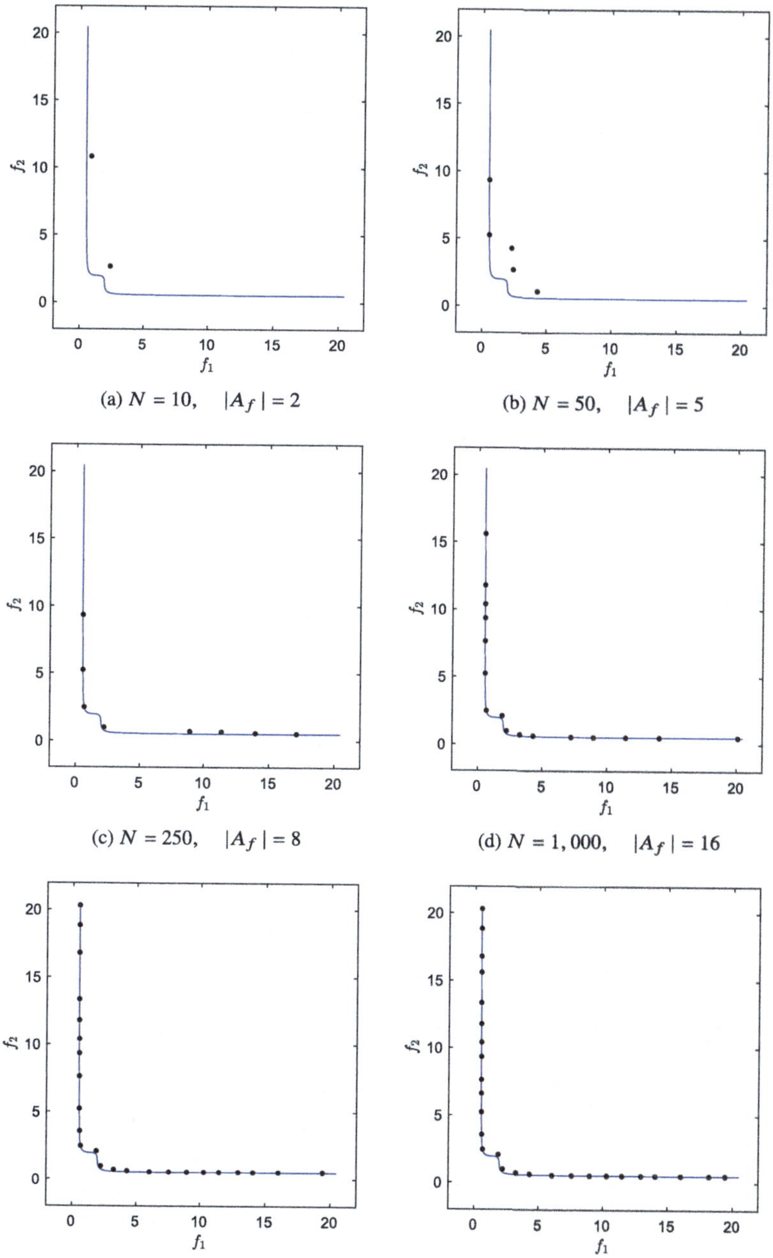

(a) $N = 10, \quad |A_f| = 2$

(b) $N = 50, \quad |A_f| = 5$

(c) $N = 250, \quad |A_f| = 8$

(d) $N = 1,000, \quad |A_f| = 16$

(e) $N = 5,000 \quad |A_f| = 24$

(f) $N = 10,000, \quad |A_f| = 27$

Fig. 7.5 Numerical results of *ArchiveUpdateTight*1 for DENT and $\epsilon = (\frac{1}{12}, \frac{1}{12})$ and $\Delta = 2$ for different numbers N of candidate solutions. A_f denotes in each case the final archive obtained by *ArchiveUpdateTight*1

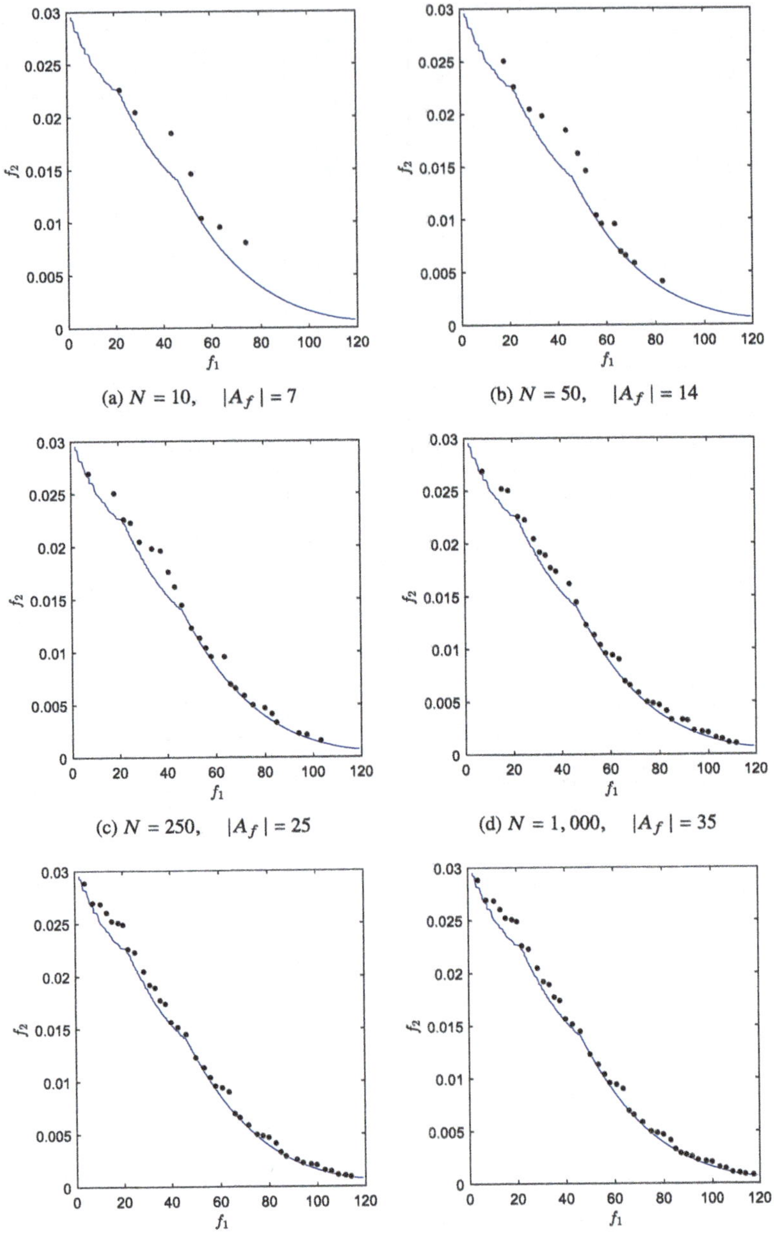

(a) $N = 10, \quad |A_f| = 7$

(b) $N = 50, \quad |A_f| = 14$

(c) $N = 250, \quad |A_f| = 25$

(d) $N = 1,000, \quad |A_f| = 35$

(e) $N = 5,000 \quad |A_f| = 41$

(f) $N = 10,000, \quad |A_f| = 43$

Fig. 7.6 Numerical results of $ArchiveUpdateTight1$ for SSW and $\epsilon = \frac{1}{3}(5, 0.005)$ and $\Delta = 1$ for different numbers N of candidate solutions. A_f denotes in each case the final archive obtained by $ArchiveUpdateTight1$

problem in Chap. 10. Another detail that is worth mentioning is the fact that in all cases the magnitudes of the final archives obtained by *ArchiveUpdateTight2* are even smaller than the respective ones obtained by *ArchiveUpdateTight1* while the (theoretical) upper bounds for both archivers indicates the opposite.

Finally, Fig. 7.8 shows the impact of the choices of ϵ and Δ on the final archiver, here for *ArchiveUpdateTight2* and problem DENT. Apparently, and as anticipated, the choices of the values do influence the final approximations. On the other hand, it is also clear that these choices are highly problem dependent. If all the entries of ϵ are in the same range, it makes sense to choose Δ also in this range since ϵ was used

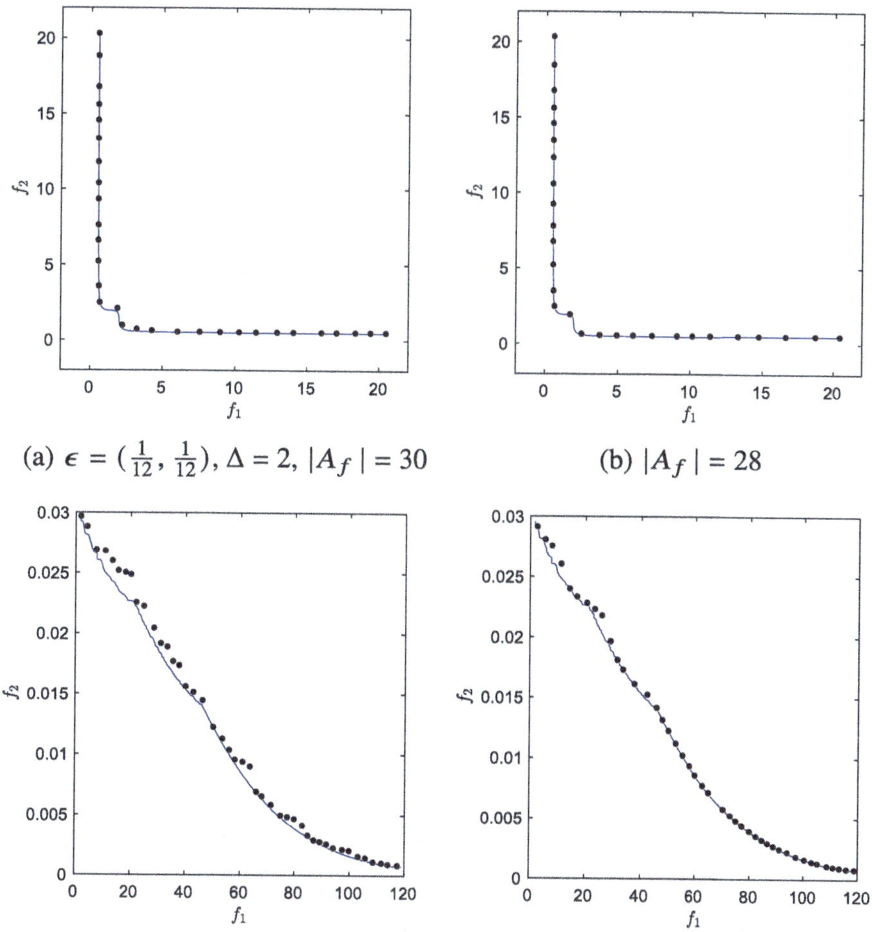

(a) $\epsilon = (\frac{1}{12}, \frac{1}{12})$, $\Delta = 2$, $|A_f| = 30$ (b) $|A_f| = 28$

(c) $\epsilon = \frac{1}{3}(5, 0.005)$, $\Delta = 4$, $|A_f| = 44$ (d) $|A_f| = 43$

Fig. 7.7 Comparison of the results of *ArchiveUpdateTight1* (left) and *ArchiveUpdateTight2* (right) on CONV2 and SSW ($N = 100{,}000$) and CONV3 and DTLZ7 ($N = 3{,}000{,}000$)

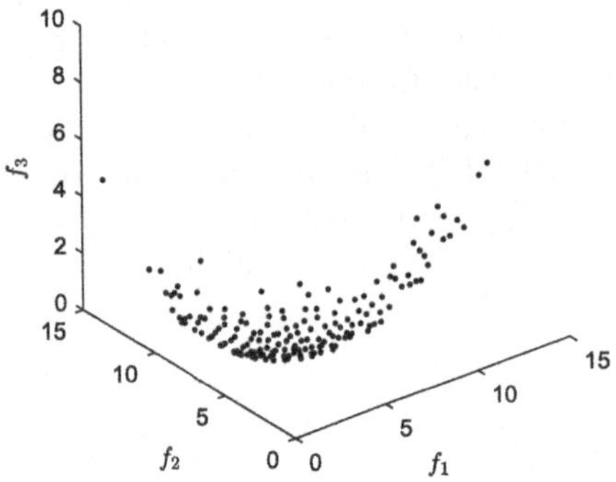

(e) $\epsilon = (\frac{1}{15}, \frac{1}{15}, \frac{1}{15}), \Delta = 0.15, |A_f| = 165$

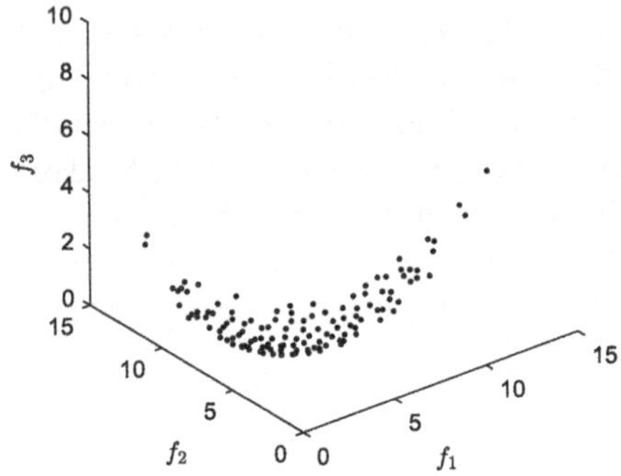

(f) $|A_f| = 131$

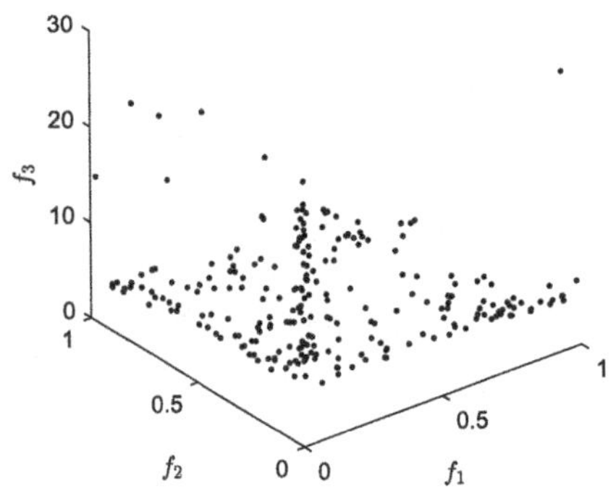

(g) $\epsilon = \frac{1}{3}(0.15, 0.15, 0.15), \quad |A_f| = 215$

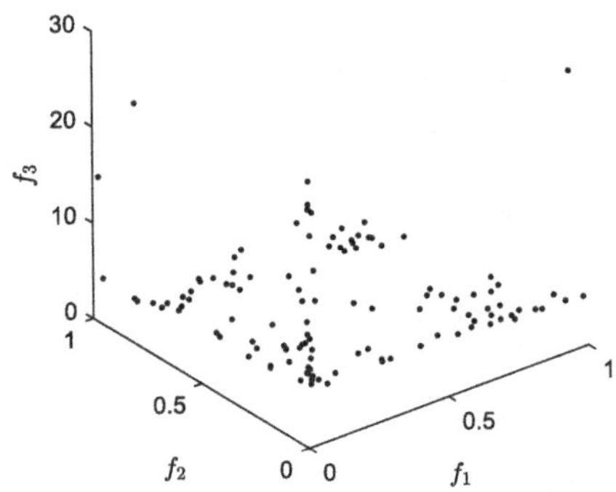

(h) $|A_f| = 114$

Fig. 7.7 (continued)

to specify the approximation quality. If the values of ϵ differ significantly (as for problem SSW) it entirely depends on preferences of the decision maker to choose a suitable value of Δ.

Next, we compare the behavior of all the five presented archiving strategies together on the test functions in order to see the particular strengths and weaknesses of all of them, and to finally give some guidance on which archiver might be suited best for a given situation.

Figures 7.9, 7.10, 7.11, 7.12 and 7.13 show the results on the archivers for the problems CONV2, DENT, SSW, CONV3, and DTLZ7, where we show the median

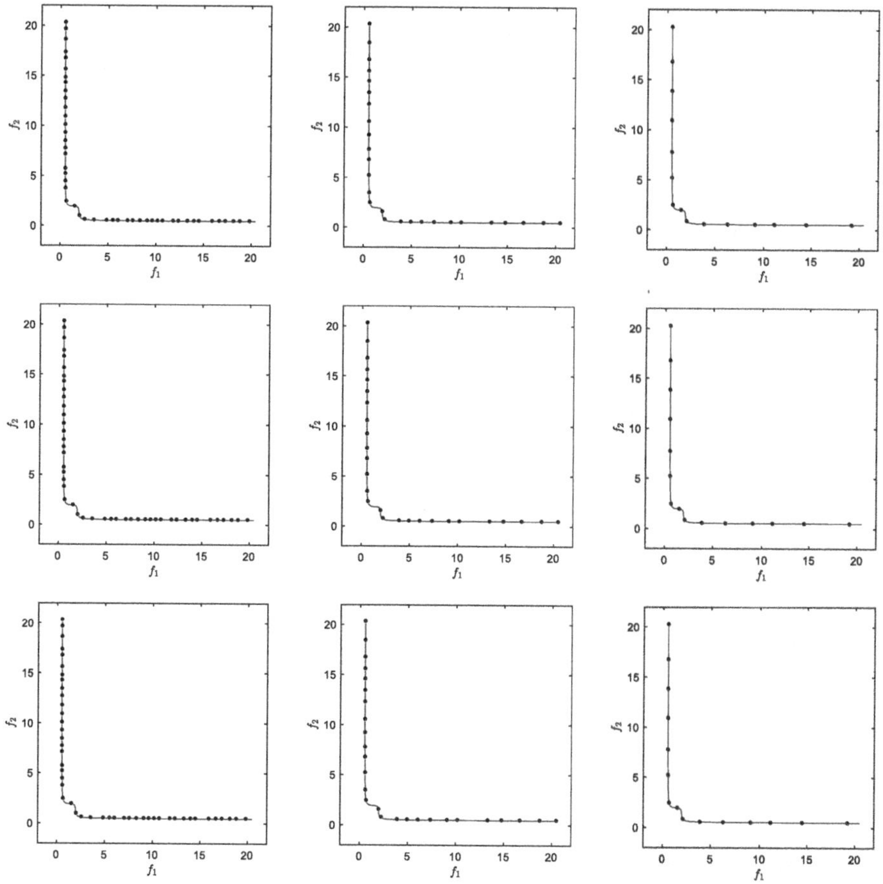

Fig. 7.8 Results of *ArchiveUpdateTight*2 on DENT for $\epsilon \in \{(0.5, 0.5), (1, 1), (2, 2)\}$ (top to bottom) and $\Delta \in \{0.5, 1, 2\}$ (left to right). All figures show the final archives using $N = 100{,}000$ candidate solutions

results out of 30 independent runs. To further elucidate the difference between the strategies, we show in Table 7.1 the computational times (in seconds) and the standard deviations for the different archivers and problems averaged over these 30 independent runs, and show in Fig. 7.14 the respective box plots. Similarly, in Table 7.2 and Fig. 7.15 we show the magnitudes of the final archives, and in Table 7.3 and Fig. 7.16 we show the approximation qualities in terms of the performance indicator Δ_2. Figure 7.17 finally shows convergence plots for all archivers for SSW and DTLZ7.

As anticipated, $ArchiveUpdateP_Q$ yields the best approximation qualities for all test problems. This, however, comes with the highest running times which results from the tremendous archive sizes which is the price one has to pay when all non-dominated solutions are stored. Apparently, all values would be even more extreme if

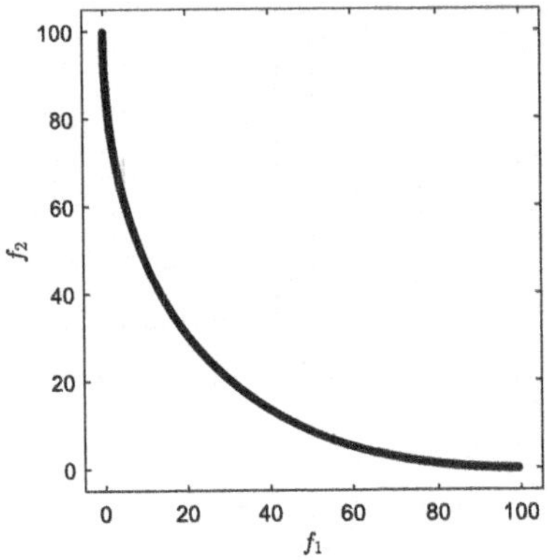

(a) $ArchiveUpdateP_Q$

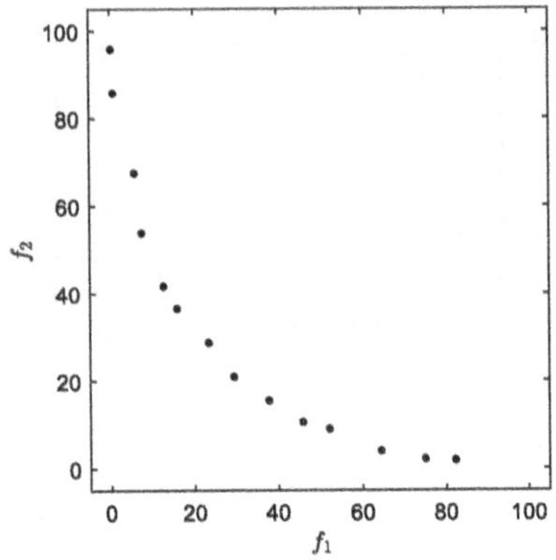

(b) $ArchiveUpdateEps1$

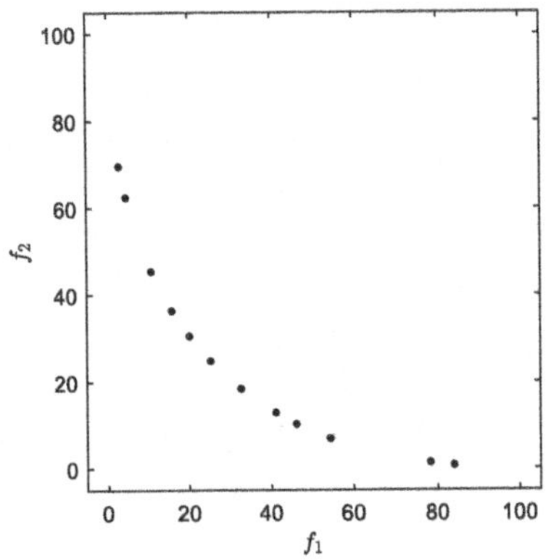

(c) $ArchiveUpdateEps2$

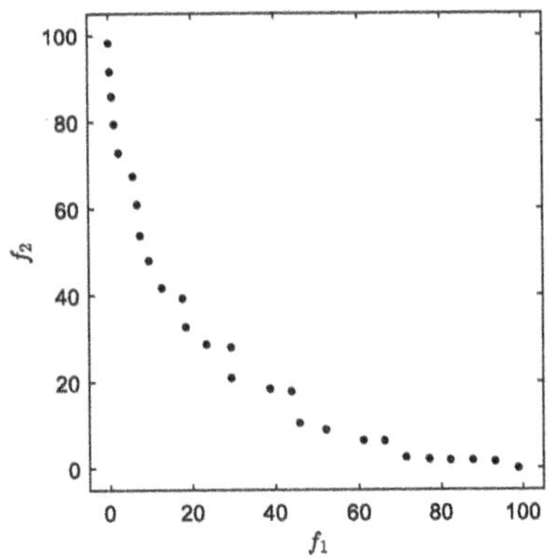

(d) $ArchiveUpdateTight1$

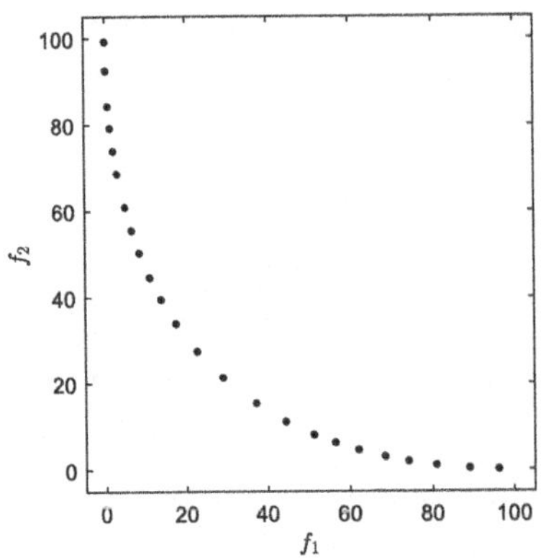

(e) $ArchiveUpdateTight2$

Fig. 7.9 Results of the five archivers on CONV2 for $\epsilon = (5, 5)$, $\Delta = 5$ and $N = 100{,}000$

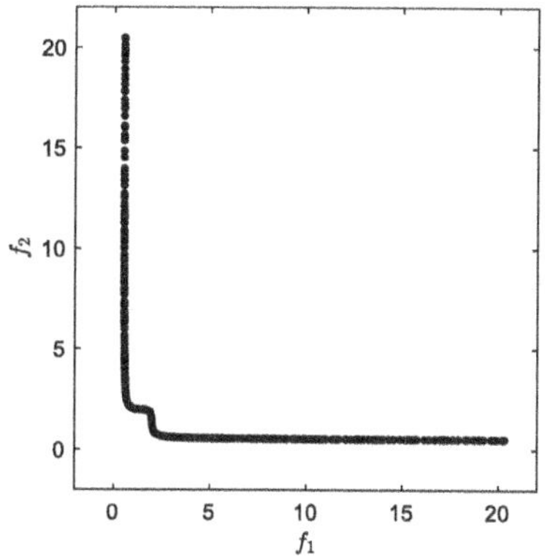

(a) $ArchiveUpdateP_Q$

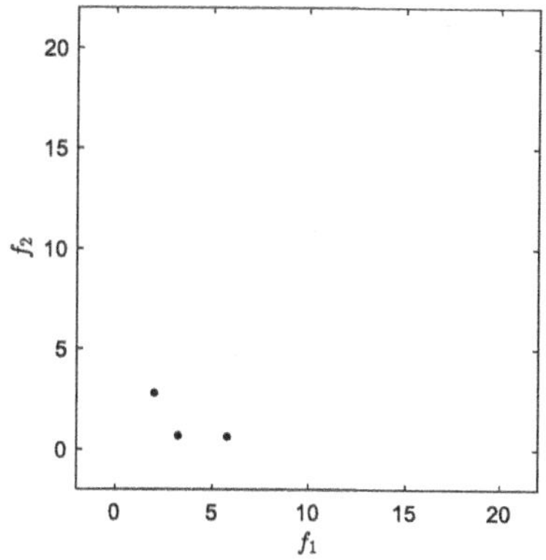

(b) $ArchiveUpdateEps1$

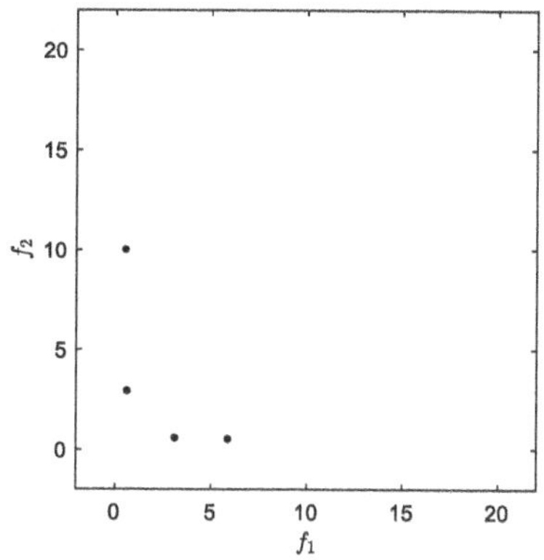

(c) $ArchiveUpdateEps2$

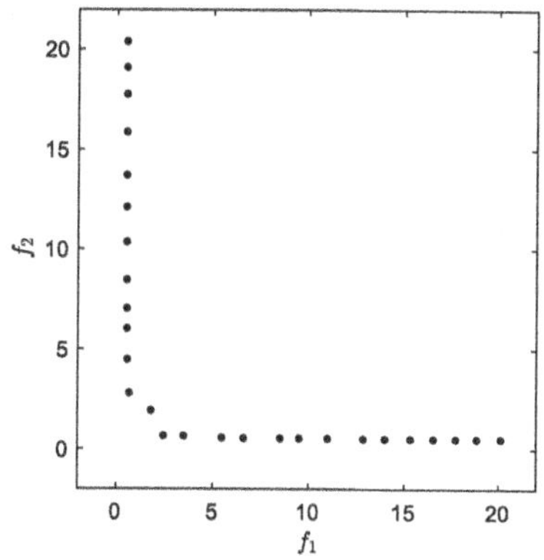

(d) $ArchiveUpdateTight1$

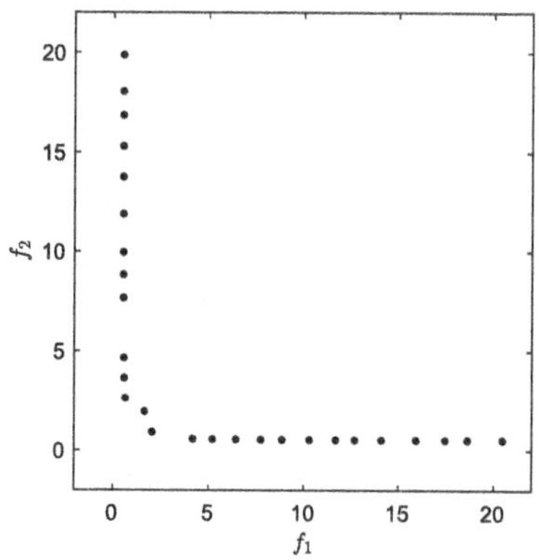

(e) $ArchiveUpdateTight2$

Fig. 7.10 Results of the five archivers on DENT for $\epsilon = (2, 2)$, $\Delta = 1$ and $N = 100{,}000$

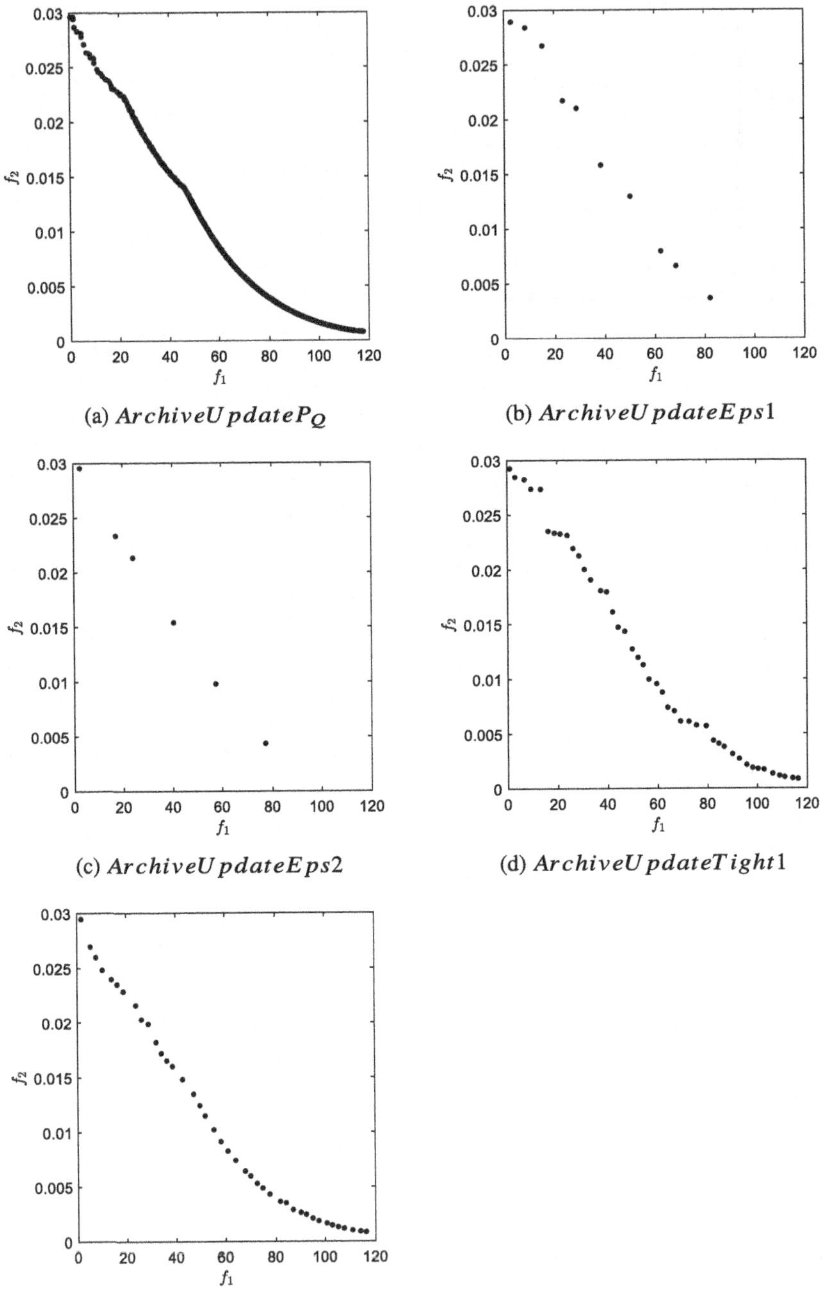

(a) $ArchiveUpdateP_Q$

(b) $ArchiveUpdateEps1$

(c) $ArchiveUpdateEps2$

(d) $ArchiveUpdateTight1$

(e) $ArchiveUpdateTight2$

Fig. 7.11 Results of the five archivers on SSW for $\epsilon = (5, 0.005)$, $\Delta = 2$ and $N = 100{,}000$

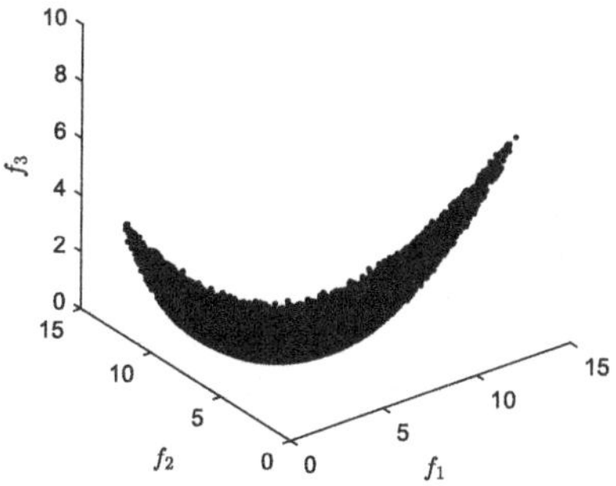

(a) $ArchiveUpdateP_Q$

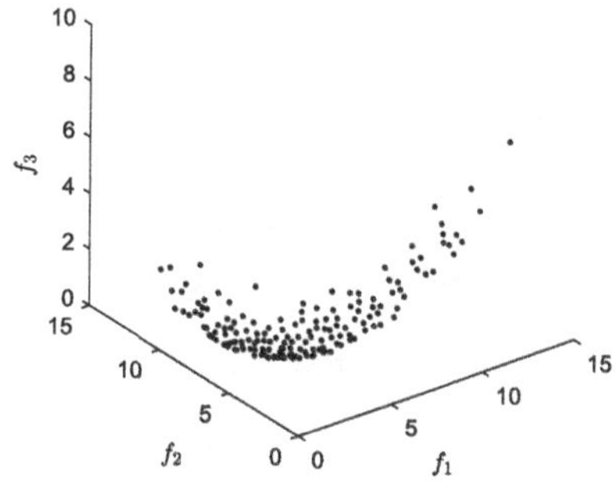

(b) $ArchiveUpdateEps1$

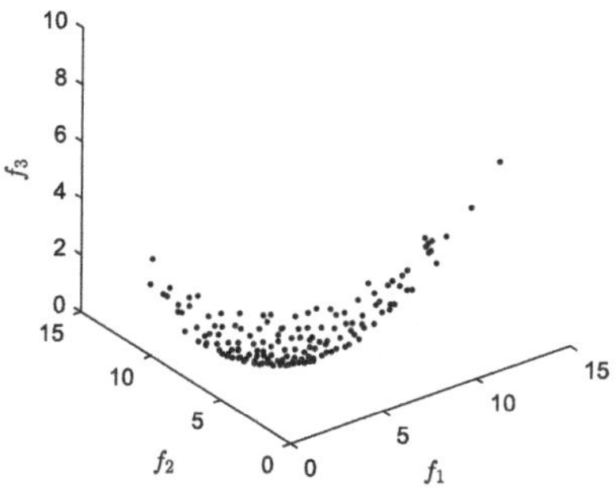

(c) $ArchiveUpdateEps2$

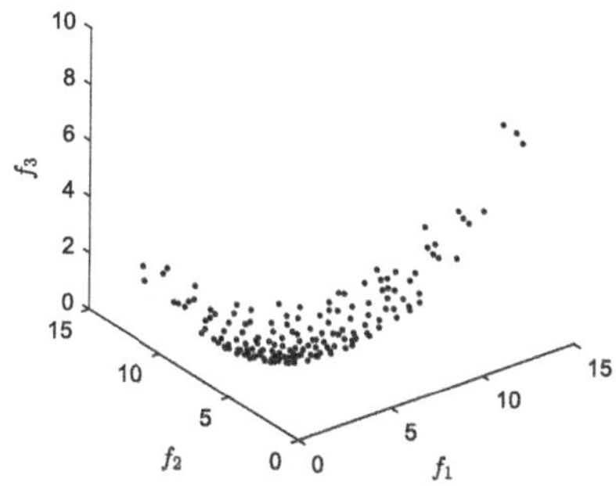

(d) $ArchiveUpdateTight1$

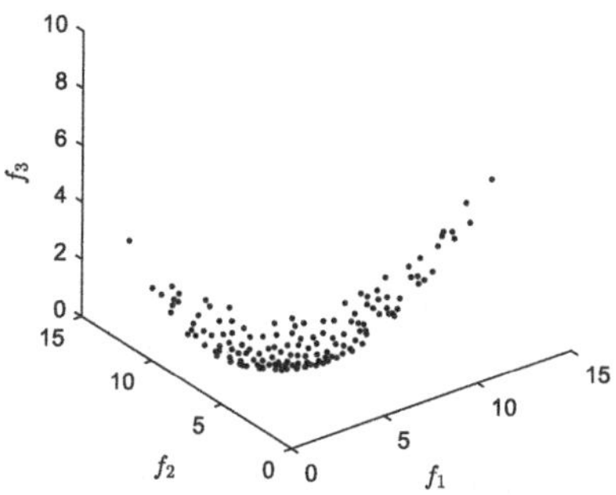

(e) $ArchiveUpdateTight2$

Fig. 7.12 Results of the five archivers on CONV3 for $\epsilon = (0.25, 0.25, 0, 25)$, $\Delta = 0.25$ and $N = 3{,}000{,}000$

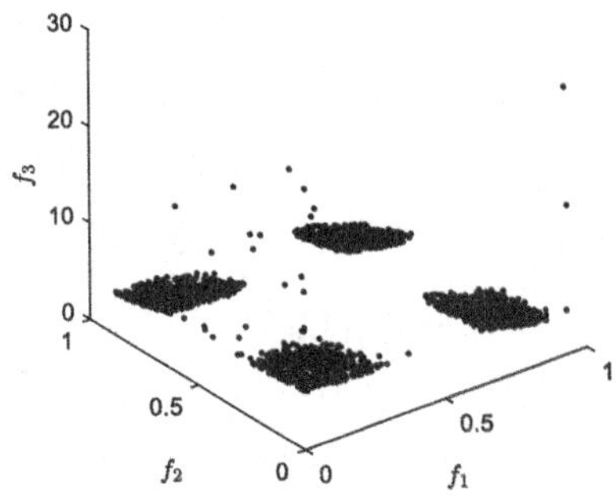

(a) $ArchiveUpdateP_Q$

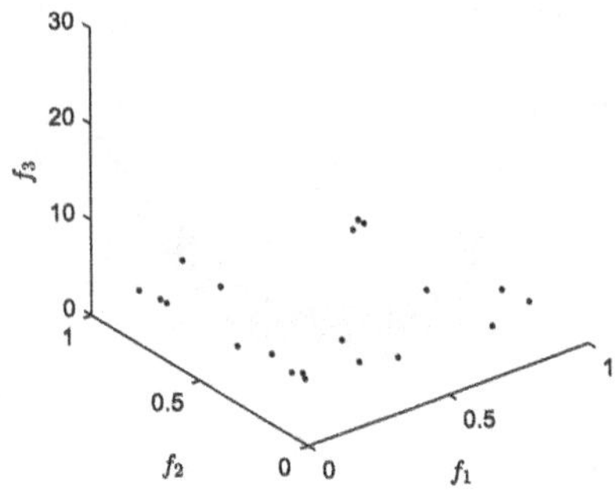

(b) $ArchiveUpdateEps1$

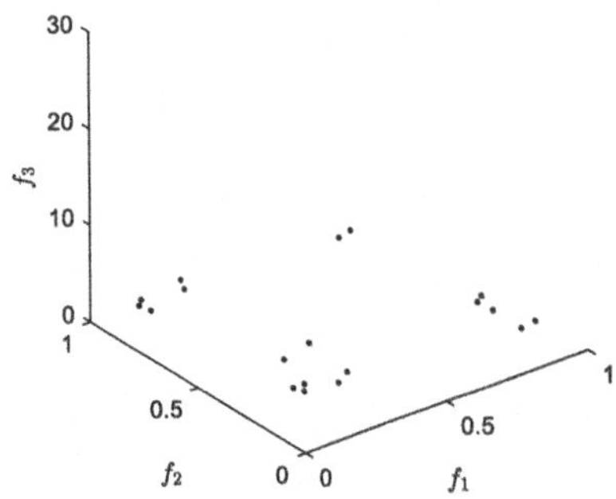

(c) $ArchiveUpdateEps2$

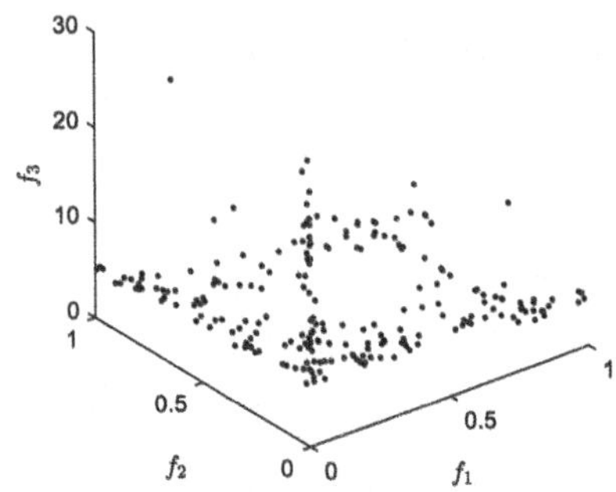

(d) $ArchiveUpdateTight1$

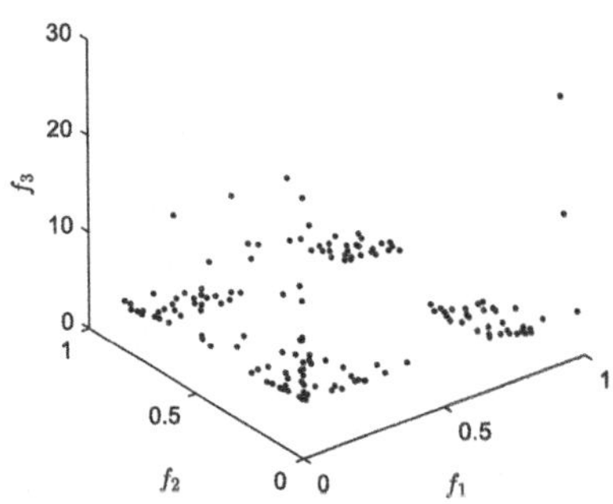

(e) $ArchiveUpdateTight2$

Fig. 7.13 Results of the five archivers on DTLZ7 for $\epsilon = (0.2, 0.2, 0, 2)$, $\Delta = 0.15$ and $N = 3,000,000$

Table 7.1 Computational times (in seconds) and standard deviations for the different archivers and problems averaged over 30 independent runs, see also Figs. 7.9, 7.10, 7.11, 7.12 and 7.13. For CONV2, SSW and DENT we have used $N = 100{,}000$, and for CONV3 and DTLZ $N = 3{,}000{,}000$ candidate solutions

Problem	P_Q	$Eps1$	$Eps2$	$Tight1$	$Tight2$
CONV2	39.0984(5.212)	1.4329(0.1889)	3.3474(0.6519)	1.7311(0.2208)	4.4797(0.5090)
DENT	13.4339(3.700)	0.9424(0.329)	1.8961(0.9987)	1.0426(0.3563)	3.2163(1.076)
SSW	17.8111(9.512)	1.0516(0.4465)	2.3169(0.9614)	1.4534(0.625)	3.4117(1.470)
CONV3	1962.31(155.093)	119.547(11.85)	282.884(148.44)	115.1945(10.425)	359.463(38.959)
DTLZ7	899.3545(192.403)	26.3836(10.345)	53.0184(14.020)	141.289(46.067)	407.114(152.095)

more candidate solutions would be chosen. This archiver is hence only a good choice if either the number N of candidate solutions is relatively low, and/or if the time for the computation of a function evaluation is relatively high. In that case, it makes sense to spend some more time for the archiving process, but in turn not to lose even one single non-dominated solution. If this is not the case, any of the other four archivers seems to be better suited. If an approximation of the Pareto front in the sense of ϵ-dominance is sufficient, either $ArchiveUpdateEps1$ or $ArchiveUpdateEps2$ can be taken. Looking at the approximation qualities reveals that $ArchiveUpdateEps1$ yields better Δ_2 values. $ArchiveUpdateEps2$ aims for Pareto optimal archive entries which results in the best GD value, which, however, results in a sacrifice in the IGD value that also negatively affects the Δ_2 value. Hence, if the entries of ϵ are large, then one may choose $ArchiveUpdateEps2$ in order to prevent that the archive entries are too far from the Pareto front, else $ArchiveUpdateEps1$ seems to be the better choice.

If approximations are desired that are ideally evenly spread along the Pareto front, any of the archivers $ArchiveUpdateTight1$ or $ArchiveUpdateTight2$ may be chosen. Note that the running times are only slightly larger as for the archivers that aim for ϵ-(approximate) Pareto fronts. The difference in the approximation quality between these two archivers does not differ which is due to the aim of both archivers to obtain a gap free approximation (which, depends, however, on the choice of Δ). Also here, $ArchiveUpdateTight2$ might be the better choice for large entries of ϵ, else both archivers can be taken. Note that Table 7.2 and Fig. 7.15 give some further numerical evidence that the archive sizes obtained by $ArchiveUpdateTight2$ are in general not higher than the ones for $ArchiveUpdateTight1$, even if this might be expected when looking at the (theoretical) upper bounds for both strategies. Both gap free archivers, however, may have some troubles to accurately approximate the borders of the different connected components of the Pareto front in case it is disconnected.

We summarize this comparison with some rough guidelines in which situation which of the archiving strategies may be suited best:

Table 7.2 Magnitudes of the final archives and standard deviations for the different archivers and problems averaged over 30 independent runs, see also Figs. 7.9, 7.10, 7.11, 7.12 and 7.13. For CONV2, SSW and DENT we have used $N = 100{,}000$, and for CONV3 and DTLZ $N = 3{,}000{,}000$ candidate solutions

Problem	P_Q	$Eps1$	$Eps2$	$Tight1$	$Tight2$
CONV2	2353.15(27.7171)	13.45(1.5381)	11.1(0.85224)	25.65(1.5985)	23.7(1.1743)
DENT	503.4(17.2517)	2.85(0.9333)	2.05(1.0501	29.1(1.2096)	27.45(1.3169)
SSW	743.4(12.9428)	10(1.4868)	7.4(1.1425)	43.6(1.0954)	42.25(2.2213)
CONV3	10804.65(77.034)	165.9(6.1379)	128.25(6.4797)	166.1(6.0254)	128.3(6.4326)
DTLZ7	2188.3(39.6141)	20.4(2.6833)	11(1.8353)	213.95(26.4664)	116.3(10.1416)

Table 7.3 Δ_2 values of the Pareto front approximations and standard deviations for the different archivers and problems averaged over 30 independent runs, see also Figs. 7.9, 7.10, 7.11, 7.12 and 7.13. For CONV2, SSW and DENT we have used $N = 100{,}000$, and for CONV3 and DTLZ $N = 3{,}000{,}000$ candidate solutions

Problem	P_Q	$Eps1$	$Eps2$	$Tight1$	$Tight2$
CONV2	0.04031(0.001)	5.7751(1.69)	6.6868(1.227)	2.5466(0.1687)	2.209(0.1362)
DENT	0.05139(0.005)	6.5011(1.156)	7.2067(1.459)	0.47112(0.043)	0.4557(0.032)
SSW	0.1155(0.021)	8.0664(3.002)	9.8299(3.526)	0.84194(0.032)	0.89215(0.075)
CONV3	0.03836(0.0002)	0.32161(0.013)	0.36693(0.02)	0.32161(0.014)	0.3669(0.02)
DTLZ7	0.08757(0.021)	0.48053(0.170)	0.8341(0.031)	0.36577(0.053)	0.3663(0.073)

- If the number of candidate solutions N considered during the algorithm and/or the computational time for a function evaluation is high, $ArchiveUpdateP_Q$ may be the best option. Otherwise,
- If approximations of the Pareto front in terms of ϵ-dominance are sufficient and the entries of ϵ are small or moderate, $ArchiveUpdateEps1$ may be the best option
- If approximations are preferred that evenly spread around the Pareto front, and the entries of ϵ are small or moderate, $ArchiveUpdateTight1$ or $ArchiveUpdateTight2$ may be used.
- If the entries of ϵ are large, then either $ArchiveUpdateEps2$ or $ArchiveUpdateTight2$ might be preferred.

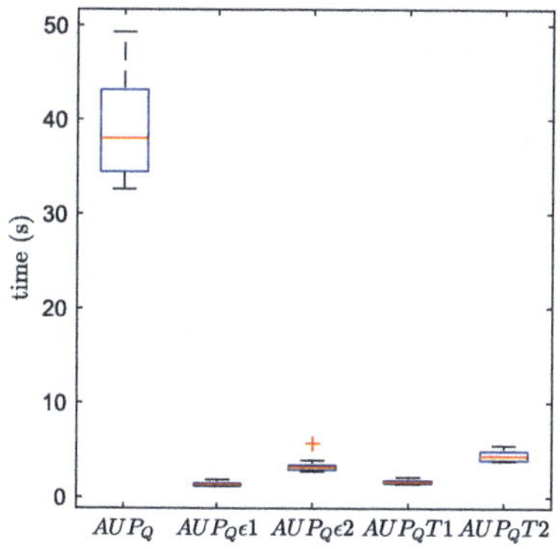

(a) $CONV2$

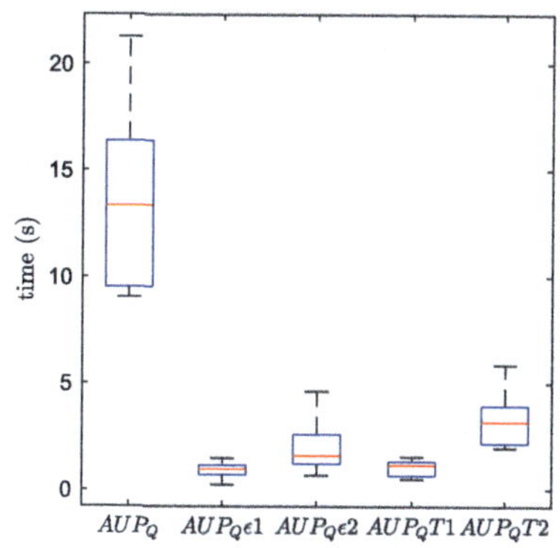

(b) $DENT$

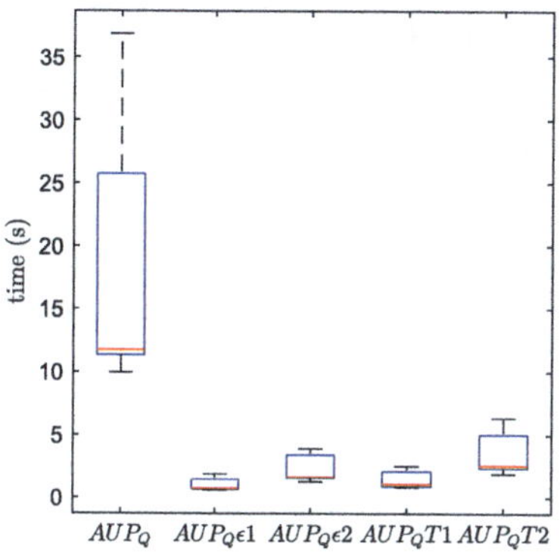

(c) SSW

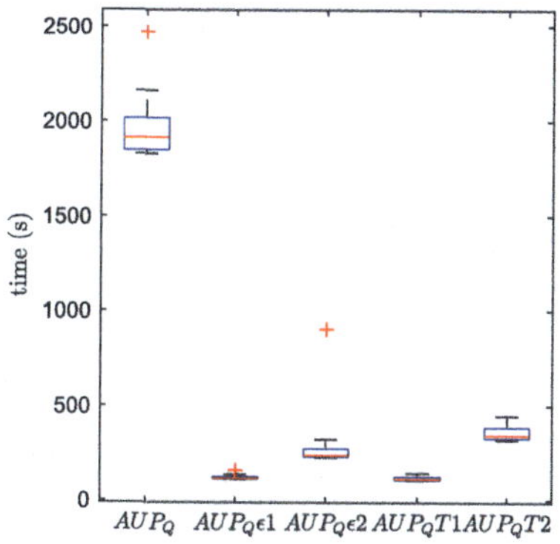

(d) $CONV3$

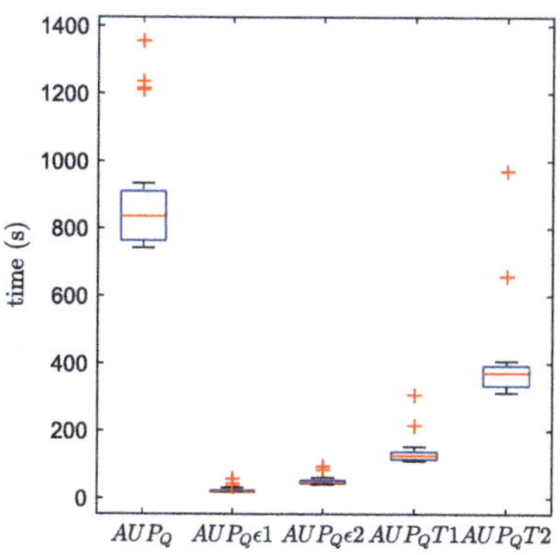

(e) $DTLZ7$

Fig. 7.14 Box plots for the running times of the different archivers for the different test problems, based on 30 independent runs

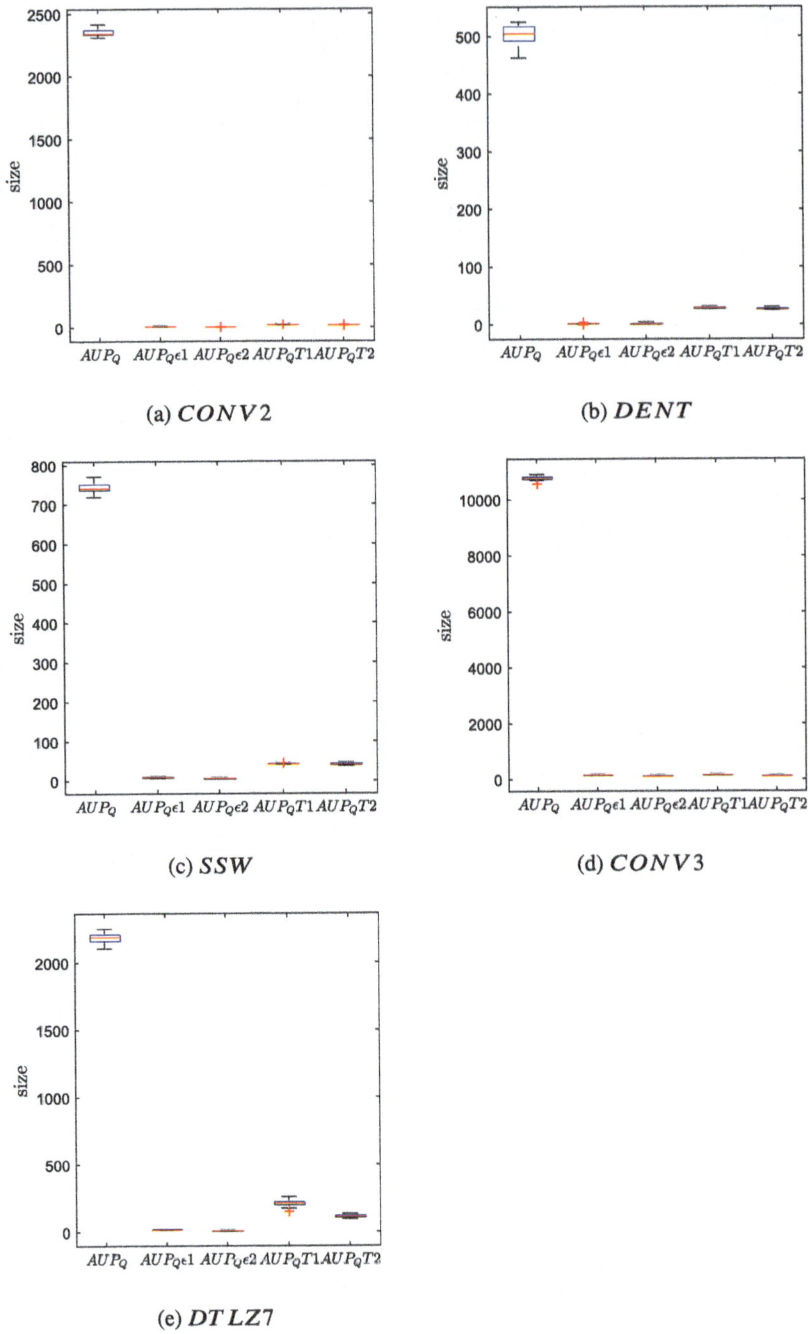

(a) $CONV2$ (b) $DENT$ (c) SSW (d) $CONV3$ (e) $DTLZ7$

Fig. 7.15 Box plots for the final archive sizes of the different archivers for the different test problems, based on 30 independent runs

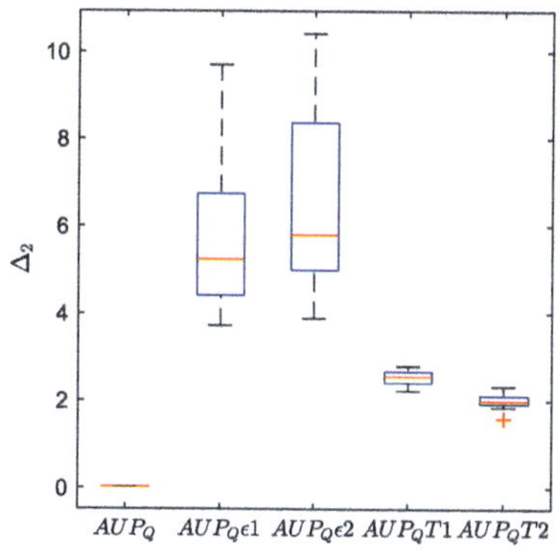

(a) $CONV2$

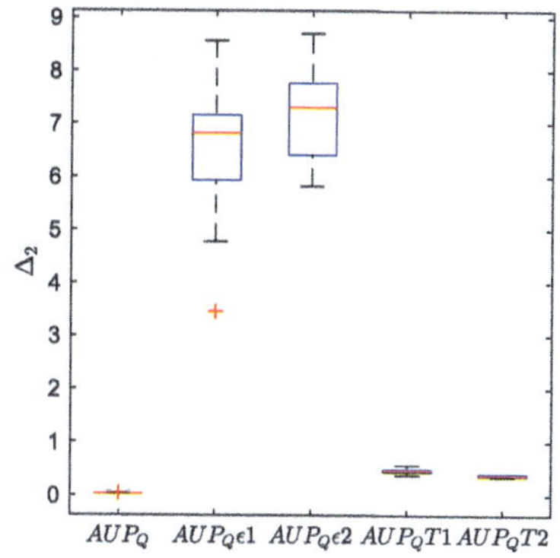

(b) $DENT$

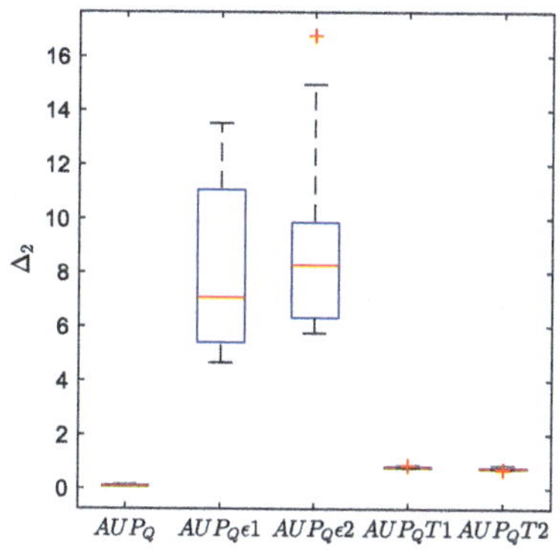

(c) SSW

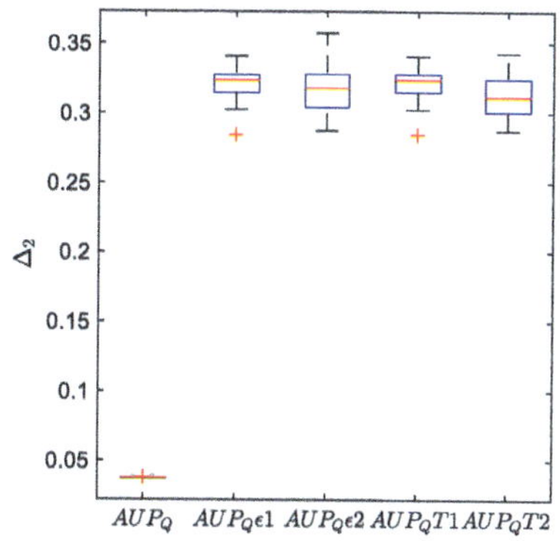

(d) $CONV3$

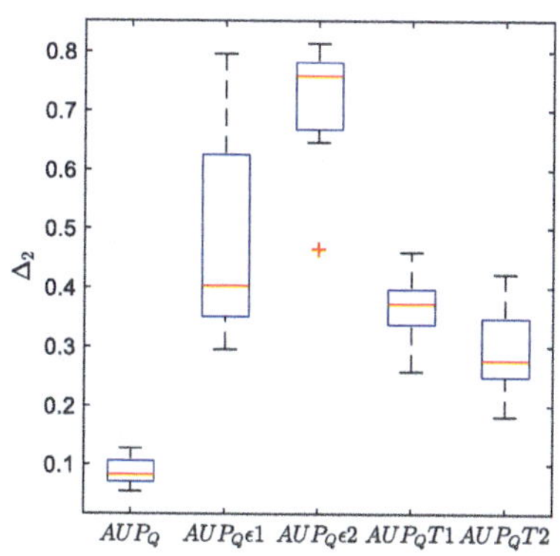

(e) $DTLZ7$

Fig. 7.16 Box plots for Pareto front approximations of the final archives for the different archivers for the different problems, based on 30 independent runs

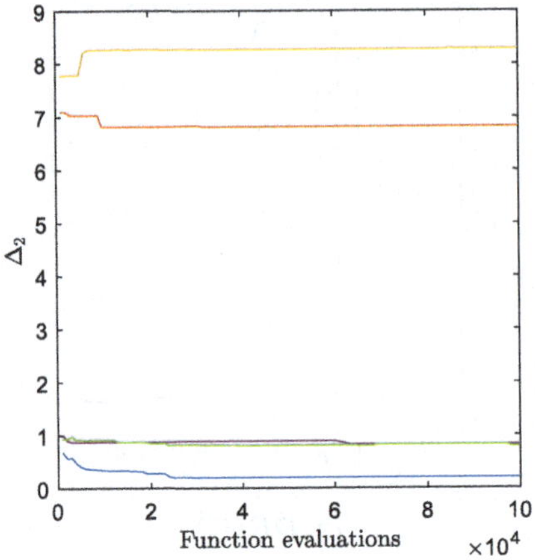

(a) *SSW*

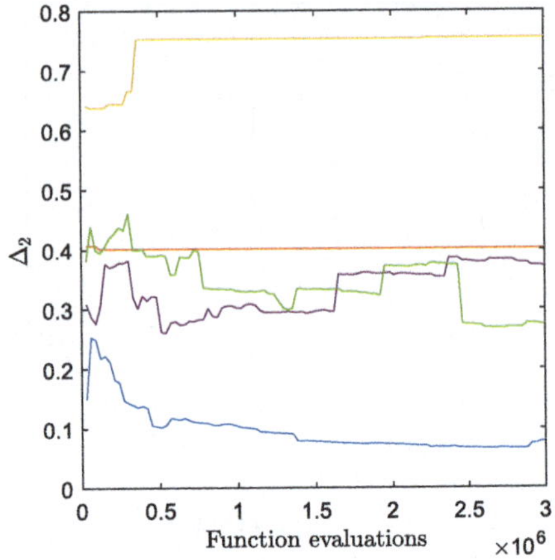

(b) *DTLZ7*

Fig. 7.17 Convergence plots for the problems SSW and DTLZ and the archivers $ArchiveUpdateP_Q$ (blue), $ArchiveUpdateEps1$ (orange), $ArchiveUpdateEps2$ (yellow), $ArchiveUpdateTight1$ (green), and $ArchiveUpdateTight2$ (purple), based on 30 independent runs

Chapter 8
Computing the Set of Approximate Solutions

In the previous chapters, we have investigated archivers that used the concept of ϵ-dominance to discretize the Pareto set/front of a given MOP. As a consequence, even the limit archives for certain archivers could contain solutions that are not optimal, but only nearly optimal. This fact, however, does not represent a problem at least from the practical point of view, as in many cases we are satisfied with such nearly optimal solutions. More precisely, for a given ϵ (which is of course again problem dependent) we may not mind if the performances of a given product (measured by the objectives f_i) differ by the entries of ϵ. As an example, consider the design of a vehicle where one objective, say f_1, is its maximal speed. In this case, we may not care if the maximal speed of this object differs by $\epsilon_1 = 0.5$ (km/h). That is, if x^* is vector in decision variable space for those f_1 takes its maximum, we will hence accept all vectors x with $f_1(x) - f_1(x^*) \leq \epsilon_1$.

So far, we have used nearly optimal (or ϵ-approximate) solutions to obtain a finite size representation of the Pareto sets/fronts. In this chapter, we will go one step further and investigate the approximation of the entire set of approximate solutions of a given multi-objective optimization problem. The reason for this is that this entire set—which we will call $P_{Q,\epsilon}$ in the sequel—might be of potential interest for the decision maker. One reason for this is that if two points x and y are similar in objective space, i.e., $F(x) \approx F(y)$, respectively $\|F(x) - F(y)\| \leq \epsilon$ for a given vector ϵ, this does not have to hold for the entries of x or y. In fact, x and y could be very distinct from each other. Hence, if both x and y are nearly optimal solutions, both could represent different realizations of a given project which practically yield the same performance. As an example for the latter, consider the bi-objective problem reported in Schütze et al. (2008e): this problem is related to (preliminary) space mission design, where the task is to identify "optimal" trajectories for a spacecraft from Earth to Mercury. The two objectives were the flight time and the fuel consumption. For this problem, the following two trajectories were identified: the first one, called x_1, was optimal and required 1002 days for the transfer, and needed a 46.2% of the available fuel budget.

O. Schütze and C. Hernández, *Archiving Strategies for Evolutionary Multi-objective Optimization Algorithms*, Studies in Computational Intelligence 938,
https://doi.org/10.1007/978-3-030-63773-6_8

The second one, called x_2, required 1005 days for the transfer and needed 46.3% of the fuel budget. Hence, x_2 is dominated by x_1 and thus not optimal. However, if one accepts a deterioration of 3 days of the flight time (in a trip that takes over two and a half years) and 0.1% difference of the fuel consumption, then the performance of x_2 can be seen nearly equal to the one of x_1. Further, it is important to note that x_1 and x_2 differ significantly in decision variable space. One of the design parameters of this BOP is the launch date, and this value differs by approximately 440 days for the two trajectories. Hence, via considering x_1 *and* x_2, the decision maker is given two possible and distinct realizations for his/her project.

8.1 The Set of Interest

Now will define the set of approximate solutions of a MOP, $P_{Q,\epsilon}$, in detail. The basis for this will be the notion of $-\epsilon$-dominance (Schütze et al. 2007a) which we will define in the following.

Definition 8.1 (*$-\epsilon$-dominance*) Let $\epsilon \in \mathbb{R}^k_+$ and $x, y \in \mathbb{R}^n$. x is said to $-\epsilon$-dominate y ($x \prec_{-\epsilon} y$) with respect to (MOP) if

$$F(x) + \epsilon \leq_p F(y) \quad \text{and} \quad F(x) + \epsilon \neq F(y). \tag{8.1}$$

Apparently, $-\epsilon$-dominance is identical to the notion of ϵ-dominance as defined earlier in this book with the crucial difference that we are now considering *negative* values. While ϵ-dominance can be considered to be a weaker concept of dominance, $-\epsilon$-dominance can, in turn, be considered to be a *stronger* concept. Figure 8.1 shows the regions of dominance for one point of a bi-objective problem for all three dominance relations which illustrates this fact.

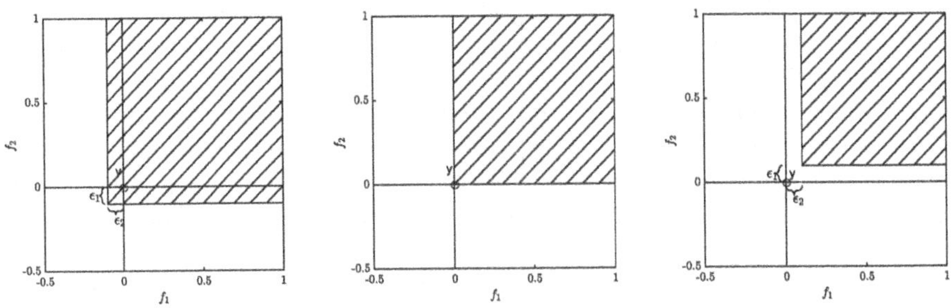

(a) ϵ - dominance (b) classical dominance (c) $-\epsilon$ - dominance

Fig. 8.1 Regions of dominance for the different dominance relations: Points those images lie in the shaded region are $(-)(\epsilon$-)dominated by points x with $F(x) = y$

Note that $-\epsilon$-dominance is unlike the classical dominance relation not transitive. In the following, we state two calculation rules for the new dominance relation that we will use later on.

Lemma 8.1 *Let (MOP) be given, and let $x, y, z \in \mathbb{R}^n$ and $\epsilon_1, \epsilon_2 \in \mathbb{R}^k_+$, then*

(a) If $x \prec_{-\epsilon_1} y$, then $x \prec_{-\epsilon_2} y, \quad \forall \epsilon_2 \leq_p \epsilon_1$
(b) If $x \prec_{-\epsilon_1} y$ and $y \prec_{-\epsilon_2} z$, then $x \prec_{-(\epsilon_1+\epsilon_2)} z$

Proof Follows directly by definition of the relation. □

Next, we define $-\epsilon$ weak Pareto points that are analogous to weak Pareto points and that will play an important role in the subsequent convergence analysis.

Definition 8.2 (*$-\epsilon$ weak Pareto point*) Let $\epsilon \in \mathbb{R}^k_+$. A point $x \in Q$ is called *$-\epsilon$ weak Pareto point* if there exists no point $y \in Q$ such that

$$F(y) + \epsilon <_p F(x). \tag{8.2}$$

Using $-\epsilon$, we can now define the set of interest. The "ideal" set of points that contain all approximate solutions of a MOP might be given by

$$P^c_{Q,\epsilon} := \{x \in Q | \exists p \in P_Q : x \prec_\epsilon p\}. \tag{8.3}$$

That is, every point $x \in P^c_{Q,\epsilon}$ is "close" (in terms of ϵ) to at least one efficient solution, measured in objective space. Note, however, that this set is not easy to catch since for its formulation Pareto optimal solutions are used. We will in the following consider a (slightly) enlarged set of interest, which we will call $P_{Q,\epsilon}$.

Definition 8.3 (*Set of approximate solutions $P_{Q,\epsilon}$*) Denote by $P_{Q,\epsilon}$ the set of points in $Q \subset \mathbb{R}^n$ which are not $-\epsilon$-dominated by any other point in Q, i.e.,

$$P_{Q,\epsilon} := \{x \in Q | \nexists y \in Q : y \prec_{-\epsilon} x\}. \tag{8.4}$$

Note that the definition of $P_{Q,\epsilon}$ is analog to the definition of the Pareto set P_Q with the difference that the new dominance relation has been used. One important difference of P_Q and $P_{Q,\epsilon}$, however, is that the latter typically forms an n-dimensional set, where P_Q can be expected to be of dimension $(k-1)$. To see this, let $x_0 \in P_Q$ be given. Hence, by definition there exists no $y \in Q$ such that $y \prec x_0$. Since F is continuous and $\epsilon \in \mathbb{R}^k_+$ there exists a neighborhood U of x_0 such that

$$\nexists y \in Q : \; y \prec_{-\epsilon} u \quad \forall u \in U \cap Q, \tag{8.5}$$

and thus, $U \cap Q \subset P_{Q,\epsilon}$, and we are done since U is n-dimensional.

Remark 8.1 The sets $P^c_{Q,\epsilon}$ and $P_{Q,\epsilon}$ defined above are closely related to the sets E^1 and E^5 considered in White (1986). The only difference is a (slightly) alternative use of the ϵ-dominance relation.

Example 8.1 (a) For the special case of SOPs (i.e., $k = 1$) with objective function f and $\epsilon > 0$, $-\epsilon$-dominance reads as

$$x \prec_{-\epsilon} y \quad \Leftrightarrow \quad f(x) + \epsilon < f(y). \tag{8.6}$$

For instance, when choosing $f : Q \to \mathbb{R}$

$$f(x) = x^2, \tag{8.7}$$

where $Q \supset [-1, 1]$ and $\epsilon = 1$, we obtain (see also Fig. 8.2)

$$P_{Q,\epsilon} = [-1, 1]. \tag{8.8}$$

(b) Figure 8.3 shows another SOP for which the set of approximate solutions is given by $P_{Q,\epsilon} = [a, b] \cup [c, d]$, i.e., by two disconnected components.
(c) Consider the unconstrained BOP $F : \mathbb{R}^2 \to \mathbb{R}^2$, where

$$F(x) = \begin{pmatrix} (x_1 - 1)^2 + (x_2 - 1)^2 \\ (x_1 + 1)^2 + (x_2 + 1)^2 \end{pmatrix}. \tag{8.9}$$

Figure 8.4 shows the sets P_Q and $P_{Q,\epsilon}$ as well as their images for $\epsilon = (1, 1)^T$.

To further understand the set of interest $P_{Q,\epsilon}$ we will in the following study its topology. We are in particularly interested in robust sets that will be required for the upcoming convergence analysis. Further, we show that $P_{Q,\epsilon}$ is connected for convex problems.

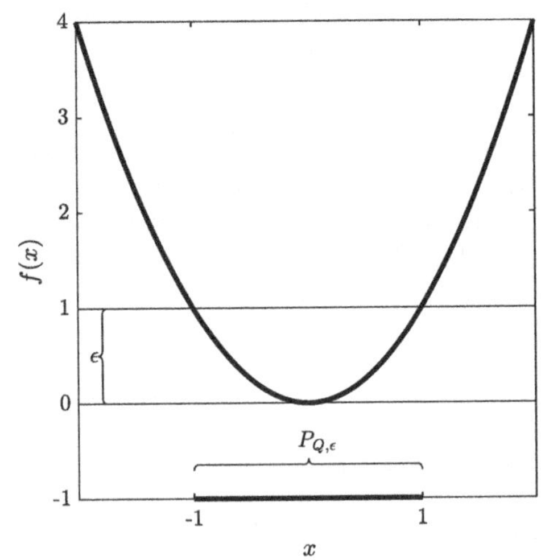

Fig. 8.2 The set of approximate solutions for $f(x) = x^2$ and $\epsilon = 1$ is given by $P_{Q,\epsilon} = [-1, 1]$

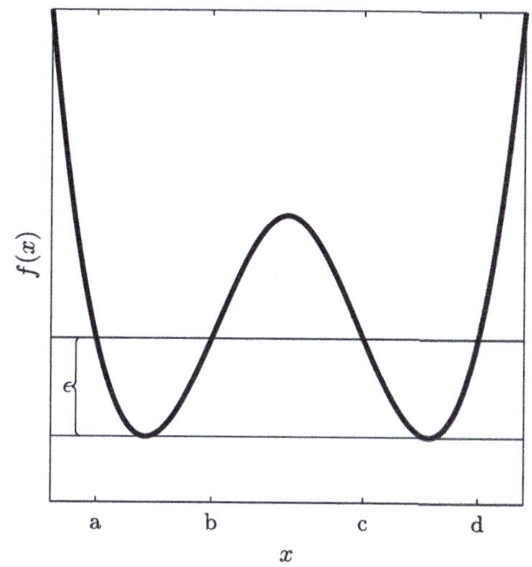

Fig. 8.3 Example of a SOP where the set of approximate solutions is $P_{Q,\epsilon} = [a, b] \cup [c, d]$, i.e., disconnected

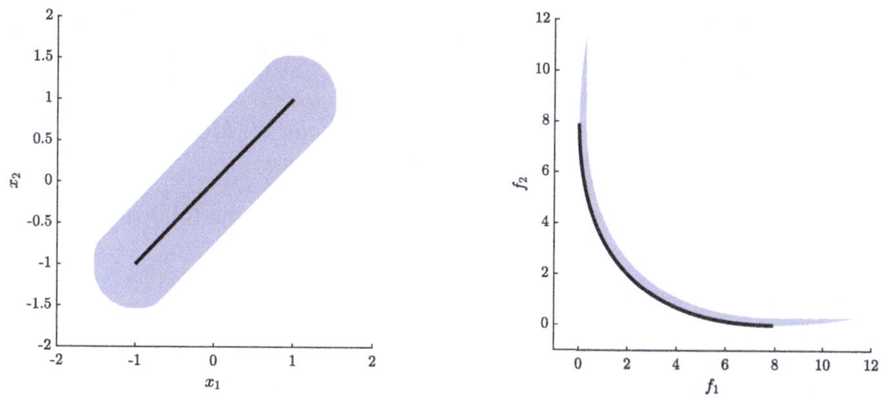

(a) decision space

(b) objective space

Fig. 8.4 Left: Pareto set (black) and approximate solutions $P_{Q,\epsilon}$ (purple) for $\epsilon = (1, 1)^T$ and MOP (8.9). Right: Pareto front (black) and $F(P_{Q,\epsilon})$ (purple)

We first recall some notations from topology. A point $x_0 \in A \subset \mathbb{R}^n$ is called an *interior point* in A if there exists a value $\delta > 0$ such that the δ-ball around x_0, $B_\epsilon(x_0)$, lies entirely in A. A point $x_0 \in A$ is called a *boundary point* of A if any δ-ball $B_\delta(x_0)$ has non-empty intersections with both A and its complement $\mathbb{R}^n \backslash A$. The set of interior points in A constitutes its *interior*, denoted by $int(A)$, and the set of boundary points its *boundary*, denoted by ∂A. A is said to be *open* if any point in A is an interior point and it is *closed* if its boundary ∂A is contained in A. The *closure* of

A, denoted by $clos(A)$, is the union of A and its boundary, i.e., $clos(A) = A \cup \partial A$. A set $A \subset \mathbb{R}^n$ is called *robust* if

$$clos(int(A)) = clos(A), \tag{8.10}$$

which means that every boundary point of A can be approached from the interior. A set $S \subset \mathbb{R}^n$ is *disconnected* if there exist open sets O_1, O_2 such that $S \subset O_1 \cup O_2$, $S \cap O_1 \neq \emptyset$, $S \cap O_2 \neq \emptyset$, and $S \cap O_1 \cap O_2 = \emptyset$. Otherwise, S is *connected.*

We will also need the following results, which yields a descent direction for unconstrained problems, in case all objectives are continuously differentiable.

Theorem 8.1 (Schaeffler et al. 2002) *Let (MOP) be given and* $q : \mathbb{R}^n \to \mathbb{R}^n$ *be defined by*

$$q(x) = \sum_{i=1}^{k} \hat{\alpha}_i \nabla f_i(x), \tag{8.11}$$

where $\hat{\alpha}$ *is a solution of*

$$\min_{\alpha \in \mathbb{R}^k} \left\{ \left\| \sum_{i=1}^{k} \alpha_i \nabla f_i(x) \right\|_2^2 ; \alpha_i \geq 0, i = 1, \ldots, k, \sum_{i=1}^{k} \alpha_i = 1 \right\}. \tag{8.12}$$

Then it is either $q(x) = 0$*, and hence,* x *a KKT point of (MOP), or* $-q(x)$ *is a descent direction for all objective functions* $f_1, \ldots, f_k$ *in* x*.*

For the subsequent considerations we will have to make the following assumptions:

(A1) let there be no weak Pareto point in $Q \backslash P_Q$,
(A2) let there be no $-\epsilon$ weak Pareto point in $Q \backslash clos(P_{Q,\epsilon})$, and
(A3) let $\mathcal{B} \subset int(Q)$ and $q(x) \neq 0$ for all $x \in \mathcal{B}$, where q is as defined in Theorem 8.1 and $\mathcal{B} := \{x \in Q | \exists y \in P_Q : F(y) + \epsilon = F(x)\}$.
(A4) let $q(x) \neq 0 \; \forall x \in \partial P_{Q,\epsilon}$, where q is as defined in Theorem 8.1.

In the following, we express the closure, the interior and the boundary of $P_{Q,\epsilon}$.

Theorem 8.2 *Let (MOP) be given, where the domain* $Q \subset \mathbb{R}^n$ *is compact. Under the assumptions (A1)–(A3) it holds:*

$$\begin{aligned} clos(P_{Q,\epsilon}) &= \{x \in Q | \nexists y \in Q : F(y) + \epsilon <_p F(x)\}, \\ int(P_{Q,\epsilon}) &= \{x \in Q | \nexists y \in Q : F(y) + \epsilon \leq_p F(x)\}, \\ \partial P_{Q,\epsilon} &= \{x \in Q | \exists y_1 \in P_Q : F(y_1) + \epsilon \leq_p F(x) \; \wedge \nexists y_2 \in Q : F(y_2) + \epsilon <_p F(x)\}. \end{aligned} \tag{8.13}$$

Proof Define $W := \{x \in Q | \nexists y \in Q : F(y) + \epsilon <_p F(x)\}$. We show $clos(P_{Q,\epsilon}) = W$ by mutual inclusion. $W \subset clos(P_{Q,\epsilon})$ follows by (A2). For the other direction assume that there exists $x \in clos(P_{Q,\epsilon}) \backslash W$. Since $x \notin W$ there exists $y \in Q$ with

$F(y) + \epsilon <_p F(x)$. By continuity of F there exists a neighborhood U of x such that $F(y) + \epsilon <_p F(u), \quad \forall u \in U$. Thus, y is $-\epsilon$-dominating all $u \in U$ (i.e., $U \cap P_{Q,\epsilon} = \emptyset$), a contradiction to the assumption that $x \in clos(P_{Q,\epsilon})$, and we obtain $clos(P_{Q,\epsilon}) = W$ as claimed. To see that the interior of $P_{Q,\epsilon}$ is equal to

$$I := \{x \in Q | \nexists y \in Q : F(y) + \epsilon \leq_p F(x)\}, \tag{8.14}$$

we use again mutual inclusion. For $int(P_{Q,\epsilon}) \subset I$ assume there exists $x \in int(P_{Q,\epsilon}) \backslash I$. Since $x \notin I$

$$\exists y_1 \in Q : \quad F(y_1) + \epsilon \leq_p F(x). \tag{8.15}$$

Since $x \in int(P_{Q,\epsilon})$ there exists no $y \in Q$ that $-\epsilon$-dominates x. Thus, equality holds in Eq. (8.15). Further, by (A1) it follows that $y_1 \in P_Q$, and we can state (8.15) as

$$\exists y_1 \in P_Q : \quad F(y_1) + \epsilon = F(x). \tag{8.16}$$

Since $x \in int(P_{Q,\epsilon})$ there exists a neighborhood $\tilde{U}$ of x with $\tilde{U} \subset int(P_{Q,\epsilon})$. Next, since by (A1) $q(x) \neq 0$, there exists $\tilde{x} \in \tilde{U}$ with $F(\tilde{x}) >_p F(x)$. This together with (8.16) leads to

$$F(y_1) + \epsilon = F(x) <_p F(\tilde{x}), \tag{8.17}$$

and hence, $y_1 \prec_{-\epsilon} \tilde{x} \in \tilde{U} \subset int(P_{Q,\epsilon})$, a contradiction. It remains to show that $I \subset int(P_{Q,\epsilon})$: assume there exists $x \in I \backslash int(P_{Q,\epsilon})$. Since $x \notin int(P_{Q,\epsilon})$ there exists a sequence $x_i \in Q \backslash P_{Q,\epsilon}, i \in \mathbb{N}$, with $\lim_{i \to \infty} x_i = x$. Thus, there exists a sequence $y_i \in Q$ with $y_i \prec_{-\epsilon} x_i$ for all $i \in \mathbb{N}$. Since $y_i \subset Q$, $i \in \mathbb{N}$, and Q is bounded, there exists a subsequence y_{i_j}, $j \in \mathbb{N}$, and $y \in Q$ with $\lim_{j \to \infty} y_{i_j} = y$ (Bolzano-Weierstrass). Since $F(y_{i_j}) + \epsilon \leq_p F(x_{i_j}), \quad \forall j \in \mathbb{N}$, it follows for the limit points that also $F(y) + \epsilon \leq_p F(x)$, which is a contradiction to $x \in I$. Thus, the interior of $P_{Q,\epsilon}$ is equal to I as desired. For the boundary we obtain

$$\partial P_{Q,\epsilon} = \{x \in Q | \exists y_1 \in Q : F(y_1) + \epsilon \leq_p F(x) \text{ and } \nexists y_2 \in Q : F(y_2) + \epsilon <_p F(x)\} \tag{8.18}$$

Since by (A1) the point y_1 in (8.18) must be in P_Q, we obtain

$$\partial P_{Q,\epsilon} = \{x \in Q | \exists y_1 \in P_Q : F(y_1) + \epsilon \leq_p F(x) \text{ and } \nexists y_2 \in Q : F(y_2) + \epsilon <_p F(x)\}, \tag{8.19}$$

and we are done. □

From the result it follows that (unless for $n = 1$ decision variable) $P_{Q,\epsilon}$ is generally neither a closed nor an open set. Further, the closure of $P_{Q,\epsilon}$ contains the $-\epsilon$ weak Pareto points.

Next, we can use Eq. (8.19) to characterize the boundary of our set of interest: since there exists an index $j \in \{1, \ldots, k\}$ with the property

$$f_j(y_1) + \epsilon_j = f_j(x),$$

points on the boundary are the $-\epsilon$ weak Pareto points that are bounded away from $F(P_Q)$ by ϵ.

We now further investigate the relation between the sets $P^c_{Q,\epsilon}$ and $P_{Q,\epsilon}$. In general, $P^c_{Q,\epsilon}$ is not a subset of $P_{Q,\epsilon}$ (see White 1986 for such an example which also holds for the notion of $-\epsilon$-dominance we are using here), however, it is a subset of the interior of $P_{Q,\epsilon}$.

Lemma 8.2 $P^c_{Q,\epsilon} \subset int(P_{Q,\epsilon})$

Proof Assume there exists $x \in P^c_{Q,\epsilon} \backslash int(P_{Q,\epsilon})$. Since $x \in P^c_{Q,\epsilon}$ there exists $p \in P_Q$ with $x \prec_\epsilon p$. Since $x \notin int(P_{Q,\epsilon})$ there exists by Theorem 8.2 $y \in Q$ with $F(y) + \epsilon \leq_p F(x)$. Thus,

$$\begin{aligned} F(y) \leq_p F(x) - \epsilon \leq F(p), \quad \text{and} \\ \exists j \in \{1, \ldots, k\} : f_j(y) \leq f_j(x) - \epsilon < f_j(p) \quad (\Rightarrow \; F(y) \neq F(p)), \end{aligned} \tag{8.20}$$

i.e., that $y \prec p$. This is a contradiction to $p \in P_Q$, and we are done. □

Example 8.2 (a) Figure 8.5 shows the different sets for the two-dimensional bi-objective problem OKA1 (Okabe et al. 2004):

$$\begin{aligned} F &: Q \subset \mathbb{R}^2 \to \mathbb{R}^2 \\ f_1(x) &= xp_1(x) \\ f_2(x) &= \sqrt{2\pi} - \sqrt{abs(xp_1(x))} + 2\sqrt{abs(xp_2(x)) - 3\cos xp_1(x) - 3} \end{aligned} \tag{8.21}$$

where

$$\begin{aligned} xp1(x) &= \cos(\pi/12)x_1 - \sin(\pi/12)x_2, \\ xp2(x) &= \sin(\pi/12)x_1 + \cos(\pi/12)x_2, \\ Q &= [6\sin(\pi/12), 6\sin(\pi/12) + 2\pi\cos(\pi/12)] \times [-2\pi\sin(\pi/12), 6\cos(\pi/12)] \end{aligned} \tag{8.22}$$

In the left figure, black marks the Pareto set, purple marks the points that are nearly optimal according to $P^c_{Q,\epsilon}$ (i.e., $P^c_{Q,\epsilon}$ is the union of the black and the purple region), and blue marks the points that are included in $P_{Q,\epsilon}$ but not in $P^c_{Q,\epsilon}$. In the right figure, the respective sets are shown in image space using the same colors.

(b) Consider the MOP $F : \mathbb{R} \to \mathbb{R}^2$, where

$$F(x) = \begin{pmatrix} (x-1)^2 \\ (x+1)^2 \end{pmatrix}. \tag{8.23}$$

Further, let $\epsilon = (1, 1)^T$ and the domain be given by $Q = [-3, 3]$ (see Fig. 8.6). For this problem we have

$$\begin{aligned} P_Q &= [-1, 1] \\ P_{Q,\epsilon} &= (-2, 2). \end{aligned} \tag{8.24}$$

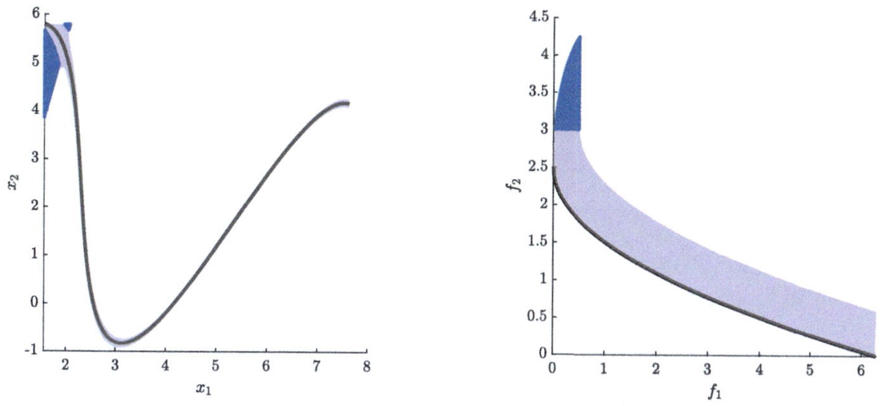

(a) decision space (b) objective space

Fig. 8.5 Left: Pareto set (black) and approximate solutions $P_{Q,\epsilon}$ (blue) for $\epsilon = (1, 1)^T$ and MOP (8.21). Right: Pareto front (black) and $F(P_{Q,\epsilon})$ (blue)

Fig. 8.6 Pareto set and set of approximate solutions for MOP (8.23)

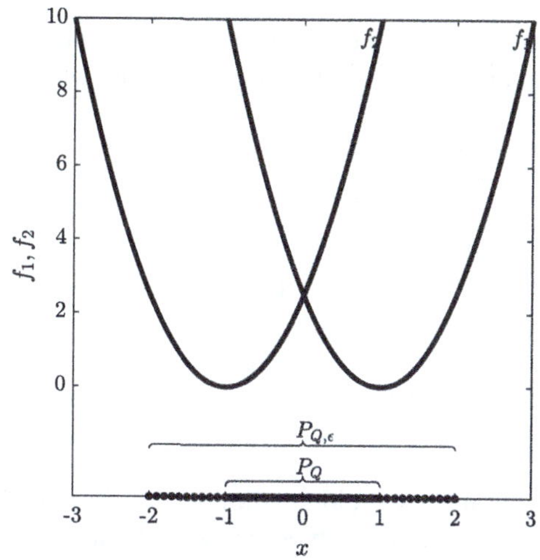

The boundary of $P_{Q,\epsilon}$ is given by $\{-2, 2\}$. Both points are $-\epsilon$-dominated since $F(-2) = (9, 1)$, $F(2) = (1, 9)$, $F(-1) = (4, 0)$ and $F(1) = (0, 4)$ and hence not included in $P_{Q,\epsilon}$. Both points are $-\epsilon$ weak Pareto points.

(c) Figure 8.7 shows an example for $k = 1$ where $P_{Q,\epsilon} = \{x^*\} \cup [c, d]$. That is, $P_{Q,\epsilon}$ is disconnected and not robust since it contains the isolated point x^*.

Example 8.2 (c) shows a case where $P_{Q,\epsilon}$ it not robust as the set contains an isolated point. The detection or approximation of such points is in many cases impossible when stochastic search methods are used. Using Lemma 8.2 we see that this is no loss since the interior of $P_{Q,\epsilon}$ can be approximated via stochastic search algorithms. For

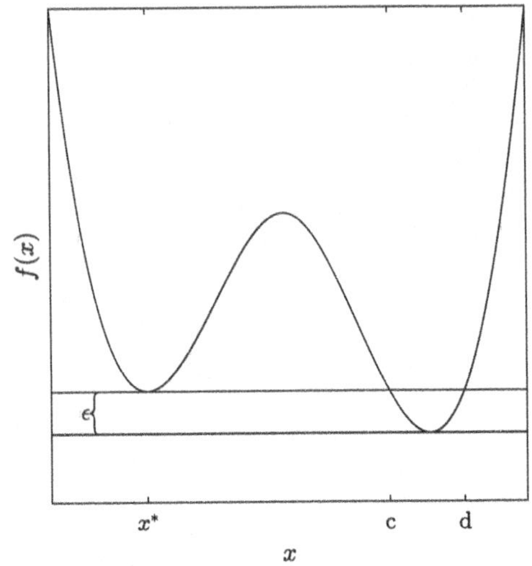

Fig. 8.7 For this problem the set of interest is $P_{Q,\epsilon} = \{x^*\} \cup [c, d]$, and hence, not robust since x^* is isolated. Note that $f'(x^*) = 0$

the subsequent convergence analysis, however, we have to make a further assumption in order to assure that $P_{Q,\epsilon}$ is robust which we will do in the following.

Lemma 8.3 *Let $Q \subset \mathbb{R}^n$ be compact. Under Assumptions (A1)–(A4) the set $P_{Q,\epsilon}$ is robust, i.e., it holds*

$$clos(int(P_{Q,\epsilon})) = clos(P_{Q,\epsilon}). \tag{8.25}$$

Proof Assume that the statement does not hold. Then there exists $x \in \partial P_{Q,\epsilon}$ and a neighborhood U of x with $U \cap int(P_{Q,\epsilon}) = \emptyset$. As $x \in \partial P_{Q,\epsilon}$ there exists by Theorem 8.2 $y \in P_Q$ with $F(y) + \epsilon \leq_p F(x)$. Since $q(x) \neq 0$ there exists an $\bar{x} \in U$ with $F(\bar{x}) <_p F(x)$. Further, as $\bar{x} \notin int(P_{Q,\epsilon})$ there exists an element $\bar{y} \in Q$ with $F(\bar{y}) + \epsilon \leq_p F(\bar{x}) <_p F(x)$, which is a contradiction to $x \in \partial P_{Q,\epsilon}$, and the claim follows. □

Finally, we address the connectedness of $P_{Q,\epsilon}$. Connectedness is a desired property e.g. when the problem is tackled via the help of local search strategies since in this case the entire set can in principle be detected when starting with a single solution $x \in P_{Q,\epsilon}$. We have already seen in the last example that connectedness of $P_{Q,\epsilon}$ is not given in general. This property, however, holds for convex problems as the following results show.

Theorem 8.3 *Let $\epsilon \in \mathbb{R}^k_+$. If Q is convex and all f_i, $i = 1, \ldots, k$, are convex, then $P_{Q,\epsilon}$ and $F(P_{Q,\epsilon})$ are connected.*

Proof The proof is based on the fact that in this case P_Q is connected (e.g., Ehrgott 2005). Assume that $P_{Q,\epsilon}$ is not connected, that is, there exist open sets $O_1, O_2 \subset \mathbb{R}^n$ such that $P_{Q,\epsilon} \subset O_1 \cup O_2$, $P_{Q,\epsilon} \cap O_1 \neq \emptyset$, $P_{Q,\epsilon} \cap O_2 \neq \emptyset$, and $P_{Q,\epsilon} \cap O_1 \cap O_2 = \emptyset$. Since P_Q is connected, it must be contained in one of these sets, without loss of generality we assume that $P_Q \subset O_1$. By assumption there exists a point $x \in$

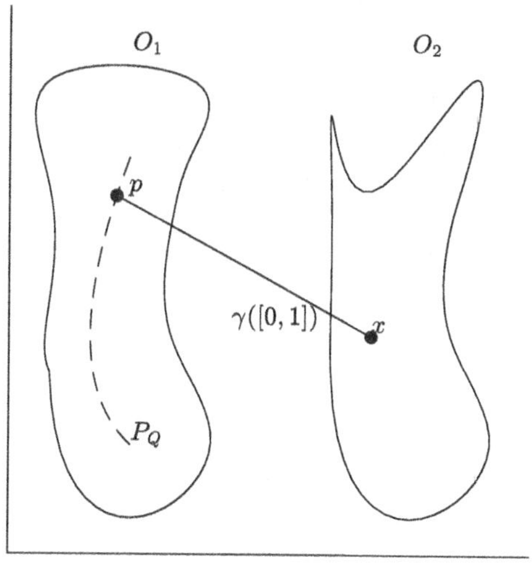

Fig. 8.8 Path connecting x and p

$P_{Q,\epsilon} \cap O_2$, and hence, $x \notin P_Q$. Further, there exists an element $p \in P_Q$ such that $F(p) \leq_p F(x)$. Since Q is convex, the following connecting path (see Fig. 8.8)

$$\gamma : [0, 1] \to Q, \quad \gamma(\lambda) = \lambda x + (1 - \lambda)p \tag{8.26}$$

is well-defined, i.e., it holds $\gamma([0, 1]) \subset Q$. Since F is convex it holds for all $\lambda \in [0, 1]$:

$$F(\gamma(\lambda)) = F(\lambda x + (1 - \lambda)p) \leq \lambda F(x) + (1 - \lambda)F(p) \leq F(x), \tag{8.27}$$

and thus that $\gamma(\lambda) \in P_{Q,\epsilon}$, which is a contradiction to the choice of O_1 and O_2. The connectedness of $F(P_{Q,\epsilon})$ follows immediately since images of connected sets under continuous mappings are connected, and the proof is complete. □

8.2 $ArchiveUpdateP_{Q,\epsilon}$

We first investigate $ArchiveUpdateP_{Q,\epsilon}$, see Algorithm 7 for the pseudo code of this archiver. It aims to maintain all $-\epsilon$-dominated solutions out of the candidate solutions that are considered during the run of the algorithm. The archiver is analog to the archiver $ArchiveUpdateP_Q$ with the only difference that the new dominance concept is used in $ArchiveUpdateP_{Q,\epsilon}$. The results for both archivers are also analog as we will see in the following.

Algorithm 7 $A := ArchiveUpdateP_{Q,\epsilon}\ (P, A_0)$

Require: P: population, A_0: current archive, $\epsilon \in \mathbb{R}^k_+$
Ensure: updated archive A
1: $A := A_0$
2: **for all** $p \in P$ **do**
3:    **if** $\nexists a \in A : a \prec_{-\epsilon} p$ **then**
4:       $A := A \cup \{p\}$
5:    **end if**
6:    **for all** $a \in A$ **do**
7:       **if** $p \prec_{-\epsilon} a$ **then**
8:          $A := A \backslash \{a\}$
9:       **end if**
10:    **end for**
11: **end for**
12: **return** A

We first consider the monotonic behavior of the archives when using $ArchiveUpdateP_{Q,\epsilon}$ within a generic stochastic search algorithm. The following result shows that for finitely many iterations the final archive A_l contains all $-\epsilon$-dominated solutions out of the candidate solutions that have been considered during the run of the algorithm.

Theorem 8.4 *Let (MOP) be given, let $l \in \mathbb{N}$, $\epsilon \in \mathbb{R}^k_+$, $P_0, \ldots, P_l \subset \mathbb{R}^n$ be finite sets, and A_i, $i = 0, \ldots, l$, be obtained by $ArchiveUpdateP_{Q,\epsilon}$ as in Algorithm 1, and $C_l = \bigcup_{i=0}^{l} P_i$. Then*

$$A_l = P_{C_l,\epsilon} = \{x \in C_l \ : \ \nexists y \in C_l : y \prec_{-\epsilon} x\}. \tag{8.28}$$

Proof Let $C_l = \{x_1, \ldots, x_m\}$, where the x_i's are considered by the archiver in this order. Let $x \in C_l$, then there exists an index $j \in \{1, \ldots, m\}$ such that $x = x_j$.

Case A: let $x \in P_{C_l,\epsilon}$. Then x will be added to the archive in the j-th step (line 3 of Algorithm 7) and later on not discarded (line 7).

Case B: let $x \notin P_{C_l,\epsilon}$. That is, there exists a point $x_i \in P_{C_l,\epsilon}$ such that $x_i \prec_{-\epsilon} x$. Hence, x is either not added to the archive (for instance, if $i < j$), or discarded (for $i > j$). □

Next, we consider the limit behavior of the respective sequence of archives. As it will be shown, the images of the archives $F(A_l)$ will come arbitrarily close to the set $F(P_{Q,\epsilon})$ during the run of the algorithm.

Theorem 8.5 *Let (MOP) be given, where F is continuous and $Q \subset \mathbb{R}^n$ is compact, and $\epsilon \in \mathbb{R}^k_+$. Then, under the assumptions (A1)–(A4) as well as (4.1), an application of Algorithm 1, where $ArchiveUpdateP_{Q,\epsilon}$ (Algorithm 7) is used to update the archive, leads to a sequence of archives A_l, $l \in \mathbb{N}$, with*

$$\lim_{l \to \infty} d_H(P_{Q,\epsilon}, A_l) = 0, \quad \textit{with probability one.} \tag{8.29}$$

Proof Since $dist(A, B) = dist(clos(A), B)$ for all sets $A, B \subset \mathbb{R}^n$ and by Lemma 8.3

$$clos(int(P_{Q,\epsilon})) = clos(P_{Q,\epsilon}),$$

is sufficient to show that the Hausdorff distance between A_l and $int(P_{Q,\epsilon})$ vanishes in the limit with probability one. First, we show that $dist(A_l, int(P_{Q,\epsilon})) \to 0$ with probability one for $l \to \infty$. It is

$$dist(A_l, int(P_{Q,\epsilon})) = \max_{a \in A_l} \inf_{p \in int(P_{Q,\epsilon})} \|a - p\|. \tag{8.30}$$

We have to show that every $x \in Q \backslash clos(P_{Q,\epsilon})$ will be discarded (if added before) from the archive after finitely many steps, and that this point will never be added further on. Let $x \in Q \backslash clos(P_{Q,\epsilon})$. Since x is by (A2) not a $-\epsilon$-weak Pareto point, there exists $p \in P_{Q,\epsilon}$ with $F(p) + \epsilon <_p F(x)$. Further, since F is continuous there exists a neighborhood U of x with $F(p) + \epsilon <_p F(u)$, $\forall u \in U$. By (4.1) it follows that there exists with probability one a number $l_0 \in \mathbb{N}$ such that there exists $x_{l_0} \in P_{l_0} \cap U \cap Q$. Thus, by construction of $ArchiveUpdateP_{Q,\epsilon}$, the point x will be discarded from the archive if it is a member of A_{l_0}, and never be added to the archive further on.

Now we consider the limit behavior of $dist(int(P_{Q,\epsilon}), A_l)$. It is

$$dist(int(P_{Q,\epsilon}), A_l) = \sup_{p \in int(P_{Q,\epsilon})} \min_{a \in A_l} \|p - a\|. \tag{8.31}$$

Let $\bar{p} \in int(P_{Q,\epsilon})$. For $i \in \mathbb{N}$ there exists by (4.1) a number l_i and a point $p_i \in P_{l_i} \cap B_{1/i}(\bar{p}) \cap Q$, where $B_\delta(p)$ denotes the open ball with center p and radius $\delta \in \mathbb{R}_+$. Since $\lim_{i \to \infty} p_i = \bar{p}$ and since $\bar{p} \in int(P_{Q,\epsilon})$ there exists an $i_0 \in \mathbb{N}$ such that $p_i \in int(P_{Q,\epsilon})$ for all $i \geq i_0$. By construction of $ArchiveUpdateP_{Q,\epsilon}$, all the points $p_i, i \geq i_0$, will be added to the archive (and never discarded further on). Thus, we have

$$dist(\bar{p}, A_l) \to 0 \quad \text{for} \quad l \to \infty$$

as desired, which completes the proof. □

Since the archiver $ArchiveUpdateP_{Q,\epsilon}$ maintains all $-\epsilon$-dominated solutions found during the run of the algorithm and since $P_{Q,\epsilon}$ can be expected to be n-dimensional, the archiver is unbounded in general. That is, we have to expect that

$$\lim_{i \to \infty} A_i = \infty \tag{8.32}$$

which means that similar to $ArchiveUpdateP_Q$, we will have to find ways to discretize the set of interest. In the following, we will discuss three such archivers. More precisely, we will investigate archivers that aim for suitable discretizations of the set of interest, where we target (i) $F(P_{Q,\epsilon})$, (ii) $P_{Q,\epsilon}$, and (iii) a combination of both.

8.3 $ArchiveUpdateP_{Q,\epsilon}D_y$

We first aim for uniform approximations in image space, i.e., of the set $F(P_{Q,\epsilon})$. Algorithm 8 shows the pseudo code of $ArchiveUpdateP_{Q,\epsilon}D_y$ which is a modification of the archive $ArchiveUpdateP_{Q,\epsilon}$. The vector $1\Delta := (\Delta, \ldots, \Delta) \in \mathbb{R}^k_+$ with a given value $\Delta \in \mathbb{R}_+$ is used for the discretization in objective space.

We will in the following proceed as for the other archives. One remarkable result is that the upper bound of the archive size when using $ArchiveUpdateP_{Q,\epsilon}D_y$ is essentially the same as for the above discussed archives that aim for gap free approximations of the Pareto front $F(P_Q)$.

Algorithm 8 $A := ArchiveUpdateP_{Q,\epsilon}D_y\ (P, A_0, \Delta)$

Require: population P, archive A_0, $\epsilon \in \mathbb{R}^k_+$, $\Delta \in \mathbb{R}_+$, $\Delta^* \in (0, \Delta)$
Ensure: updated archive A

$A := A_0$
for all $p \in P$ **do**
    if $\nexists a_1 \in A : a \prec_{-\epsilon} p$ and $\nexists a_2 \in A : d_H(F(a_2), F(p)) \leq \Delta^*$ **then**
        $A := A \cup \{p\}$
        for all $a \in A$ **do**
            if $p \prec_{-(\epsilon+1\Delta)} a$ **then**
                $A := A \backslash \{a\}$
            end if
        end for
    end if
end for
return A

First, we address the monotonic behavior of the archiver. Due to the discretization we cannot expect an analog statement to Theorem 8.4. The next example shows that $ArchiveUpdateP_{Q,\epsilon}D_y$ can generate archives that contain elements $x \in P_{Q,\epsilon+2\Delta} \backslash P_{Q,\epsilon}$, even in the limit archive. Theorem 8.6 will show that this deterioration of 2Δ (measured in objective space) is already the upper bound.

Example 8.3 Assume we are given the MOP defined by the map $F : \mathbb{R} \to \mathbb{R}$, $F(x) = x$, and the domain $Q = [0, 5]$. Let $\epsilon = 1$ and $\Delta^* = \Delta = 0.1$. Then, the set of interest is given by $P_{Q,\epsilon} = [0, 1]$.

Now assume that we are given the one-element archive $A_0 = \{a_1\}$, where $a_1 = 1.2$. If next the point $a_2 = 0.1$ is considered by $ArchiveUpdateP_{Q,\epsilon}D_y$, it will be added to A since $d_H(F(a_1), F(a_2)) > \Delta$ and since $a_2 \in P_{Q,\epsilon}$ is not $-\epsilon$-dominated by a_1 nor by any other point $x \in Q$. Hence, a_2 will not be discarded from the archiver in any following archiving step. Further, since a_2 is not $-(\epsilon + \Delta)$-dominating a_1 it holds for the updated archive that $A_1 = \{a_1, a_2\}$. In further steps, no point $a \in [0, \Delta]$ will be added to the archive since for all of these points it is $d_H(F(a), F(a_2)) \leq \Delta^*$, and hence, $a_2 \in P_{Q,\epsilon+2\Delta} \backslash P_{Q,\epsilon}$ will remain an element of all further archives during the run of the algorithm.

If, on the other hand, the point $a_1 = 0$ is added to the archive, no element $x \in Q\backslash P_{Q,\epsilon}$ will ever be added to the archive in subsequent steps by $ArchiveUpdateP_{Q,\epsilon}D_y$.

Theorem 8.6 *Let $l \in \mathbb{N}$, $\epsilon \in \mathbb{R}^k_+$, $P_0, \ldots, P_l \subset \mathbb{R}^n$ be finite sets, and $A_i, i = 1, \ldots, l$, be obtained by $ArchiveUpdateP_{Q,\epsilon}D_y$ as in Algorithm 1. Then, for $C_l = \bigcup_{i=0}^l P_i$ it holds:*

$$\begin{aligned} &(i) \quad dist(F(P_{C_l,\epsilon}), F(A_l)) \leq \Delta \\ &(ii) \quad dist(F(A_l), F(P_{C_l,\epsilon})) \leq dist(F(P_{C_l,\epsilon+2\Delta}), F(P_{C_l,\epsilon})) \\ &(iii) \quad d_H(F(P_{C_l,\epsilon}), F(A_l)) \leq \max(\Delta, dist(F(P_{C_l,\epsilon+2\Delta}), F(P_{C_l,\epsilon}))) \end{aligned} \tag{8.33}$$

Proof *Ad (i)*: It is

$$dist(F(P_{C_l,\epsilon}), F(A_l)) = \max_{p\in P_{C_l,\epsilon}} \min_{a\in A_l} \|F(p) - F(a)\|_\infty.$$

Let $p \in P_{C_l,\epsilon}$. Then p was either added to the archive during the update process (say in the i-th step), or it was discarded by an archive element $a_i \in A_i$ with $\|F(a_i) - F(p)\|_\infty \leq \Delta^*$ (line 3). In the former case p is not discarded further on since $p \in P_{C_l,\epsilon}$ and due to the removal strategy in line 6. In the latter case a_i has to be inside $P_{C_l,\epsilon+1\Delta}$ and will also be a member of A_l. Hence, the maximal distance is given by Δ.
Ad (ii): It is

$$dist(F(A_l), F(P_{C_l,\epsilon})) = \max_{a\in A_l} \min_{p\in P_{C_l,\epsilon}} \|F(a) - F(p)\|_\infty.$$

Assume there exists $a \in A_l$ which is not in $P_{C_l,\epsilon+2\Delta}$, i.e., an element $x \in P_{C_l,\epsilon}$ that $-(\epsilon + 2\Delta)$ dominates a. If x has been considered by the archiver before a, then either x has been added to the archive or it has been discarded due to an element $a_2 \in A_i$ with $\|F(a_2) - F(x)\|_\infty \leq \Delta^*$. Since $x \in P_{C_l,\epsilon}$ and analog to the above discussion $a_2 \in P_{C_l,\epsilon+1\Delta}$ it follows that none of the elements will be discarded further on which implies that a would not be added to the archive (line 3). If on the other side x is being considered after a by the archiver (say in step j), x is either added to the archive or discarded due to an element $\tilde{a}_2 \in A_j$ with $\|F(\tilde{a}_2) - F(p)\|_\infty \leq \Delta^*$. Since both points at least $-(\epsilon + 1\Delta)$ dominate a, also in this case a will be removed from the archive which is a contradiction to the assumption on a. Hence, every archive entry of A_l has to be inside $P_{C_l,\epsilon+2\Delta}$.
Ad (iii): follows immediately by (i) and (ii). □

The results show that the obtained approximation error depends on (i) Δ and (ii) the distance of of $F(P_{Q,\epsilon+2\Delta})$ to $F(P_{Q,\epsilon})$. While Δ can be set as it is the approximation parameter, the latter is given by the considered MOP. As we have seen above, this distance can get very large (in pathological cases).

The next theorem is a result on the limit behavior of the archiver.

Theorem 8.7 *Let (MOP) be given, where F is continuous and $Q \subset \mathbb{R}^n$ is compact, and $\epsilon \in \mathbb{R}^k_+$. Then, under the assumptions (A1)–(A4) as well as (4.1), where $AchiveUpdateP_{Q,\epsilon}D_y(P, A, \Delta)$ is used to update the archive, leads to a sequence $A_l, l \in \mathbb{N}$, of archives with*

(a)

$$\|F(a_1) - F(a_2)\|_\infty \geq \Delta^*, \qquad \forall l \in \mathbb{N}$$

(b) There exists with probability one an integer $l_0 \in \mathbb{N}$ such that for all $l \geq l_0$ it holds:

(b1) $dist(F(P_{Q,\epsilon}), F(A_l)) < \Delta$
(b2) $dist(F(A_l), F(P_{Q,\epsilon})) \leq dist(F(P_{Q,\epsilon+2\Delta}), F(P_{Q,\epsilon}))$
(b3) $d_H(F(P_{Q,\epsilon}), F(A_l)) \leq D$, *where* $D = max(\Delta, dist(F(P_{Q,\epsilon+2\Delta}), F(P_{Q,\epsilon})))$

Proof Before we come to the proof we make some remarks: a point p is discarded from an existing archive A in two cases (see line 3 of Algorithm 8):

$$\begin{aligned} &(D1) \quad \exists a_1 \in A : \ a_1 \prec_{-\epsilon} p, \quad \text{or} \\ &(D2) \quad \exists a_2 \in A : \ \|F(a_2) - F(p)\|_\infty \leq \Delta^*. \end{aligned} \tag{8.34}$$

Further, let

$$B_\delta^\infty(x) := \{y \in \mathbb{R}^k \ : \ \|x - y\|_\infty < \delta\}$$

the k-dimensional open box around $x \in \mathbb{R}^k$.

Now we are in the position to state the proof.

Claim (a): follows immediately by construction of the algorithm and by an inductive argument.

Claim (b1): By *(a)* it follows that for an element a from a given archive A it holds

$$F(\tilde{a}) \notin B_{\Delta^*}^\infty(F(a)), \quad \forall \tilde{a} \in A \backslash \{a\}, \tag{8.35}$$

Since further Q is compact and F is continuous it follows that $F(Q)$ is bounded, and thus, there exists an upper bound for the number of entries in the archive for a given MOP, denote by $n_0 = n_0(\Delta^*, F(Q))$.

Since $clos(P_{Q,\epsilon})$ is compact and $dist(F(P_{Q,\epsilon}), F(A_l)) = dist(F(clos(P_{Q,\epsilon})), F(A_l))$, and since $A_l, l \in \mathbb{N}$, is finite it follows that

$$dist(F(P_{Q,\epsilon}), F(A_l)) = \max_{p \in P_{Q,\epsilon}} \min_{a \in A_l} \|F(p) - F(a)\|_\infty$$

That is, the claim is right for an archive $A_l, l \in \mathbb{N}$, if for every $p \in P_{Q,\epsilon}$ there exists an element $a \in A_l$ such that $\|F(p) - F(a)\|_\infty < \Delta$. Thus, $F(P_{Q,\epsilon})$ must be contained in $C_{A_l,\Delta}$, where

$$C_{A,\Delta} := \bigcup_{a \in A} B_\Delta^\infty(F(a)).$$

First, we show that if there exists an $l_0 \in \mathbb{N}$ with $dist(F(P_{Q,\epsilon}), F(A_l)) < \Delta$, this property holds for all $l \geq l_0$.
Assume that such an l_0 is given. Define

$$\tilde{A} := \left\{a \in A_{l_0} | \exists p \in P_{Q,\epsilon} : \|F(p) - F(a)\|_\infty < \Delta\right\}, \tag{8.36}$$

since it holds that

$$p \in P_{Q,\epsilon} \text{ and } a \in Q : \|F(p) - F(a)\| \leq \Delta \Rightarrow a \in P_{Q,\epsilon+1\Delta},$$

it follows that $\tilde{A} \subset P_{Q,\epsilon+1\Delta}$, and thus, no element $a \in \tilde{A}$ will be discarded further on due to the construction of $ArchiveUpdateP_{Q,\epsilon}$. Since $dist(F(P_{Q,\epsilon}), F(A_l)) < \Delta$ it follows that for all $p \in P_{Q,\epsilon}$ there exists an element $a \in \tilde{A}$ such that $\|F(p) - F(a)\|_\infty < \Delta$. By the above discussion this holds for all $l \geq l_0$, and since no element $a \in \tilde{A}$ is discarded during the run of the algorithm, and the claim follows.

It remains to show the existence of such an integer l_0, which we will do by contradiction: first, we show that by using $ArchiveUpdateP_{Q,\epsilon}$ and under the assumptions made above only finitely many replacements can be done during the run of the algorithm. Then we construct a contradiction by showing that under the assumptions made above infinitely many replacements have to be done during the run of the algorithm with the given setting.

Let a finite archive A_0 be given. If a point $p \in \mathbb{R}^n$ replaces a point $a \in A_0$ (see lines 4 and 7 of Algorithm 8) it follows by construction of $ArchiveUpdateP_{Q,\epsilon}$ that

$$F(p) <_p F(a) - \Delta. \tag{8.37}$$

Since the relation $\prec$ is transitive, there exists for every $a \in A$ a "history" of replaced points $a_i \in A_{l_i}$ where Eq. (8.37) holds for a_i and a_{i-1}. Since $F(Q)$ is bounded there exist $l_i, u_i \in \mathbb{R}, i = 1, \ldots, k$, such that $F(Q) \subset [l_1, u_1] \times \cdots \times [l_k, u_k]$. After r replacements there exists at least one $a \in A_{l(r)}$ such that the length h of the history of a is at least $h \geq \lceil r/n_0 \rceil$, where n_0 is the maximal number of entries in the archive (see above). Denote by $a_0 \in A_0$ the root of the history of a. For a, a_0 it follows that

$$F(a) \leq F(a_0) - h\Delta,$$

for $\tilde{h} > h_{max} := \Delta^{-1} \max_{i=1,\dots,k} u_i - l_i$ we obtain a contradiction since in that case there exists $i \in \{1, \dots, n\}$ with $f_i(a) < l_i$ and thus $F(a) \notin F(Q)$. Hence it follows that there can be done only finitely many such replacements during the run of an algorithm.

Assume that such an integer l_0 as claimed above does not exist, that is, that $F(P_{Q,\epsilon}) \not\subset C_{A_l,\Delta}$ for all $l \in \mathbb{N}$. Hence there exists a sequence of points

$$p_i \in P_{Q,\epsilon}: \quad y_i = F(p_i) \in F(P_{Q,\epsilon}) \backslash C_{A_i,\Delta} \quad \forall i \in \mathbb{N}. \tag{8.38}$$

Since $P_{Q,\epsilon} \subset Q$ and Q is compact there exists an accumulation point $p^* \in clos(P_{Q,\epsilon})$, that is, there exists a subsequence $\{i_j\}_{j\in\mathbb{N}}$ with

$$p_{i_j} \to p^* \in clos(P_{Q,\epsilon}) \text{ for } j \to \infty. \tag{8.39}$$

From Lemma 8.3 we know that $P_{Q,\epsilon}$ is robust under the assumptions (A1)–(A3), i.e., that

$$clos(int(P_{Q,\epsilon})) = clos(P_{Q,\epsilon}). \tag{8.40}$$

Hence, the set

$$\tilde{U}_1 := B^{\infty}_{(\Delta-\tilde{\Delta})/2}(y^*) \cap \overset{\circ}{P}_{Q,\epsilon}, \tag{8.41}$$

where $y^* := F(p^*)$, is not empty. By (4.1) it follows that there exists with probability one an $l_1 \in \mathbb{N}$ and an element $\tilde{x}_1 \in P_{l_0+l_1}$ generated by *Generate()* with $\tilde{y}_1 = F(\tilde{x}_1) \in \tilde{U}_1$. There are two cases for the archive $A_{l_0+l_1}$: (a) x_1 can be discarded from the archive, or (b) x_1 is added to it.

Assume first that x_1 is discarded. Since $x_1 \in P_{Q,\epsilon}$ there exists no $\bar{x} \in Q$ such that $\bar{x}$ $-\epsilon$-dominates x_1. Hence, $(D1)$ can not occur (see (8.34)), and thus, there must exist an $a_2 \in A_{l_0+l_1}$ such that $\|F(a_2) - F(x_1)\|_\infty \leq \Delta^*$ (see $(D2)$). Thus, whether x_1 is added to the archive or not there exists an $\tilde{a}_1 \in A_{l_0+l_1}$ such that $\|F(\tilde{a}_1 - y^*\|_\infty \leq \Delta$ (since in case x_1 is added to the archive $\tilde{a}_1 = x_1$ can be chosen), and we obtain

$$\|F(\tilde{a}_1) - \tilde{y}\|_\infty \leq \|F(\tilde{a}_1) - F(x_1)\|_\infty + \|F(x_1) - \tilde{y}\|_\infty < \Delta \quad \forall \tilde{y} \in U_1. \tag{8.42}$$

By (8.38) and (8.39) there exist integers $j_1, \tilde{l}_1 \in \mathbb{N}$ with

$$y_{i_{j_1}} \in \tilde{U}_1 \backslash C_{l_0+l_1+\tilde{l}_1,\Delta}. \tag{8.43}$$

Since by (8.42) it holds that $\|y_{i_{j_1}} - F(a_1)\|_\infty < \Delta$ it follows that $a_1 \notin A_{l_0+l_1+\tilde{l}_1}$, which is only possible via a replacement in Algorithm 8 (lines 4 and 7).

In an analogous way, a sequence $\{a_i\}_{i\in\mathbb{N}}$ of elements can be constructed which have to be replaced by other elements. Since this leads to a sequence of infinitely many replacements. This is a contradiction to the assumption, and the proof is complete.

Claim (b2): Let $\tilde{A}$ and l_0 as above, and let $l \geq l_0$. Further, let $x \in Q \backslash P_{Q,\epsilon+2\Delta}$, that is, there exists a $p \in P_{Q,\epsilon}$ such that $p \prec_{-(\epsilon+2\Delta)} x$. Since $l \geq l_0$ there exists an $a \in \tilde{A} \subset A_l$ such that $\|F(p) - F(a)\|_\infty < \Delta$. Combining both facts we see that $a \prec_{-(\epsilon+1\Delta)} x$. Thus, no element $x \in Q \backslash P_{Q,\epsilon+2\Delta}$ is contained in A_l, $l \geq l_0$, or will ever be added to the archive further on. The claim follows since the archive can only contain elements in $P_{Q,\epsilon+2\Delta}$ (see also Example 8.3).

Claim (b3): follows immediately by (b1) and (b2). □

Remark 8.2 (a) For the discretization parameters $\Delta = \Delta^* = 0$ the archiver is identical $ArchiveUpdateP_{Q,\epsilon}$.

(b) We have used one scalar $\Delta_0 \in \mathbb{R}_+$ to discretize $F(P_{Q,\epsilon})$. It is of course also possible to use a vector $\Delta \in \mathbb{R}^k_+$ (with different entries for each objective) which will lead to analog results. Given the k-dimensional vectors Δ and Δ^* one can e.g. change the second condition of line 3 of $ArchiveUpdateP_{Q,\epsilon}D_y$ ($\nexists a_2 \in A: d_H(F(a_2), F(p)) \leq \Delta^*$) by

$$\nexists a_2 \in A: |f_i(a_2) - f_i(p)| \leq \Delta_i^*, \quad i = 1, \ldots, k. \tag{8.44}$$

(c) Similar to the above discretization strategies, the value $0 < \Delta^* < \Delta$ is used to guarantee convergence. For the implementations, it is fine to use $\Delta^* = \Delta$.

Next, we investigate the bounds on the archive sizes. Similar to the above methods that have a discretization parameter, lower bound is given by one (if Δ is chosen sufficiently large). The next theorem is a result on the upper bounds.

Theorem 8.8 *Let $\epsilon \in \mathbb{R}^k_+$, $\Delta^*, \Delta \in \mathbb{R}_+$ with $\Delta^* < \Delta$ be given. Further, let $m_i = \min_{x \in Q} f_i(x)$, $M_i = \max_{x \in Q} f_i(x)$, $1 \leq i \leq k$, and l_0 as in Theorem 8.7. Then, when using ArchiveUpdate$P_{Q,\epsilon}$, the archive size maintained in Algorithm 1 for all $l \geq l_0$ is bounded by*

$$|A_l| \leq \left(\frac{1}{\Delta^*}\right)^k \sum_{i=1}^k (\epsilon_i + 2\Delta + \Delta^*) \prod_{\substack{j=1 \\ j \neq i}}^k (M_j - m_j + \Delta^*). \tag{8.45}$$

Proof Let A_l with $l \geq l_0$ be given. Since $A_l \subset P_{Q,\epsilon+2\Delta}$ (see Theorem 8.7) we are interested in an upper bound on the volume of $F(P_{Q,\epsilon+2\Delta})$. For this, we consider first the $(k-1)$-dimensional volume of the Pareto front $F(P_Q)$. Due to the nature of nondominance we can assume that $F(P_Q)$ is located in the graph of a map

$$\begin{aligned} \Phi_f &: K \to \mathbb{R}^k \\ \Phi_f(u_1, \ldots, u_{k-1}) &= (u_1, \ldots, u_{k-1}, f(u_1, \ldots, u_{k-1})), \end{aligned} \tag{8.46}$$

where $K := [m_1, M_1] \times \ldots \times [m_{k-1}, M_{k-1}]$ and $f: K \to [m_k, M_k]$ which satisfies the monotonicity conditions

$$f(u_1, \ldots, u_{i-1}, x_i, u_{i+1}, \ldots, u_{k-1}) \leq f(u_1, \ldots, u_{i-1}, y_i, u_{i+1}, \ldots, u_{k-1})$$
$$\forall i = 1, \ldots, k-1,$$
$$u_j \in [m_j, M_j], j = 1, \ldots, i-1, i+1, \ldots, k-1,$$
$$x_i, y_i \in [m_i, M_i], x_i \leq y_i. \tag{8.47}$$

Further, we can assume that f is sufficiently smooth. If not, we can replace f by a smooth function $\tilde{f}$ such that the volume of $\Phi_{\tilde{f}}$ is larger than the volume of Φ_f: if $f(m_i, \ldots, m_{k-1}) = m_k$, the Pareto front consists of one point, $F(P_Q) = \{(m_1, \ldots, m_k)\}$, and has minimal volume 0. Since we are interested in upper bounds on the volume we can omit this case. Doing so, a smooth function $\tilde{f} : K \rightarrow [m_k, M_k]$ exists with $\tilde{f}(x) \leq f(x)$, $\forall x \in K$, that fulfills conditions (8.47). The $(k-1)$-dimensional volume of $\Phi_{\tilde{f}}$ is obviously larger than the volume of Φ_f. Under the smoothness assumption we can replace condition (8.47) by

$$\frac{\partial f}{\partial u_i} u \leq 0, \quad \forall u \in K, \ \forall i = 1, \ldots, k-1. \tag{8.48}$$

The $(k-1)$-dimensional volume of Φ_f within range K is given by (see Forster 1984):

$$Vol_{k-1}(\Phi_f) = \int_K \sqrt{||\nabla f||^2 + 1} du, \tag{8.49}$$

Analog to the proof of Theorem 6.2 the volume can be estimated by using partial integration and the monotonicity conditions (8.48) as follows:

$$Vol_{k-1}(\Phi_f) \leq \sum_{i=1}^{k} \prod_{\substack{j=1 \\ j \neq i}}^{k} (M_j - m_j). \tag{8.50}$$

Considering this and the nature of $-\epsilon$-dominance we can bound the k-dimensional volume of $F(P_{Q,\epsilon+2\Delta})$ by:

$$Vol_k(F(P_{Q,\epsilon+2\Delta})) \leq \sum_{i=1}^{k} (\epsilon_i + 2\Delta) \prod_{\substack{j=1 \\ j \neq i}}^{k} (M_j - m_j), \tag{8.51}$$

Since $\|F(a_1) - F(a_2)\| \geq \Delta^*$ for all $a_1, a_2 \in A_l$ it follows that the boxes

$$B^{\infty}_{\frac{1}{2}\Delta^*}(F(a)), \quad a \in A_l, \tag{8.52}$$

are mutually non overlapping. Further, if $F(a) \in F(P_{Q,\epsilon+2\Delta})$, then $B^{\infty}_{\frac{1}{2}\Delta^*}(F(a))$ is included in a $\Delta^*/2$-neighborhood $\tilde{F}$ of $F(P_{Q,\epsilon+2\Delta})$ with

$$Vol_k(\tilde{F}) \le \sum_{i=1}^{k}(\epsilon_i + 2\Delta + \Delta^*)\prod_{\substack{j=1\\ j\neq i}}^{k}(M_j - m_j + \Delta^*). \tag{8.53}$$

The maximal number of entries in A_l can now be estimated by

$$|A_l| \le \frac{Vol_k(\tilde{F})}{Vol_k(B^{\infty}_{\frac{1}{2}\Delta^*}(F(a)))}, \tag{8.54}$$

and the claim follows since the volume of $B^{\infty}_{\frac{1}{2}\Delta^*}(F(a))$ is obviously given by $(\Delta^*)^k$. □

Using the above result, we can now consider the growth of the archives with respect to the value of Δ which we have used for the discretization. If Δ is smaller than every entry of ϵ, i.e., if we have

$$\epsilon_i = c_i\Delta, \quad \text{where} \quad c_i > 1, i = 1, \ldots, k,$$

and we set for simplicity $\Delta = \Delta^*$ we get for the upper bound

$$|A_l(D_y)| \le \left(\frac{1}{\Delta}\right)^{k-1}\sum_{i=1}^{k}(c_i + 3)\prod_{\substack{j=1\\ j\neq i}}^{k}(M_j - m_j + \Delta^*). \tag{8.55}$$

Here, we denote by $A_l(D_y)$ the archive obtained by $ArchiveUpdateP_{Q,\epsilon}D_y$ for sake of distinction. We hence obtain

$$|A_l(D_y)| \in O\left(\left(\frac{1}{\Delta}\right)^{k-1}\right) \quad \text{for} \quad \Delta \to 0 \text{ and } l \to \infty. \tag{8.56}$$

That is, we obtain—up to constants—the same upper bounds as for the archivers described above that aim for Pareto front approximations. Considering the fact that the Pareto fronts typically form $(k-1)$-dimensional objects we can hence say that the order of the magnitudes of $A_l(D_y)$ is already optimal.

8.4 $ArchiveUpdateP_{Q,\epsilon}D_x$

Another strategy to obtain a finite size approximation of the set of interest wold be to make the disretization in decision variable space. Algorithm 9 shows the peudo code of $ArchiveUpdateP_{Q,\epsilon}D_x$ which does such a discretization. The archiver is identical to $ArchiveUpdateP_{Q,\epsilon}$ except for the condition in line 3: a point p is added to the archive if there exists no element a_1 in the current archive A that $-\epsilon$-dominates

p, *and* if there does not exist another point a_2 from A those Hausdorff distance to p is less or equal to a value Δ^*, which is a value less than a given discretization parameter Δ (which is now used in decision variable space).

Algorithm 9 $A := ArchiveUpdateP_{Q,\epsilon}D_x\ (P, A_0, \Delta)$

Require: population P, archive A_0, $\epsilon \in \mathbb{R}^k_+$, $\Delta \in \mathbb{R}_+$, $\Delta^* \in (0, \Delta)$
Ensure: updated archive A

$A := A_0$
for all $p \in P$ **do**
  if $\nexists a_1 \in A : a \prec_{-\epsilon} p$ and $\nexists a_2 \in A : d_H(a_2, p) \leq \Delta^*$ **then**
    $A := A \cup \{p\}$
    for all $a \in A$ **do**
      if $p \prec_{-\epsilon} a$ **then**
        $A := A \backslash \{a\}$
      end if
    end for
  end if
end for
return A

We leave out a detailed analysis of this archiver since its use is restricted to very low dimensional problems. Since we can expect that $P_{Q,\epsilon}$ is n-dimensional, we also expect that for the archive sizes it holds

$$|A_l(D_x)| \in O\left(\left(\frac{1}{\Delta}\right)^n\right) \quad \text{for} \quad \Delta \to 0 \text{ and } l \to \infty, \tag{8.57}$$

where $A_l(D_y)$ denotes the archives obtained by $ArchiveUpdateP_{Q,\epsilon}D_x$. We will see that this is indeed the case in the next section where we discuss a more promising discretization strategy.

8.5 $ArchiveUpdateP_{Q,\epsilon}D_{xy}$

The two archivers presented above realize the discretization of the set of interest either in decision variable space *or* in objective space. In order to provide the decision maker with the best overview of his/her possibilities, it would be desired to have an archiver that aims for a suitable distribution in both spaces. The archiver $ArchiveUpdateP_{Q,\epsilon}D_{xy}$ has been designed for this purpose, see Algorithm 10 for a pseudo code of this method. This algorithm has two discretization parameters, Δ_x and Δ_y. The aim of $ArchiveUpdateP_{Q,\epsilon}D_{xy}$ is to keep solutions that have similar values in objective space (measured by Δ_y which is relatively small) but that have significantly different values in decision variable space (measured by Δ_x which is relatively large). Note that $ArchiveUpdateP_{Q,\epsilon}D_{xy}$ is identical to $ArchiveUpdateP_{Q,\epsilon}D_y$ for $\Delta_x = \infty$ (respectively, if Δ_x is chosen large enough for a given problem).

Algorithm 10 $A := ArchiveUpdateP_{Q,\epsilon}D_{xy}\ (P, A_0, \epsilon, \Delta_x, \Delta_y)$

Require: population P, archive A_0, $\epsilon \in \mathbb{R}^k_+$, $\Delta_x \in \mathbb{R}_+$, $\Delta_y \in \mathbb{R}_+$, $\Delta_x^* \in (0, \Delta_x)$, $\Delta_y^* \in (0, \Delta_y)$
Ensure: updated archive A
1: $A := A_0$
2: **for all** $p \in P$ **do**
3:   **if** $\nexists a_1 \in A : a_1 \prec_{-\epsilon} p$ and $\nexists a_2 \in A : (d_H(F(a_2), F(p)) \leq \Delta_y^*$ and $d_H(a_2, p) \leq \Delta_x^*)$ **then**
4:     $A \leftarrow A \cup \{p\}$
5:     $\hat{A} = \{a_1 \in A | \nexists a_2 \in A : a_2 \prec_{-(\epsilon+\Delta_y)} a_1\}$
6:     **for all** $a \in A \backslash \hat{A}$ **do**
7:       **if** $p \prec_{-(\epsilon+\Delta_y)} a$ and $dist(a, \hat{A}) \geq 2\Delta_x$ **then**
8:         $A \leftarrow A \backslash \{a\}$
9:       **end if**
10:     **end for**
11:   **end if**
12: **end for**
13: **return** A

The limit behavior of $ArchiveUpdateP_{Q,\epsilon}D_{xy}$ is investigated next.

Theorem 8.9 *Let (MOP) be given, where F is continuous and Q is compact, $\epsilon \in \mathbb{R}^k_+$, $\Delta_x, \Delta_x^*, \Delta_y, \Delta_y^* \in \mathbb{R}_+$ with $\Delta_y^* < \Delta_y$ and $\Delta_x^* < \Delta_x$. Then, under the assumptions (A1)–(A3) and (4.1), an application of Algorithm 1, where $AchiveUpdateP_{Q,\epsilon}D_{xy}$ is used to update the archive, leads to a sequence of archives $A_l, l \in \mathbb{N}$, such that there exists an $l_0 \in \mathbb{N}$ such that for all $l \geq l_0$:*

(i) $dist(P_{Q,\epsilon}, A_l) < \Delta_x$ *and* $dist(F(P_{Q,\epsilon}), F(A_l)) < \Delta_y$
(ii) $dist(A_l, P_{Q,\epsilon}) \leq dist(B_{2\Delta_x}(P_{Q,\epsilon+2\Delta_y}), P_{Q,\epsilon})$ *and* $dist(F(A_l), F(P_{Q,\epsilon})) \leq dist(F(B_{2\Delta_x}(P_{Q,\epsilon+2\Delta_y})), F(P_{Q,\epsilon}))$
(iii) $d_H(P_{Q,\epsilon}, A_l) \leq \max(\Delta_x, dist(B_{2\Delta_x}(P_{Q,\epsilon+2\Delta_y}), P_{Q,\epsilon})$ *and* $d_H(F(P_{Q,\epsilon}), F(A_l)) \leq \max(\Delta_y, dist(F(B_{2\Delta_x}(P_{Q,\epsilon+2\Delta_y})), F(P_{Q,\epsilon}))$

Proof Since the results in objective space are analog to the results in Theorem 8.7, we only prove here the results in decision variable space.
Note that a point p is discarded from an existing archive A in two cases (line 2):

$$\begin{aligned} &(D1)\quad \exists a_1 \in A :\ a_1 \prec_{-\epsilon} p, \quad \text{or} \\ &(D2)\quad \exists a_2 \in A :\ \|F(a_2) - F(p)\|_\infty \leq \Delta_y^* \text{ and } \|a_2 - p\|_\infty \leq \Delta_x^*. \end{aligned} \tag{8.58}$$

(i) since $clos(P_{Q,\epsilon})$ is compact, $dist(P_{Q,\epsilon}, A_l) = dist(clos(P_{Q,\epsilon}), A_l)$ and $A_l, l \in \mathbb{N}$, is finite it follows that

$$dist(P_{Q,\epsilon}, A_l) = \max_{p \in clos(P_{Q,\epsilon})} \min_{a \in A_l} \|p - a\|_\infty. \tag{8.59}$$

The claim is right for an archive $A_l, l \in \mathbb{N}$, if for every $p \in P_{Q,\epsilon}$ there exists an element $a \in A_l$ such that $\|p - a\|_\infty < \Delta_x$. Thus, we need to prove that $P_{Q,\epsilon}$ is contained in

$$C_{A,\Delta_x} := \bigcup_{a \in A} B^\infty_{\Delta_x}(a), \tag{8.60}$$

where $B^\infty_\delta(x) := \{y \in \mathbb{R}^n \; : \; \|x - y\|_\infty < \delta\}$.

First we show that if there exists an $l_0 \in \mathbb{N}$ with $dist(P_{Q,\epsilon}, A_l) < \Delta_x$ then this inequality holds for all $l \geq l_0$. Assume that such an l_0 is given. Let

$$\tilde{A} := \left\{a \in A_{l_0} | \exists p_1 \in P_{Q,\epsilon} : \|p_1 - a\|_\infty < \Delta_x \text{ or } \exists p_2 \in P_{Q,\epsilon} : \|F(p_2) - F(a)\|_\infty < \Delta_y\right\}. \tag{8.61}$$

The following discussion shows that no element $a \in \tilde{A}$ will be discarded further on due to the construction of the archiver. It holds

$$p \in P_{Q,\epsilon} \text{ and } a \in Q : \|p - a\| \leq \Delta_x \;\; \Rightarrow a \in B_{\Delta_x}(P_{Q,\epsilon}), \tag{8.62}$$

Assume an element $a \in \tilde{A}$ is discarded. For this to happen the following has to hold: (i) $a \notin P_{Q,\epsilon+\Delta_y}$, (ii) an element $a_2 \in A_l$ is added to the archiver such that $\nexists a_3 \in \tilde{A} : dist(F(a_2), F(a_3)) < \Delta_y$ and $a_2 \prec_{-(\epsilon+\Delta_y)} a$, and (iii) $\nexists b \in \tilde{A} : dist(a, b) \leq 2\Delta_x$, and hence, $dist(a, \hat{A}) > 2\Delta_x$. For all $p \in P_{Q,\epsilon}$ there exists an element $a \in \tilde{A}$ such that $\|p - a\|_\infty < \Delta_x$. Thus, there exists an element b such that $b \in \tilde{A} : dist(a, b) \leq 2\Delta_x$ and $b \in \hat{A}$. Otherwise, there would exists elements in $P_{Q,\epsilon}$ such that their distance to $\tilde{A}$ is bigger than Δ_x which is a contradiction to the assumption. Further, since by assumption $dist(P_{Q,\epsilon}, A_{l_0}) < \Delta_x$, there exists for all $p \in P_{Q,\epsilon}$ an element $a \in \tilde{A}$ with $\|p - a\|_\infty < \Delta_x$, and by the above consideration this holds for all archives A_l with $l \geq l_0$.
It remains to show the existence of such an integer l_0, which we will do by contradiction: first we show that by using $ArchiveUpdateP_{Q,\epsilon}D_{xy}$ and under the assumptions made above only finitely many replacements can be done during the run of the algorithm. Then, we construct a contradiction by showing that under the assumptions made above infinitely many replacements have to be done during the run of the algorithm with the given setting.
Let a finite archive A_0 be given. If a point $p \in \mathbb{R}^n$ replaces a point $a \in A_0$ (lines 5 to 7) it follows by construction of $ArchiveUpdateP_{Q,\epsilon}D_{xy}$ that

$$F(p) <_p F(a) - \Delta_y \text{ and } \|p - a\| \geq \Delta_y. \tag{8.63}$$

Since the relation $\prec$ is transitive, there exists for every $a \in A$ a "history" of replaced points $a_i \in A_{l_i}$ where Equation (8.63) holds for a_i and a_{i-1} which has to be finite as $F(Q)$ is bounded.

Assume that such an integer l_0 as claimed above does not exist, that is, that $P_{Q,\epsilon} \not\subset C_{A_l,\Delta}$ for all $l \in \mathbb{N}$. Hence, there exists a sequence of points

$$p_i \in P_{Q,\epsilon}: \quad p_i \in P_{Q,\epsilon} \backslash C_{A_i,\Delta} \quad \forall i \in \mathbb{N}. \tag{8.64}$$

Since $P_{Q,\epsilon} \subset Q$ and Q is compact there exists an accumulation point $p^* \in clos(P_{Q,\epsilon})$, that is, there exists a subsequence $\{i_j\}_{j\in\mathbb{N}}$ with

$$p_{i_j} \to p^* \in clos(P_{Q,\epsilon}) \text{ for } j \to \infty. \tag{8.65}$$

By (8.3) it follows that the set

$$\tilde{U}_1 := B^{\infty}_{(\Delta_x - \Delta_x^*)/2}(p^*) \cap int(P_{Q,\epsilon}), \tag{8.66}$$

is not empty. By (4.1) it follows that there exists with probability one an $l_1 \in \mathbb{N}$ and an element $\tilde{x}_1 \in P_{l_0+l_1}$ generated by *Generate()* with $\tilde{x}_1 \in \tilde{U}_1$. There are two cases for the archive $A_{l_0+l_1}$: (a) x_1 can be discarded from the archive, or (b) x_1 is added to it. Assume first that x_1 is discarded. Since $x_1 \in P_{Q,\epsilon}$ there exists no $\bar{x} \in Q$ such that $\bar{x}$ $-\epsilon$-dominates x_1. Hence, $(D1)$ can not occur (see (8.58)), and thus, there must exist an $a_2 \in A_{l_0+l_1}$ such that $\|F(a_2) - F(x_1)\|_\infty \le \Delta_y^*$ and $\|a_2 - x_1\| \le \Delta_x^*$ (see $(D2)$). Thus, whether x_1 is added to the archive or not there exists an $\tilde{a}_1 \in A_{l_0+l_1}$ such that $\|\tilde{a}_1 - y^*\|_\infty \le \Delta_x$ (since in case x_1 is added to the archive $\tilde{a}_1 = x_1$ can be chosen), and we obtain

$$\|\tilde{a}_1 - \tilde{y}\|_\infty \le \|\tilde{a}_1 - x_1\|_\infty + \|x_1 - \tilde{y}\|_\infty < \Delta_x \quad \forall \tilde{y} \in \tilde{U}_1 \tag{8.67}$$

By (8.64) and (8.65) there exist integers $j_1, \tilde{l}_1 \in \mathbb{N}$ with

$$y_{i_{j_1}} \in \tilde{U}_1 \backslash C_{l_0+l_1+\tilde{l}_1,\Delta_x}. \tag{8.68}$$

Since by (8.67) it holds that $\|y_{i_{j_1}} - a_1\|_\infty < \Delta_x$ it follows that $a_1 \notin A_{l_0+l_1+\tilde{l}_1}$, which is only possible via a replacement in Algorithm 10 (lines 4 to 7). Analogously, a sequence $\{a_i\}_{i\in\mathbb{N}}$ of elements can be constructed which have to be replaced by other elements. Since this leads to a sequence of infinitely many replacements. This is a contradiction to the assumption, and the proof is complete.

(ii) let $\tilde{A}$ and l_0 be as above, and let $l \ge l_0$. Further, let $x \in Q \backslash B_{2\Delta_x}(P_{Q,\epsilon+2\Delta_y})$, that is, there exists a $p \in P_{Q,\epsilon}$ such that $p \prec_{-(\epsilon+2\Delta_y)} x$. Since $l \ge l_0$ there exists an $a \in \tilde{A} \subset A_l$ such that $\|F(p) - F(a)\|_\infty < \Delta_y$. Combining both facts we see that $a \prec_{-(\epsilon+\Delta_y)} x$ and $dist(a, x) \ge \Delta_x$ (line 6). Thus, no element $x \in Q \backslash B_{2\Delta_x}(P_{Q,\epsilon+2\Delta_y})$ is contained in A_l, $l \ge l_0$, or will ever be added to the archive further on. The claim follows since the archive can only contain elements in $B_{2\Delta_x}(P_{Q,\epsilon+2\Delta_y})$.

(iii) follows immediately by (i) and (ii). □

Finally, we discuss the upper bounds on the archive sizes generated by $ArchiveUpdateP_{Q,\epsilon}D_{xy}$. Since for $\Delta_x = \infty$ (or large enough for a given problem) the archiver coincides with $ArchiveUpdateP_{Q,\epsilon}D_y$, the bounds are equal to the ones discussed above for this case.
Let $\Delta_x < \infty$. Since the domain Q is compact, it is contained inside a n-dimensional box, i.e.,

$$Q \subset [a_1, b_1] \times \cdots \times [a_n, b_n]$$

for some $a_i, b_i \in \mathbb{R}$ for $i = 1, \ldots, n$. The set $P_{Q,\epsilon+2\Delta}$ is included in the box $[a_1, b_1] \times \cdots \times [a_n, b_n]$, and the number of points that can be added to the archive per coordinate direction is $1/\Delta_x^* + 1$. The upper bound is hence given by

$$|A_l(D_{xy})| \leq \left(\frac{1}{\Delta^*} + 1\right)^n \prod_{j=1}^{n} (b_i - a_i). \tag{8.69}$$

The next example shows that this bound is tight.

Example 8.4 Consider the MOP where each objective is constant, i.e.,

$$\begin{aligned} F &: [0, 1]^n \to \mathbb{R}^k \\ F(x) &= c_0, \end{aligned} \tag{8.70}$$

where $c_0 \in \mathbb{R}^k$, and let $\Delta_x = \frac{1}{s}$ for a $s \in \mathbb{N}$. Define

$$\begin{aligned} x_{i_1,\ldots,i_n} &= (i_1 \Delta_x, \ldots, i_n \Delta_x) \\ \mathcal{D} &:= \left\{x_{i_1,\ldots,i_n} | 0 \leq i_1, \ldots, i_n \leq s\right\}. \end{aligned} \tag{8.71}$$

It is $\mathcal{D} \subset [0, 1]^n$ and $d_H(z_1, z_2) \geq \Delta_x > \Delta_x^*$ for all $z_1, z_2 \in \mathcal{D}$, $z_1 \neq z_2$. Assume that only points $z \in \mathcal{D}$ are considered. Then, all elements in $\mathcal{D}$ will be added to the archive leading to a final archive A with

$$|A| = |\mathcal{D}| = (s + 1)^n.$$

The growth of the magnitudes of the archives can hence be expressed by

$$o\left(\left(\frac{1}{\Delta_x}\right)^n\right) \quad \text{for} \quad \Delta_x \to 0. \tag{8.72}$$

For the applicability of the archiver, the adjustment of Δ_x is hence of utmost importance.

8.6 Numerical Results

Here we compare the four archivers $ArchiveUpdateP_{Q,\epsilon}$ (which stores all $-\epsilon$-dominated solutions), $ArchiveUpdateP_{Q,\epsilon}D_x$ (which makes a disretization in decision variable space), $ArchiveUpdateP_{Q,\epsilon}D_y$ (discretization in objective space), and $ArchiveUpdateP_{Q,\epsilon}D_{xy}$ (discretization in both spaces) on some test functions. For this, we will first investigate the test problems TWO-TO-ONE (Preuss et al. 2006, $n = 2, k = 2$), DEB99 (Deb 2001, $n = 2, k = 2$), SYM-PART (Rudolph et al. 2007, $n = 2, k = 2$), SSW (Schaeffler et al. 2002, $n = 3, k = 2$), OMNI-TEST (Deb and Tiwari 2008, $n = 4, k = 2$), and LSS (Lamé Super Sphere, Emmerich and Deutz 2007, $n = 5, k = 2$). See Appendix A for a brief description of all test functions. Note that the dimensions of both decision variable space and objective space are very low for each problem. This is due to the limitations to obtain the exact sets P_Q and $F(P_Q)$ for problems with higher dimensions. We will later on shortly consider the three-objective problem CONV3 with $n = 30$ decision variables.

Table 8.1 shows the computational times and standard deviations for the four archivers on each problem. The results are taken from 30 independent runs, and for each problem and run we have taken $N = 100{,}000$ candidate solutions. Figure 8.9 shows the according box plots. Analogously, Table 8.2 and Fig. 8.10 show the magnitudes of the final archives, Table 8.3 and Fig. 8.11 show the approximation quality of the final archive in decision variable space in terms of Δ_2 (i.e., $\Delta_2(A_f, P_{Q,\epsilon})$), and Table 8.4 and Fig. 8.12 show the respective approximation qualities in objective space (i.e., $\Delta_2(F(A_f), F(P_{Q,\epsilon}))$). Finally, Figs. 8.13, 8.14, 8.15, 8.16, 8.17, 8.18 show the results of each archiver that are closest to the median in objective space.

For TWO-TO-ONE we have chosen to use the parameters $\epsilon = (0.1, 0.1)^T$, $\Delta_x = 0.08$ and $\Delta_y = 0.3$. As anticipated, the archivers obtained by $ArchiveUpdateP_{Q,\epsilon}$ yield by far the best approximation qualities both in decision variable space and objective space. In turn, however, the archive sizes and running times are significantly higher compared to the other archiving strategies. Similar to the archivers that aim for Pareto front approximations, the values will get more extreme for larger values of candidate solutions. From Fig. 8.13 it can be observed that, as desired, the solutions obtained by $ArchiveUpdateP_{Q,\epsilon}D_x$ are more uniformly distributed in decision vari-

Table 8.1 Computational times (in seconds) and standard deviations for the different archivers and problems averaged over 30 independent runs, see also Figs. 8.13, 8.14, 8.15, 8.16, 8.17, 8.18. For all test problems we have used $N = 100{,}000$ candidate solutions

Problem	$P_{Q,\epsilon}$	$P_{Q,\epsilon}D_x$	$P_{Q,\epsilon}D_y$	$P_{Q,\epsilon}D_{xy}$
DEB99	49.3647 (2.5889)	0.25499 (0.040031)	0.80745 (0.07338)	1.0466 (0.11972)
TWO-TO-ONE	4.948 (0.5917)	0.18944 (0.05489)	0.31727 (0.08765)	0.51654 (0.1265)
SYM-PART	0.37361 (0.03246)	0.094547 (0.00475)	0.064819 (0.004197)	0.28177 (0.022714)
SSW	8.8613 (1.2924)	2.6764 (0.47808)	10.3503 (1.2439)	10.9089 (1.0806)
OMNI-TEST	0.063722 (0.010763)	0.072183 (0.012132)	0.071788 (0.011692)	0.083861 (0.023796)
LSS	69.1527 (4.989)	9.3839 (0.95501)	0.54536 (0.063897)	72.4677 (4.9738)

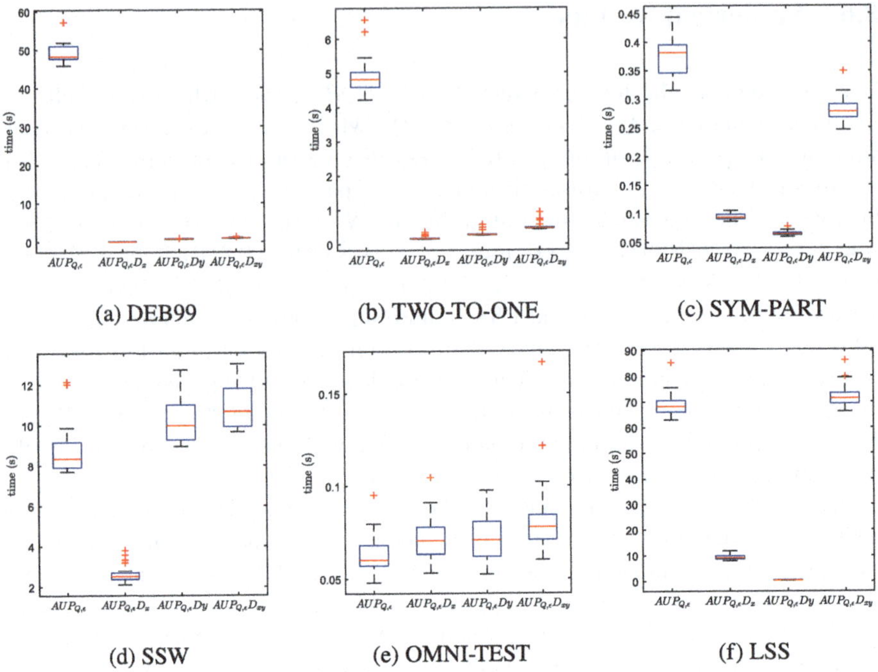

(a) DEB99 (b) TWO-TO-ONE (c) SYM-PART

(d) SSW (e) OMNI-TEST (f) LSS

Fig. 8.9 Box plots for the running times of the different archivers for the different test problems, based on 30 independent runs

able space, while $ArchiveUpdateP_{Q,\epsilon}D_y$ produces more evenly distributed solutions in objective space. The archive generated by $ArchiveUpdateP_{Q,\epsilon}D_{xy}$ yields satisfying results in both spaces for this example. These differences can not entirely be reflected by the Δ_2 values. Note that (i) the final approximation quality heavily depends on the choice of the design parameters Δ_x and Δ_y. That is, with a small enough value of Δ_y compared to Δ_x one can obtain better approximations in decision variable space for a particular problem, and vice versa. Further, (ii) this is a demonstrational example where the number of decision variables is equal to the number of objectives, while it is more usual that n is (much) larger than k. However, it can be said that for this example all three archivers that use a discretization maintain much smaller archives leading to much shorter running times, while all of them yield approximations that can be considered to be "perfect" with respect to the values of ϵ, Δ_x, and Δ_y.

For DEB99 we have chosen to use $\epsilon = (0.011, 0.011)^T$, $\Delta_x = 0.1$ and $\Delta_y = 0.01$, see Fig. 8.14 for the median plots. This example confirms the observations we have made for the previous example. We further stress two things: (i) the choices of Δ_x

Table 8.2 Archive sizes and standard deviations for the final archives for the different archivers and problems averaged over 30 independent runs, see also Figs. 8.13, 8.14, 8.15, 8.16, 8.17, 8.18. For all test problems we have used $N = 100{,}000$ candidate solutions

Problem	$P_{Q,\epsilon}$	$P_{Q,\epsilon} Dx$	$P_{Q,\epsilon} Dy$	$P_{Q,\epsilon} Dxy$
DEB99	8433.75 (162.2438)	27.75 (2.0229)	218.85 (5.451)	374.9 (8.4536)
TWO-TO-ONE	2754.45 (56.3929)	104.55 (3.6198)	251.85 (5.6314)	436.35 (7.8959)
SYM-PART	718.55 (25.3927)	95.7 (7.1606)	109.9 (4.9726)	564.6 (19.3837)
SSW	3042.35 (50.014)	638.7 (17.6757)	3122.75 (46.2452)	3267.8 (52.3456)
OMNI-TEST	71.5 (6.1601)	61.75 (5.8298)	69.85 (4.1074)	102.45 (6.9848)
LSS	9423.35 (108.9231)	1090.9 (25.4742)	159.55 (4.1228)	9501.85 (110.6974)

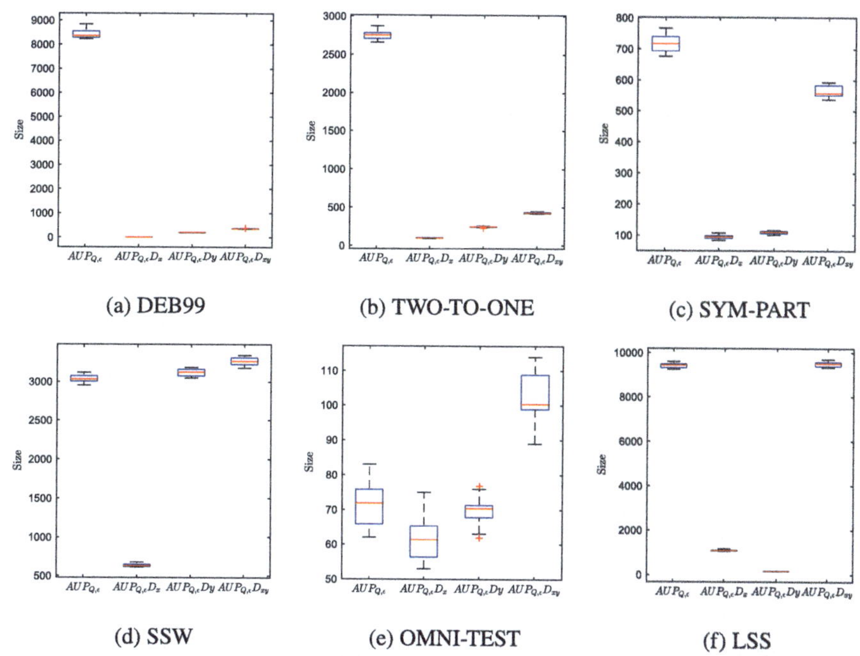

Fig. 8.10 Box plots for the archive sizes of the final archives of the different archivers for the different test problems, based on 30 independent runs

and Δ_y lead to the fact that the archiver that aims for an approximation in objective space yields a better approximation (in terms of Δ_2, but not in the distribution) in decision variable space than $ArchiveUpdateP_{Q,\epsilon}D_x$. However, as said above, other values of the design parameters will lead to different results. (ii) Note that the "approximated area" of $P_{Q,\epsilon}$ by all archivers that use discretization (in particular for $ArchiveUpdateP_{Q,\epsilon}D_x$) is larger than the actual set of interest. This is an effect that has been explained in a previous chapter (see e.g. Fig. 6.6 and related discussion), but also applies here.

Table 8.3 Approximation qualities $\Delta_2(A_f, P_{Q,\epsilon})$ and standard deviations of the final archives A_f of the different archivers and the set of approximate solutions $P_{Q,\epsilon}$, averaged over 30 independent runs. For all test problems we have used $N = 100{,}000$ candidate solutions

Problem	$P_{Q,\epsilon}$	$P_{Q,\epsilon}Dx$	$P_{Q,\epsilon}Dy$	$P_{Q,\epsilon}Dxy$
DEB99	0.003065 (0.00017457)	0.053732 (0.0040127)	0.030806 (0.0094713)	0.017528 (0.0011363)
TWO-TO-ONE	0.010884 (0.00015225)	0.049015 (0.0030624)	0.0426 (0.0013516)	0.027265 (0.00029685)
SYM-PART	0.074748 (0.0023741)	0.22923 (0.014039)	0.22317 (0.030648)	0.07796 (0.0022362)
SSW	0.55405 (0.0061432)	0.90514 (0.014737)	0.55367 (0.0055663)	0.54935 (0.0050098)
OMNI-TEST	1.3028 (0.096693)	1.3207 (0.10582)	1.2907 (0.099056)	1.2942 (0.079562)
LSS	0.37893 (0.0014296)	0.62643 (0.0050743)	1.0726 (0.018546)	0.37843 (0.0014308)

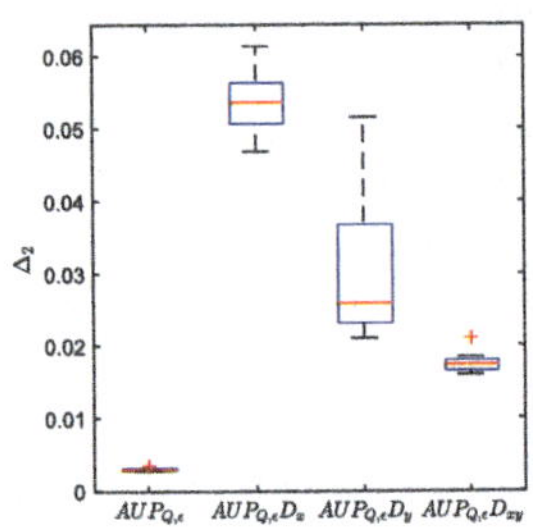

(a) DEB99

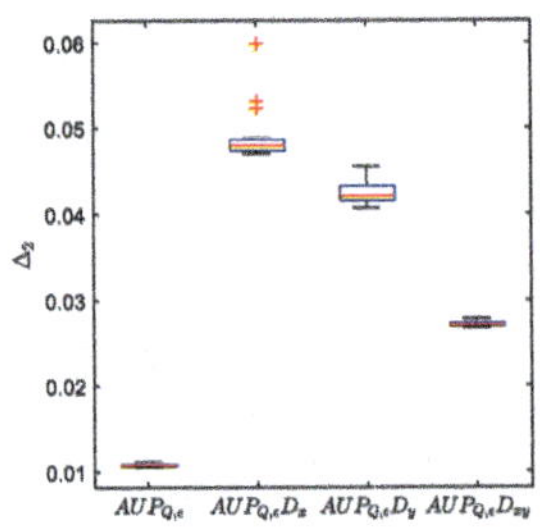

(b) TWO-TO-ONE

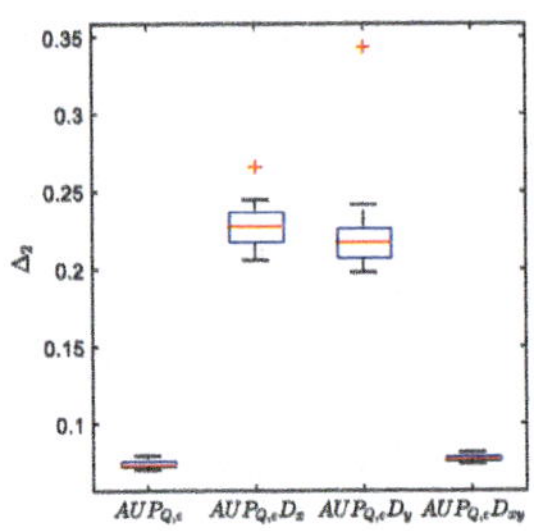

(c) SYM-PART

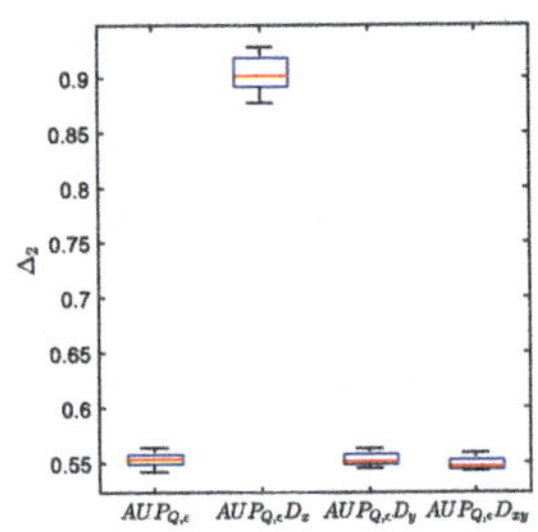

(d) SSW

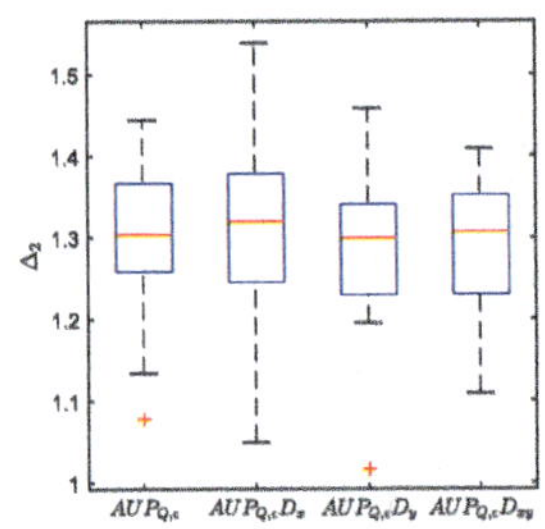

(e) OMNI-TEST

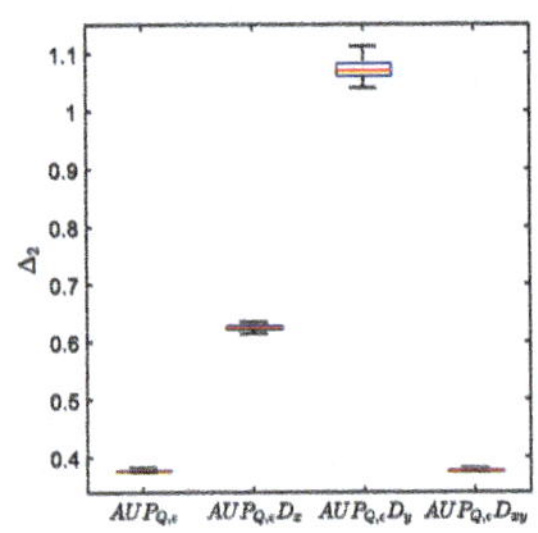

(f) LSS

Fig. 8.11 Box plots for the approximation qualities $\Delta_2(A_f, P_{Q,\epsilon})$ of the final archives A_f of the different archivers and the set of approximate solutions $P_{Q,\epsilon}$ for the different test problems, based on 30 independent runs

Table 8.4 Approximation qualities $\Delta_2(F(A_f), F(P_{Q,\epsilon}))$ and standard deviations of the images of the final archives $F(A_f)$ of the different archivers and the images of the sets of approximate solutions $F(P_{Q,\epsilon})$, averaged over 30 independent runs. For all test problems we have used $N =$ 100,000 candidate solutions

Problem	$P_{Q,\epsilon}$	$P_{Q,\epsilon}Dx$	$P_{Q,\epsilon}Dy$	$P_{Q,\epsilon}Dx_y$
DEB99	0.0053371 (0.0012126)	0.35252 (0.074002)	0.024554 (0.00077259)	0.019034 (0.00064781)
TWO-TO-ONE	0.018269 (0.00043599)	0.14981 (0.034163)	0.075747 (0.0029867)	0.052797 (0.0014648)
SYM-PART	0.018417 (0.00028506)	0.15409 (0.054131)	0.032696 (0.0025527)	0.02049 (0.0012081)
SSW	0.043693 (0.0057024)	0.10926 (0.0065502)	0.041465 (0.0054924)	0.041354 (0.0055319)
OMNI-TEST	0.081565 (0.013187)	0.085879 (0.014269)	0.093442 (0.015583)	0.084754 (0.010007)
LSS	0.0041046 (4.2734e-05)	0.099015 (0.022633)	0.028101 (0.00043943)	0.004165 (0.00010352)

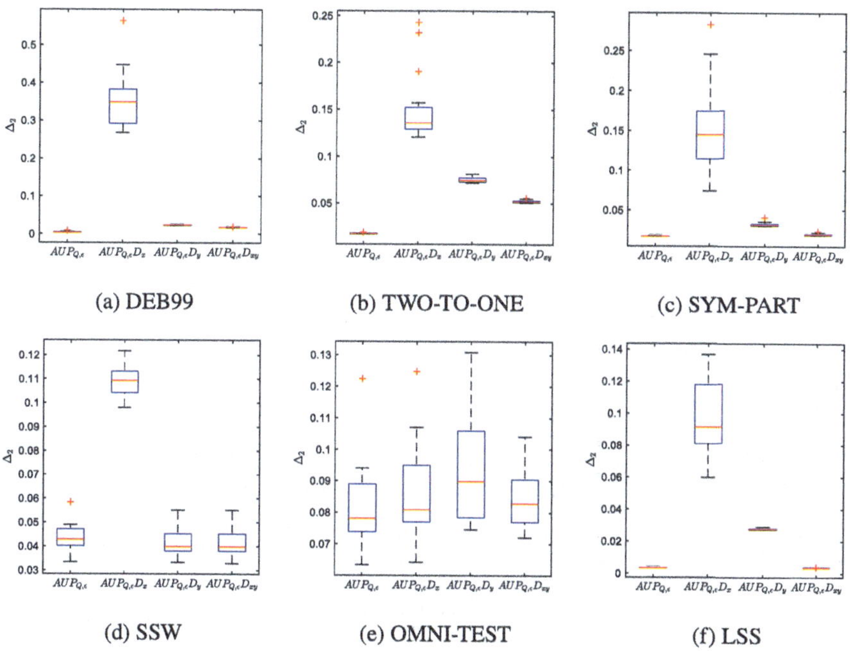

(a) DEB99 (b) TWO-TO-ONE (c) SYM-PART

(d) SSW (e) OMNI-TEST (f) LSS

Fig. 8.12 Box plots for the approximation qualities $\Delta_2(F(A_f), F(P_{Q,\epsilon}))$ of the images of the final archives $F(A_f)$ of the different archivers and the images of the sets of approximate solutions $F(P_{Q,\epsilon})$ for the different test problems, based on 30 independent runs

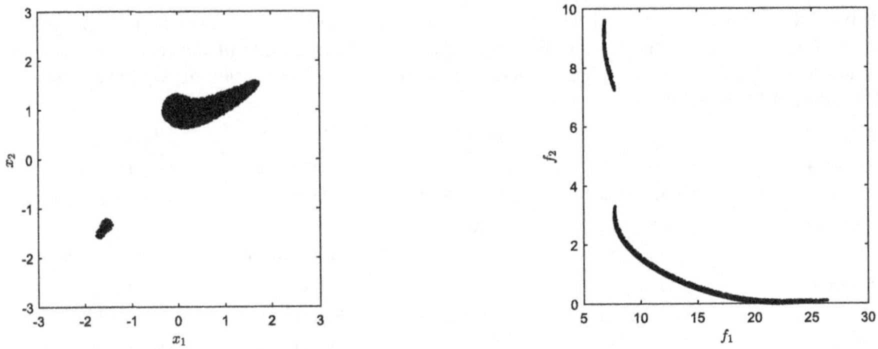

(a) $ArchiveUpdateP_{Q,\epsilon}$

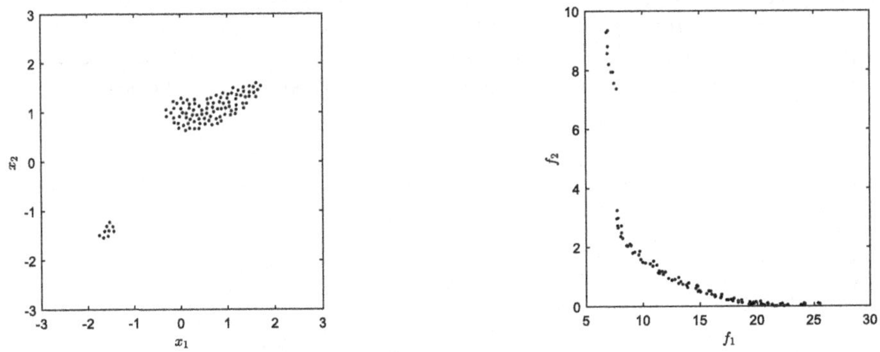

(b) $ArchiveUpdateP_{Q,\epsilon}Dx$

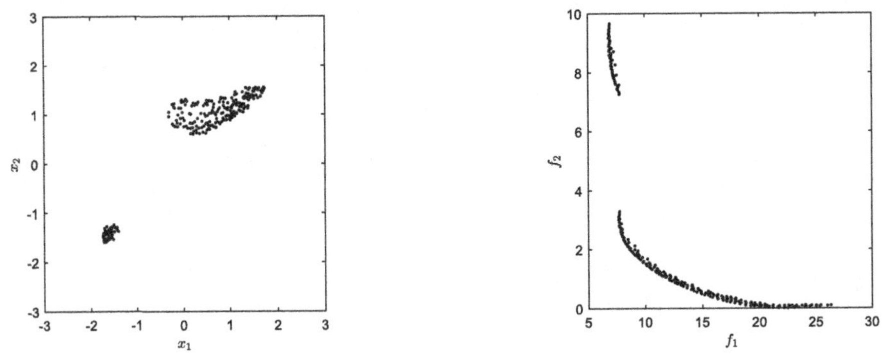

(c) $ArchiveUpdateP_{Q,\epsilon}Dy$

Fig. 8.13 Results of the four $P_{Q,\epsilon}$ archivers for TWO-TO-ONE

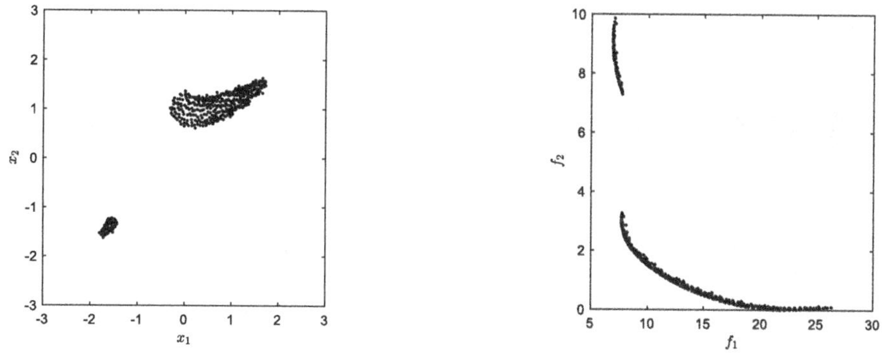

(d) $ArchiveUpdateP_{Q,\epsilon}D_{xy}$

Fig. 8.13 (continued)

SYM-PART is a 2-dimensional bi-objective problem those Pareto set falls into 9 connected components, where each of the maps to the entire Pareto front. That is, each of the points of the Pareto front has 9 pre-images. Similarly, the set of approximate solutions $P_{Q,\epsilon}$ also falls into 9 connected components, and each of the components that contain one of the components of P_Q map to the entire set $F(P_{Q,\epsilon})$. Figure 8.15 shows the median plots for the results of all four archivers, where we have used the design parameters $\epsilon = (0.15, 0.15)^T$, $\Delta_x = 0.1$ and $\Delta_y = 0.05$. In this example, $ArchiveUpdateP_{Q,\epsilon}D_y$ yields a much better distribution in objective space than $ArchiveUpdateP_{Q,\epsilon}D_x$ while it also manages to obtain an acceptable approximation of all 9 components of $P_{Q,\epsilon}$. The latter, however, is owed to the fact that the candidate solutions have been drawn uniformly at random from the domain. We will come back to this problem in Chap. 10.

Next, we consider problem SSW using the values $\epsilon = (0.01, 0.0001)^T$, $\Delta_x = 1$ and $\Delta_y = 0.01$, see Fig. 8.16. As opposed to the previous examples, the results for this test problem are not satisfying. Surprisingly, at least at first sight, the archive sizes and the running times for the two strategies $ArchiveUpdateP_{Q,\epsilon}D_y$ and $ArchiveUpdateP_{Q,\epsilon}D_{xy}$ that use a discretization are even higher than for the archives that use $ArchiveUpdateP_{Q,\epsilon}$ that stores *all* $-\epsilon$-dominated solutions. This has two reasons: first, note that the ranges in objective space differ significantly. While the Pareto front ranges in the first objective from 0 to 120, the range in the second objective is [0, 0.03]. Accordingly, there is not *one* value of Δ_y which is suitable for both objectives, and consequently, many candidate solutions are added to the archive (note that both $ArchiveUpdateP_{Q,\epsilon}D_y$ and $ArchiveUpdateP_{Q,\epsilon}D_{xy}$ use Δ_y). Second, due the effect we have already seen for DEB99 (enlarged approximated area) it can happen that even more solutions are stored than for $ArchiveUpdateP_{Q,\epsilon}$.

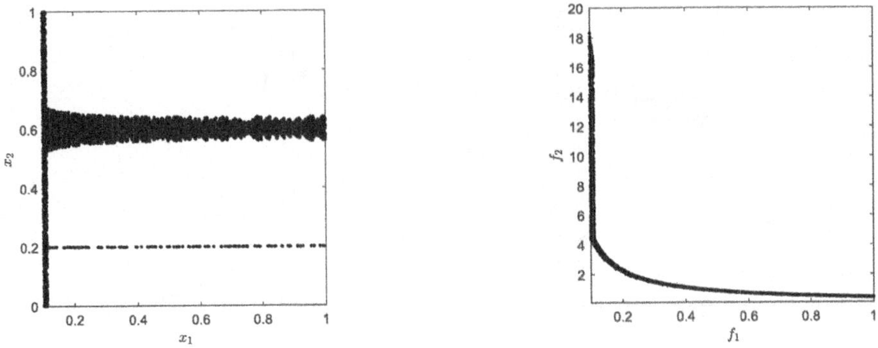

(a) $ArchiveUpdateP_{Q,\epsilon}$

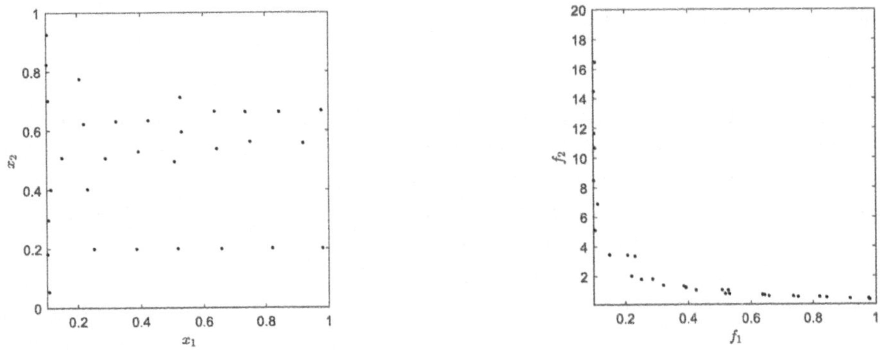

(b) $ArchiveUpdateP_{Q,\epsilon}Dx$

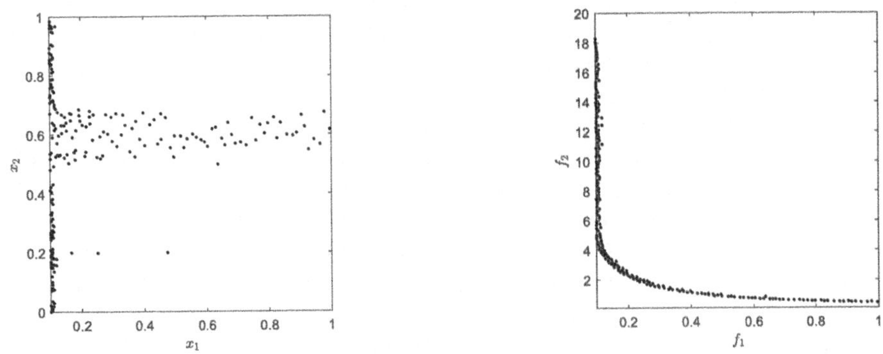

(c) $ArchiveUpdateP_{Q,\epsilon}Dy$

Fig. 8.14 Results of the four $P_{Q,\epsilon}$ archivers for DEB99

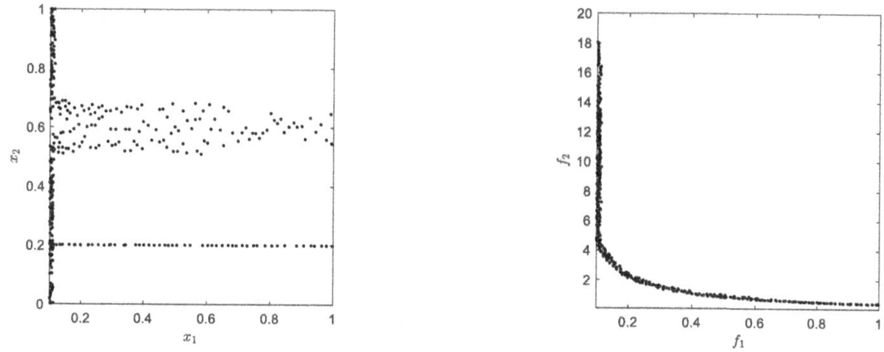

(d) $ArchiveUpdateP_{Q,\epsilon}D_{xy}$

Fig. 8.14 (continued)

A possible remedy is of course to use a k-dimensional vector for Δ_y as described in Remark 8.2 (b). Figure 8.19 shows some numerical results for the same problem but where we have modified the acceptance strategy for a candidate solution for $ArchiveUpdateP_{Q,\epsilon}D_y$ and $ArchiveUpdateP_{Q,\epsilon}D_{xy}$ as described in Remark 8.2 (b). Figure 8.20 shows the resulting box plots taken from 30 independent runs. Using this modification, the results are as "expected" and desired for each algorithm, and are as e.g. for TWO-TO-ONE. Needless to say that an analog modification can of course be done in decision variable space.

OMNI-TEST is similar to SYM-PART with the difference that OMNI-TEST has a 4-dimensional decision variable space, and that there are 25 connected components that each map to the entire Pareto front (analog for $P_{Q,\epsilon}$). Figure 8.17 shows the median results for $\epsilon = (0.01, 0.01)^T$, $\Delta_x = 0.1$ and $\Delta_y = 0.05$, where the left figure is a projection of the final archives in (x_1, x_2, x_3) space. The results are not in line with the previous ones. For instance, the shortest running times are for the archivers that are built using $ArchiveUpdateP_{Q,\epsilon}$. The reason for this is that $N = 100{,}000$ randomly chosen candidate solutions were not enough to suitably approximate the entire set of approximate solutions. However, note that all archivers were able to detect most of the connected components of $P_{Q,\epsilon}$ (which is again owed to the fact that the candidate solutions were taken uniformly at random from the domain).

Figure 8.18 shows some results on LSS using $\epsilon = (0.1, 0.1)^T$, $\Delta_x = 0.1$ and $\Delta_y = 0.05$. The results confirm the anticipated effects of each archiver, and also show the challenge to adapt the design parameters to properly adapt the set of interest. Note that for this setting, the final archive sizes when using $ArchiveUpdateP_{Q,\epsilon}D_{xy}$ are very large (more than 9,000 out of 100,000 candidate solutions). In turn, the archive sizes for $ArchiveUpdateP_{Q,\epsilon}D_x$ and $ArchiveUpdateP_{Q,\epsilon}D_y$, that focus on one space, are much lower (around 1000 and 150, respectively) while yielding good approximation qualities.

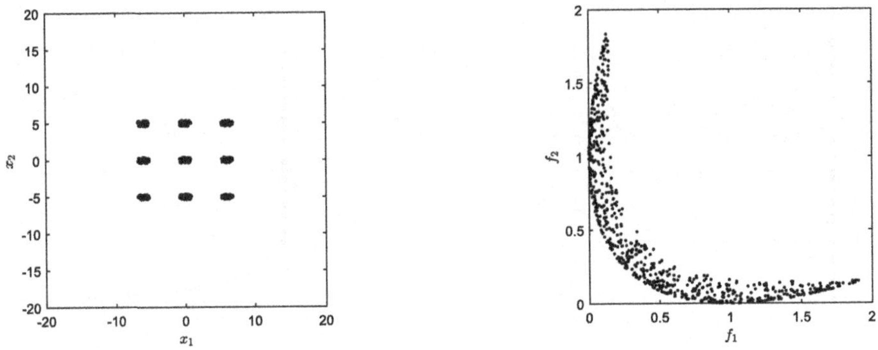

(a) $ArchiveUpdateP_{Q,\epsilon}$

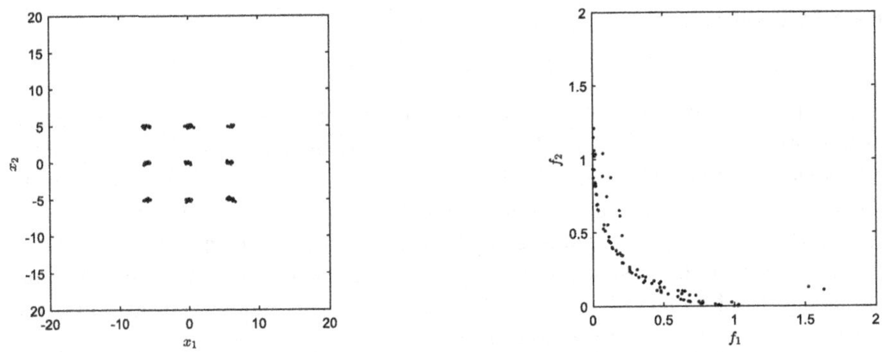

(b) $ArchiveUpdateP_{Q,\epsilon}Dx$

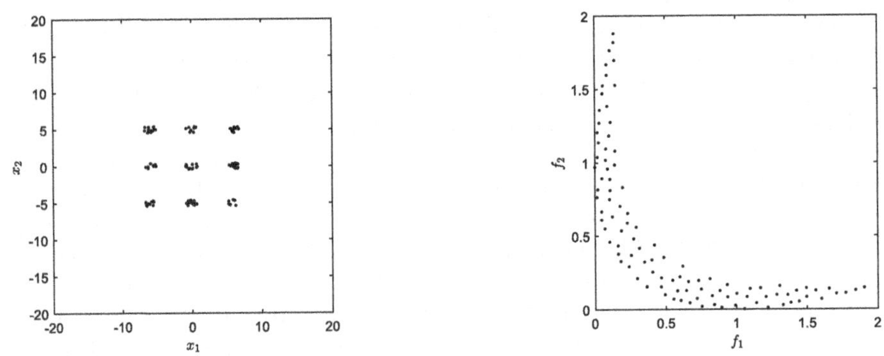

(c) $ArchiveUpdateP_{Q,\epsilon}Dy$

Fig. 8.15 Results of the four $P_{Q,\epsilon}$ archivers for SYM-PART

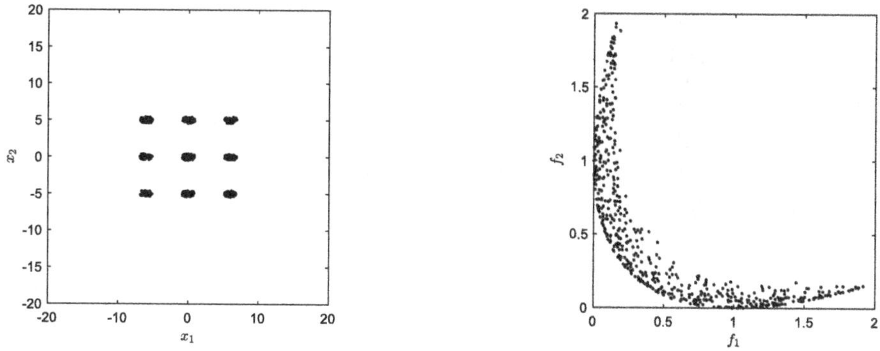

(d) $ArchiveUpdateP_{Q,\epsilon}D_{xy}$

Fig. 8.15 (continued)

So far, we have in this section only considered low dimensional problems, mainly since it is hard to obtain a nearly perfect expression for $P_{Q,\epsilon}$ and $F(P_{Q,\epsilon})$ for higher dimensional problems that are required to measure the approximation qualities. All the above archiving strategies are of course applicable to problems of any dimension (both with respect to n and k). As example for a higher dimensional problem, we have chosen to use problem CONV3, but this time using $n = 30$ decision variables. Figure 8.21 shows the median runs on this problem for the four archivers, and Fig. 8.22 shows the boxplots for the resulting running times and the obtained final archive sizes, while we had to omit the approximation qualities since $P_{Q,\epsilon}$ and $F(P_{Q,\epsilon})$ could not be determined analytically nor numerically up to the required quality.

In order to obtain sufficient approximations of the set of approximate solutions for this problem, we could not proceed as above and only take randomly chosen solutions out of the domain. Instead, we have used a steady-state MOEA as follows: first, we have fed the archives with $N = 20{,}000$ randomly chosen candidate solutions out of the domain Q. In the next step, we have used the resulting archive to obtain the new populations, and, in turn, to further update the archive (see Algorithm 1 of Laumanns et al. 2002). The overall function budget has been set to 100,000 for each run (20,000 to generate the initial archive, and another 80,000 for the steady-state MOEA). We have used $\epsilon = (3, 3, 3)^T$, $\Delta_x = 1.5$, and $\Delta_y = 3$ to discretize the sets of interest. As anticipated, storage of all promising solutions without any discretization leads to tremendously large archives: $ArchiveUpdateP_{Q,\epsilon}$ stored in average nearly every second candidate solution. The situation changes, however, when using the discretization. In our example, the use of $ArchiveUpdateP_{Q,\epsilon}D_y$ led to an (averaged) archive size of 110, and the use of $ArchiveUpdateP_{Q,\epsilon}D_{xy}$ to an archive size of around 430. It is certainly not problematic to store this amount of data.

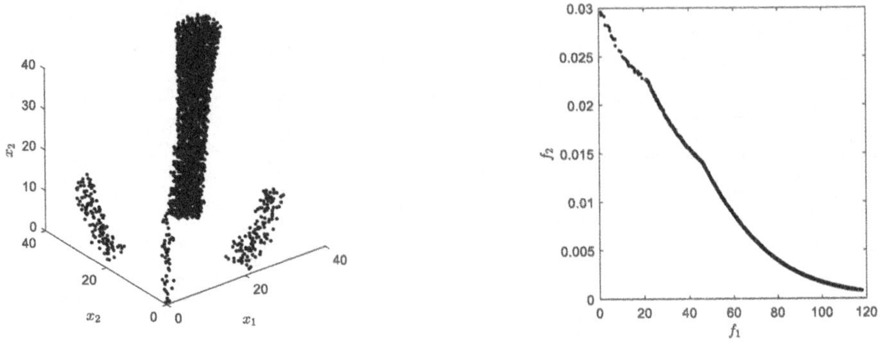

(a) $ArchiveUpdateP_{Q,\epsilon}$

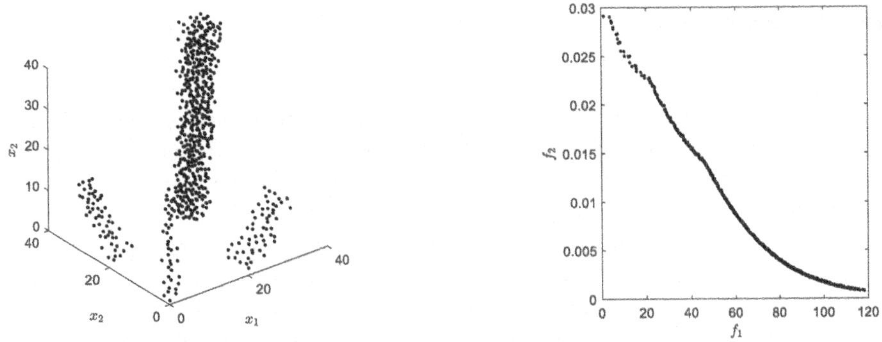

(b) $ArchiveUpdateP_{Q,\epsilon}Dx$

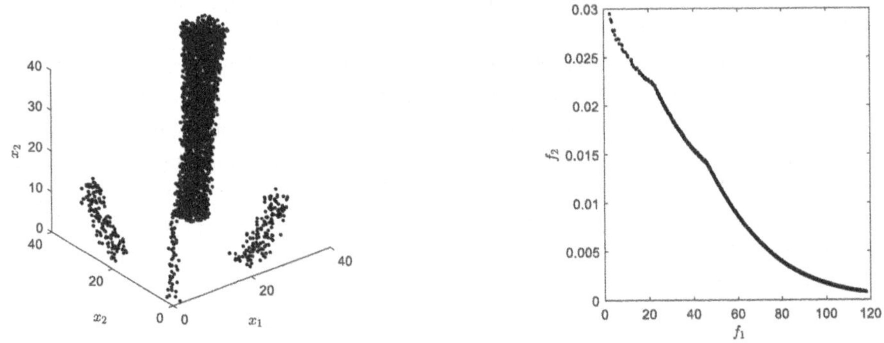

(c) $ArchiveUpdateP_{Q,\epsilon}Dy$

Fig. 8.16 Results of the four $P_{Q,\epsilon}$ archivers for SSW

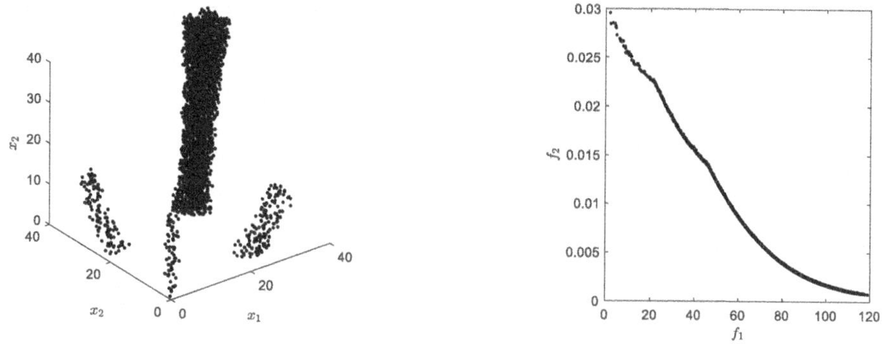

(d) $ArchiveUpdateP_{Q,\epsilon}D_{xy}$

Fig. 8.16 (continued)

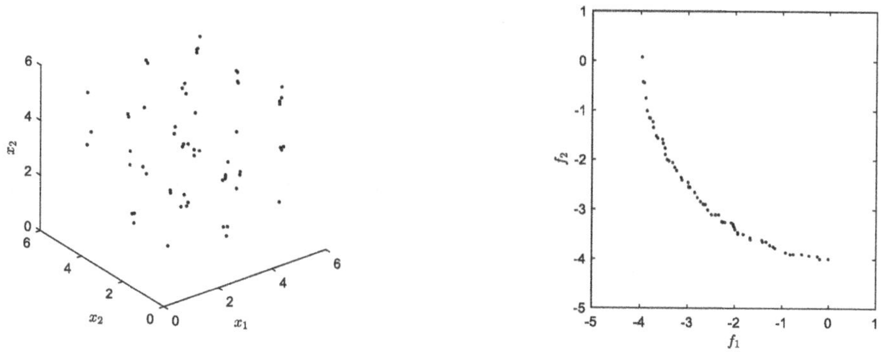

(a) $ArchiveUpdateP_{Q,\epsilon}$

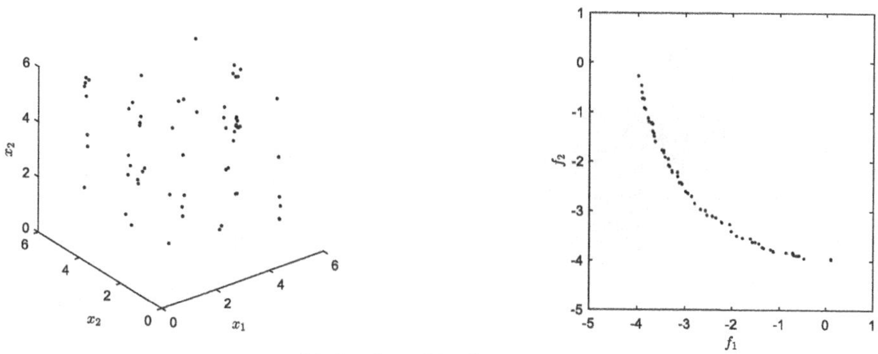

(b) $ArchiveUpdateP_{Q,\epsilon}Dx$

Fig. 8.17 Results of the four $P_{Q,\epsilon}$ archivers for OMNI-TEST

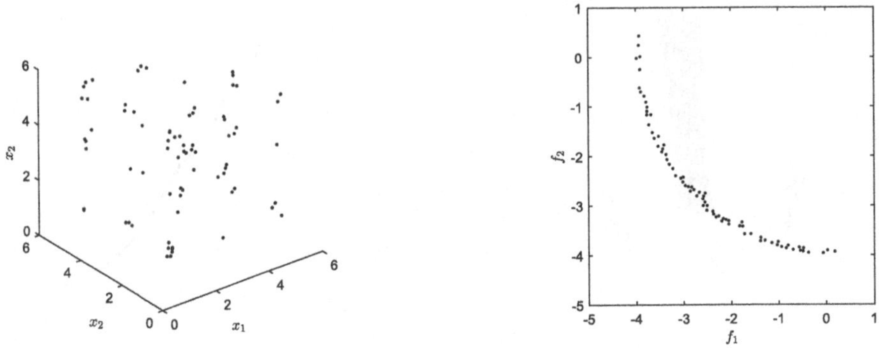

(c) $ArchiveUpdateP_{Q,\epsilon}Dy$

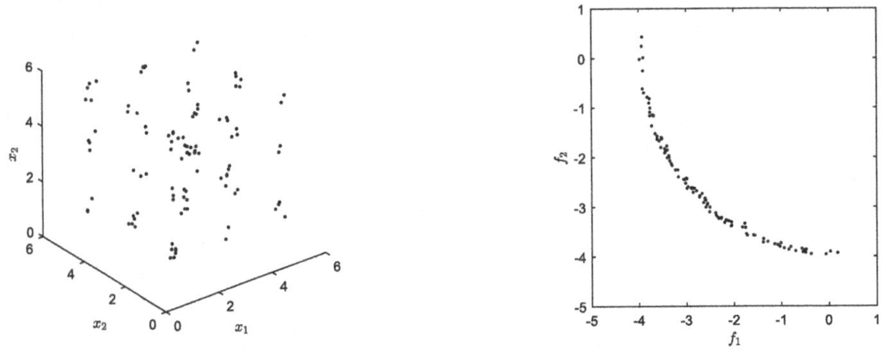

(d) $ArchiveUpdateP_{Q,\epsilon}D_{xy}$

Fig. 8.17 (continued)

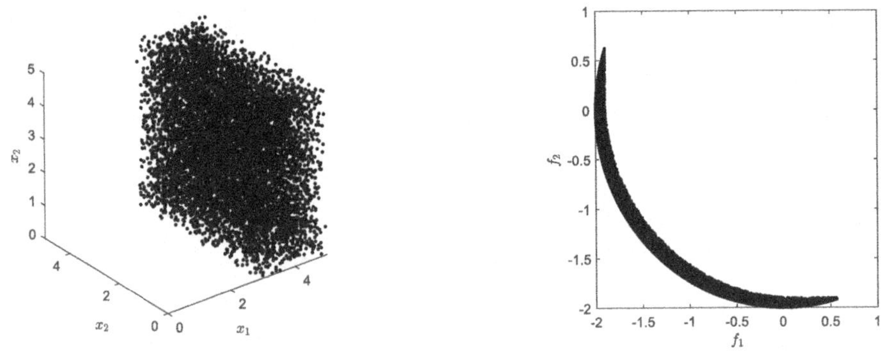

(a) $ArchiveUpdateP_{Q,\epsilon}$

Fig. 8.18 Results of the four $P_{Q,\epsilon}$ archivers for LSS

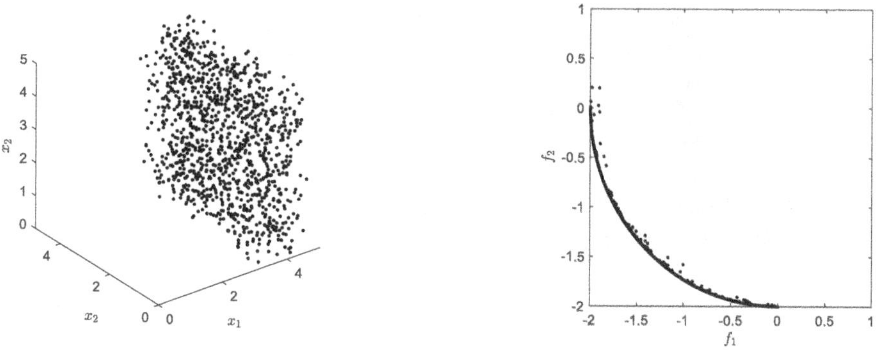

(b) $ArchiveUpdateP_{Q,\epsilon}Dx$

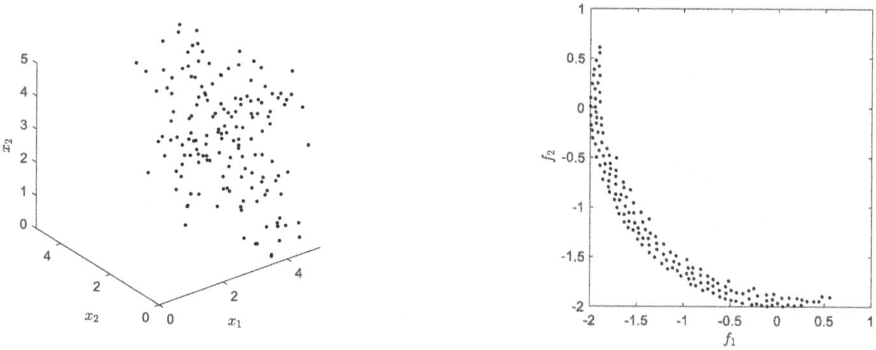

(c) $ArchiveUpdateP_{Q,\epsilon}Dy$

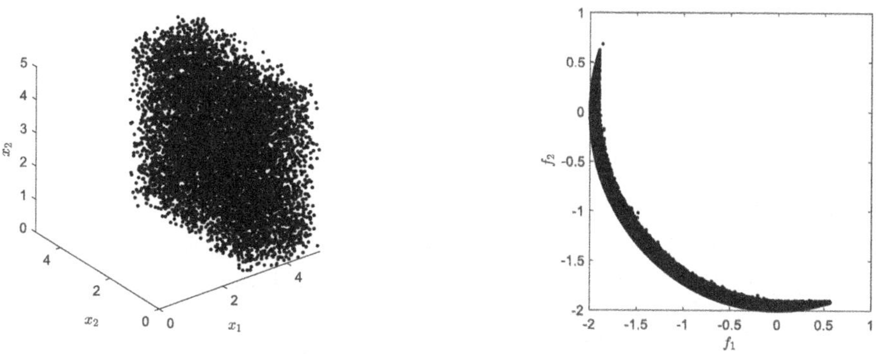

(d) $ArchiveUpdateP_{Q,\epsilon}D_{xy}$

Fig. 8.18 (continued)

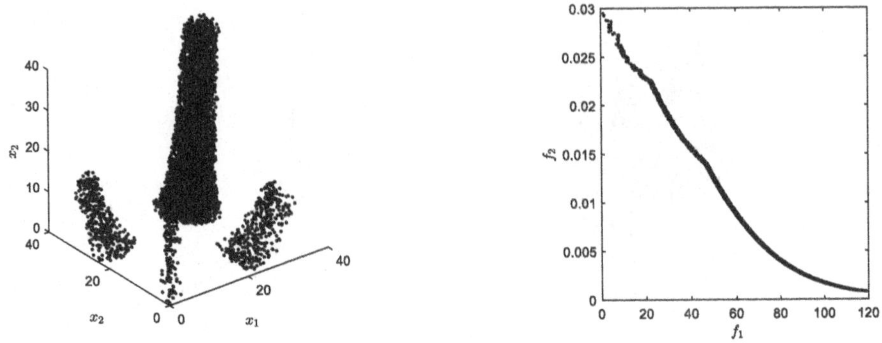

(a) $ArchiveUpdateP_{Q,\epsilon}$

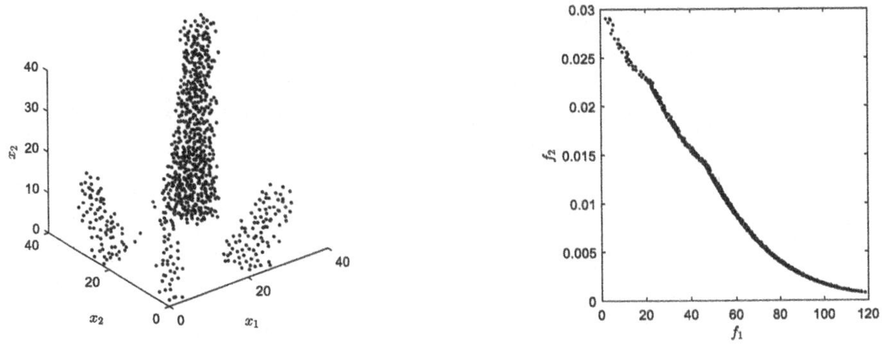

(b) $ArchiveUpdateP_{Q,\epsilon}Dx$

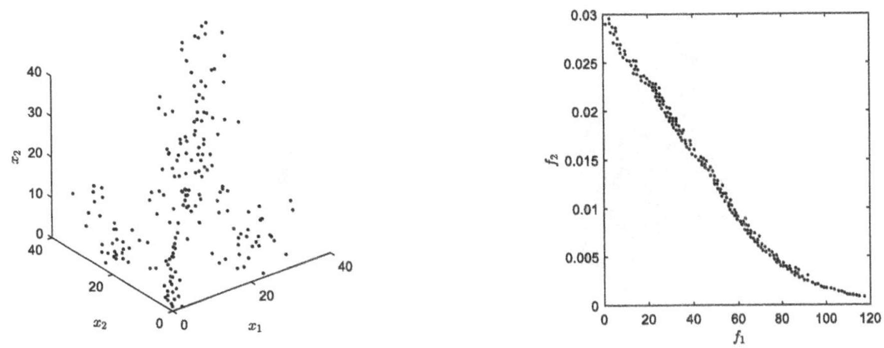

(c) $ArchiveUpdateP_{Q,\epsilon}Dy$

Fig. 8.19 Results of the four $P_{Q,\epsilon}$ archivers for SSW but using $\Delta_y = (1.2, 0.003)^T$. Compare to Fig. 8.16

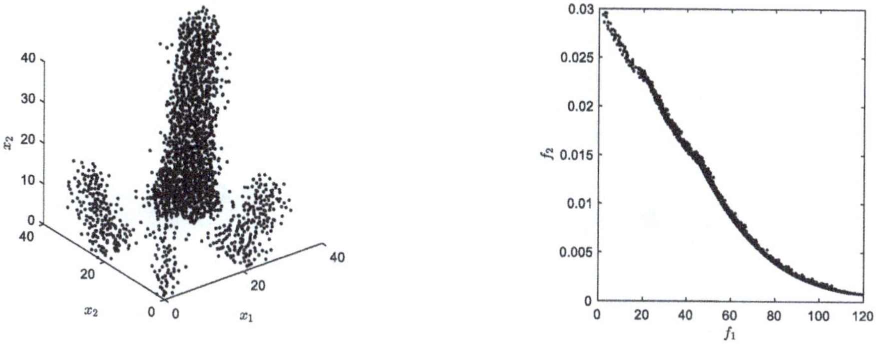

(d) $ArchiveUpdateP_{Q,\epsilon}D_{xy}$

Fig. 8.19 (continued)

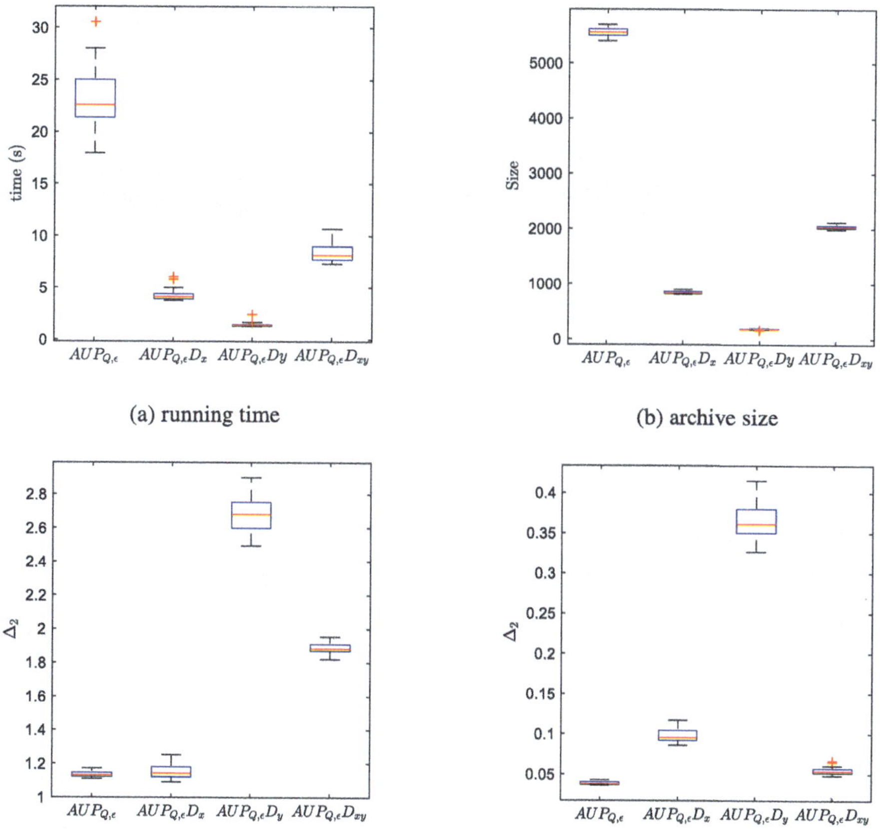

(a) running time

(b) archive size

(c) $\Delta_2(A_f, P_{Q,\epsilon})$

(d) $\Delta_2(F(A_f), F(P_{Q,\epsilon}))$

Fig. 8.20 Box plots for running time, archive size and approximation quality of the four archivers on SSW with $\Delta_x = (1.2, 0.003)^T$, averaged over 30 independent runs. Compare to Fig. 8.19

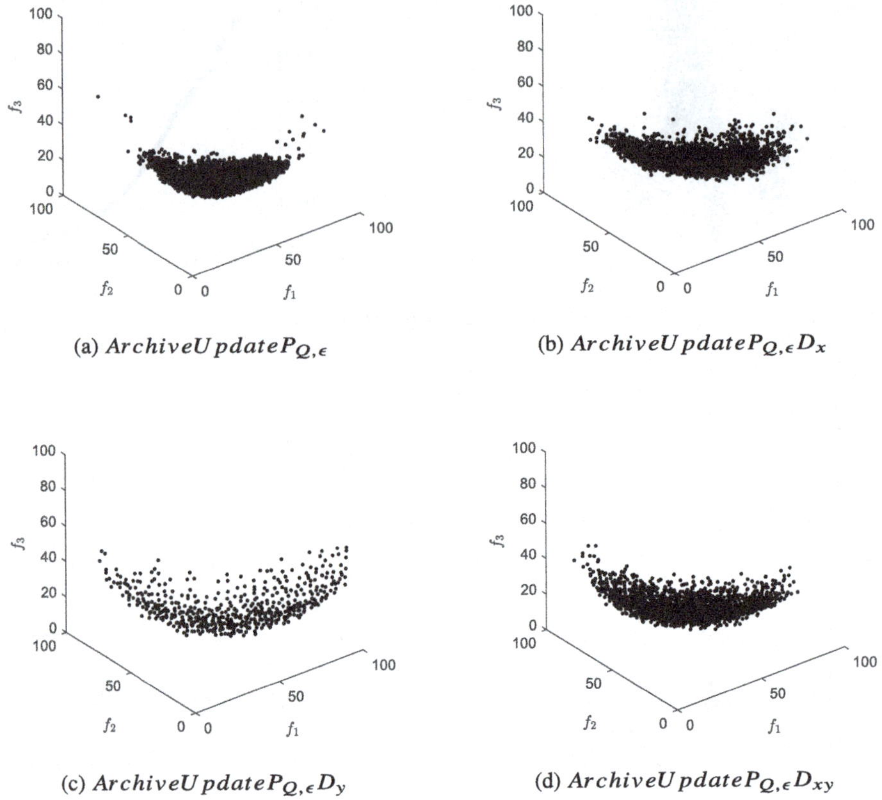

(a) $ArchiveUpdateP_{Q,\epsilon}$

(b) $ArchiveUpdateP_{Q,\epsilon}D_x$

(c) $ArchiveUpdateP_{Q,\epsilon}D_y$

(d) $ArchiveUpdateP_{Q,\epsilon}D_{xy}$

Fig. 8.21 Results of the four $P_{Q,\epsilon}$ archivers for CONV3 using $n = 30$ decision variables

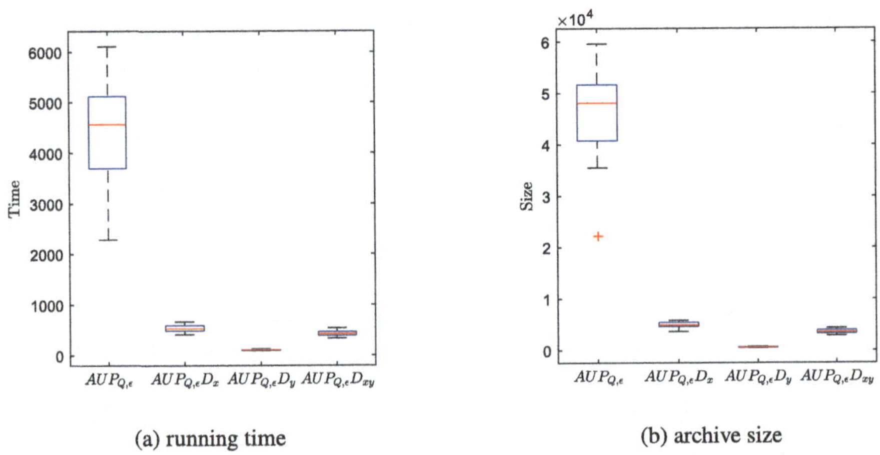

(a) running time

(b) archive size

Fig. 8.22 Box plots for running times and archive sizes for the results on CONV3 for $n = 30$. Compare to Fig. 8.21

Chapter 9
A Short Excursion to Scalar Optimization: Computing the Set of Approximate Solutions for SOPs

The concepts we have seen in the previous chapter can also be applied to scalar objective optimization problems (SOPs). Unlike for MOPs, where one expects an entire manifold of solutions, the solution of a SOP is typically given by a singleton. That is, if one is "only" interested in the best solution for a given SOP, one can simply keep and update the best found solution during the run of the algorithm. For instance, within an evolutionary algorithm, the best found solution in each step will always be kept in the current population, and there is no need for an additional archiver. This situation changes, however, if one is next to the optimum also interested in approximate solutions, e.g. as alternatives or back up solutions for the decision maker. As example, consider the hypothetical problem depicted in Fig. 9.1. The point x_1 is the unique optimum for this problem and hence the only point of interest if one seeks for the best solution. If one allows a deterioration of $\epsilon > 0$ in objective space, then one may either be as well interested in the point x_2 which is locally optimal and which has a deterioration less than ϵ in objective space. Or, depending on the problem, one may even want to compute solutions in the neighborhood of both points x_1 and x_2 since these points are nearly also optimal, while neither of them is locally (or globally) optimal. The archiving strategies we will discuss in the following will allow for both.

9.1 The Set of Interest

A continuous scalar optimization problem (SOP) can be expressed mathematically as

$$\min_{x \in Q} f(x), \qquad \text{(SOP)}$$

O. Schütze and C. Hernández, *Archiving Strategies for Evolutionary Multi-objective Optimization Algorithms*, Studies in Computational Intelligence 938,
https://doi.org/10.1007/978-3-030-63773-6_9

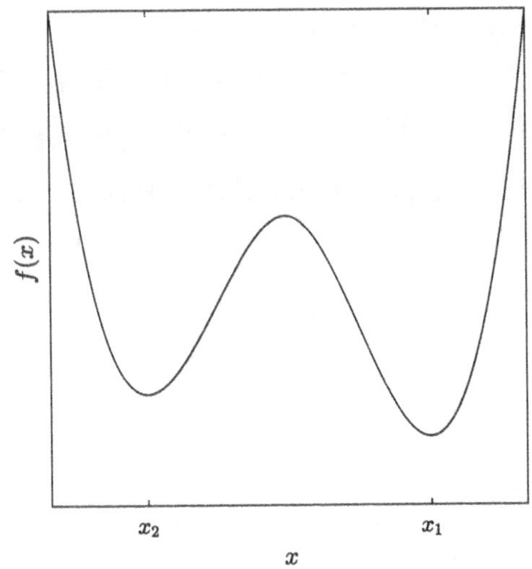

Fig. 9.1 Example of an unconstrained SOP that contains one global minimum at x_1 and another local minimum at x_2. Since x_1 and x_2 are distinct from each other (in decision variable space) but have similar objective values, both points may be of interest for the decision maker

where $f : Q \to \mathbb{R}$ is the (only) objective function and where $Q \subset \mathbb{R}^n$ is the domain. Analogous to MOPs, we call a point $x \in Q$ of problem (SOP) feasible, and points $x \notin Q$ infeasible.
A solution to (SOP) is a point $x^* \in Q$ that satisfies

$$f(x^*) \leq f(y) \qquad \forall y \in Q. \tag{9.1}$$

We denote by M_Q the set of all solutions to (SOP) within the domain Q, i.e.,

$$M_Q = \{x \in Q \; : \; f(x) \leq fy) \;\; \forall y \in Q\}. \tag{9.2}$$

This set can also consist of several isolated solutions (as for the sine function within a given interval) or even be a continuous set (as for plateau functions). For most SOPs, however, M_Q is a singleton, i.e., contains exactly one solution. This includes also the problem that is depicted in Fig. 9.1. As discussed above, next to the global optimum x_1 also the local optimum x_2 (or points near to it) may be of interest to the decision maker. In order to include such points, we consider the set of nearly optimal solutions. The definition of this set, $M_{Q,\epsilon}$, is analog to the definition of $P_{Q,\epsilon}$ for multi-objective optimization problems, which we present in the following.

Definition 9.1 ($-\epsilon$*-dominance for SOPs*) Let $\epsilon \in \mathbb{R}_+$ and $x, y \in \mathbb{R}^n$. x is said to $-\epsilon$-dominate y ($x \prec_{-\epsilon} y$) with respect to (SOP) if

$$f(x) + \epsilon < f(y). \tag{9.3}$$

Definition 9.2 (*Set of approximate solutions $M_{Q,\epsilon}$*) Denote by $M_{Q,\epsilon}$ the set of points in $Q \subset \mathbb{R}^n$ which are not $-\epsilon$-dominated by any other point in Q, i.e.,

$$M_{Q,\epsilon} := \{x \in Q \ : \ \nexists y \in Q : y \prec_{-\epsilon} x\}. \tag{9.4}$$

The above definition is done in analogy to $P_{Q,\epsilon}$. Since we are now considering only one objective, we can simplify the definition of $M_{Q,\epsilon}$ as follows: let $x^* \in M_Q$ be a solution to (SOP) and $y^* = f(x^*)$, then the set of approximate solutions is given by

$$M_{Q,\epsilon} = \{x \in Q \ : \ f(x) \leq y^* + \epsilon\}. \tag{9.5}$$

That is, if Q is compact, then also $M_{Q,\epsilon}$, and the boundary of this set is either given by boundary points of Q or by points in B, where

$$B = \{x \in Q \ : \ f(x) = y^* + \epsilon\}. \tag{9.6}$$

Points $b \in B$ are the equivalent to the $-\epsilon$-weak Pareto points for MOPs.

Example 9.1 (a) Consider the unconstrained SOP that is given by the objective

$$\begin{aligned} f :&\mathbb{R}^n \to \mathbb{R} \\ f(x) &= \sum_{i=1}^{n} x_i^2. \end{aligned} \tag{9.7}$$

Then, the point $x^* = 0$ is the only solution to (SOP). Further, for a given $\epsilon > 0$ the closed ball with centre x^* and radius $\sqrt{\epsilon}$ is equal to the set of approximate solutions, and B is given by the $(n-1)$-sphere with same centre and radius, i.e.,

$$\begin{aligned} M_Q &= \{0\} \\ M_{Q,\epsilon} &= \left\{x \in \mathbb{R}^n \ : \ \sum_{i=1}^{n} x_i^2 \leq \epsilon\right\} \\ B &= \left\{x \in \mathbb{R}^n \ : \ \sum_{i=1}^{n} x_i^2 = \epsilon\right\}. \end{aligned} \tag{9.8}$$

(b) Consider the unconstrained SOP that is depicted in Fig. 9.2. For this problem, we have

$$\begin{aligned} M_Q &= \{x^*\} \\ M_{Q,\epsilon} &= [a, b] \cup [c, d] \\ B &= \{a, b, c, d\}. \end{aligned} \tag{9.9}$$

In particular, the set of approximate solutions is disconnected for this example.

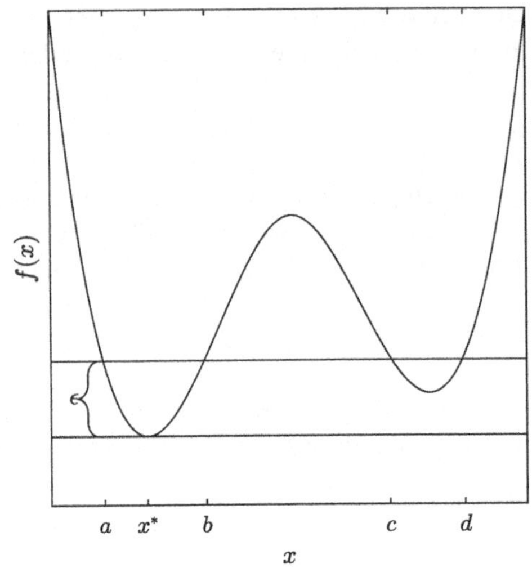

Fig. 9.2 Example of a SOP where $M_{Q,\epsilon}$ is disconnected

We further discuss the set of interest which is analog to the discussion of $P_{Q,\epsilon}$:

- (Dimensionality) We can expect that $M_{Q,\epsilon}$ forms a set of dimension n.
- (Connectedness) The set $M_{Q,\epsilon}$ does not have to be connected as the above example shows. Connectedness is given for convex problems.

Corollary 9.1 *Let $\epsilon \in \mathbb{R}_+$. If Q and the objective f of (SOP) are convex, then $M_{Q,\epsilon}$ and $F(M_{Q,\epsilon})$ are connected.*

- (Robustness) For the convergence analysis we have to assume that the set of approximate solutions does not contain isolated points. If we assume that B is contained in the interior of the domain B (and hence, that B forms the boundary of $M_{Q,\epsilon}$) and that the gradient of f does not vanish on each boundary point

$$B \subset int(Q) \quad \text{and} \quad \nabla f(x) \neq 0 \;\; \forall x \in B \tag{9.10}$$

then one can show analog to the multi-objective case that the set $M_{Q,\epsilon}$ is robust, i.e.,

$$clos(int(M_{Q,\epsilon}) = clos(M_{Q,\epsilon}). \tag{9.11}$$

Finally, we stress that the concept of approximate solutions can be used in the context of scalar optimization to determine multiple optima. This is due to the fact that each point $x^* \in M_Q$ is also contained in $M_{Q,\epsilon}$ for every $\epsilon > 0$. Even more, it holds

$$M_Q = \bigcap_{\epsilon>0} M_{Q,\epsilon}, \tag{9.12}$$

i.e., the above characteristic only holds for points $x^* \in M_Q$.
Elitist approaches, on the other hand, have strong limitations to approximate M_Q if it is not a singleton. This is due to the fact that one can expect that there is only one 'best' solution out of a given finite set of candidate solutions. We have to take this into consideration for archivers that aim for suitable discretizations of $M_{Q,\epsilon}$. That is, the discretization cannot merely be done in objective space leading to the need of an additional discretization parameter in decision variable space. If this one is chosen appropriately, however, we will be able to approximate all solutions $x^* \in M_Q$ as we will see on several examples.

In evolutionary computation, several methods exist that share some characteristics with the archivers proposed in the sequel. There exist, for instance, several mechanisms to preserve the diversity within the populations of an evolutionary algorithm (Shir et al. 2009; Crepinsek et al. 2013; Segura et al. 2015, 2016, 2017; Cuate and Schütze 2019). This, however, is mainly done in order to prevent pre-mature convergence, and nearly optimal solutions are not explicitly maintained. Further, the goal in multi-modal optimization is to ideally detect *all* locally optimal solutions of a given SOP, i.e., also local solutions that are not nearly optimal (Sebag et al. 2005; Stoean et al. 2010; Preuss 2015; Wessing and Preuss 2016; Liu et al. 2020).

9.2 $ArchiveUpdateM_{Q,\epsilon}$

We again first consider the archiver that maintains all $-\epsilon$-non-dominated solutions out of the given candidate set. See Algorithm 11 for the pseudo code of this archiver which is called $ArchiveUpdateM_{Q,\epsilon}$. The archiver coincides with $ArchiveUpdateP_{Q,\epsilon}$ for the special case that one objective is considered. The following results on the monotonic behavior and the convergence are hence analog which we state for sake of completeness.

Algorithm 11 $A := ArchiveUpdateM_{Q,\epsilon}\ (P, A_0)$

Require: P: population, A_0: current archive, $\epsilon \in \mathbb{R}_+$
Ensure: updated archive A

$A := A_0$
for all $p \in P$ **do**
  if $\nexists a \in A:\ f(a) + \epsilon \leq f(p)$ **then**
    $A := A \cup \{p\}$
  end if
  for all $a \in A$ **do**
    if $f(p) + \epsilon < f(a)$ **then**
      $A := A \backslash \{a\}$
    end if
  end for
end for
return A

Corollary 9.2 *Let (SOP) be given, let $l \in \mathbb{N}$, $\epsilon \in \mathbb{R}_+$, $P_0, \ldots, P_l \subset \mathbb{R}^n$ be finite sets, and A_i, $i = 0, \ldots, l$, be obtained by $ArchiveUpdateM_{Q,\epsilon}$ as in Algorithm 1, and $C_l = \bigcup_{i=0}^{l} P_i$. Then*

$$A_l = M_{C_l,\epsilon} = \{x \in C_l \ : \ \nexists y \in C_l : y \prec_{-\epsilon} x\}. \tag{9.13}$$

Corollary 9.3 *Let (SOP) be given, where f is continuous and $Q \subset \mathbb{R}^n$ is compact, and $\epsilon \in \mathbb{R}^k_+$. Then, under the assumptions (4.1) and (9.10), an application of Algorithm 1, where $ArchiveUpdateP_{Q,\epsilon}$ (Algorithm 11) is used to update the archive, leads to a sequence of archives A_l, $l \in \mathbb{N}$, with*

$$\lim_{l \to \infty} d_H(M_{Q,\epsilon}, A_l) = 0, \quad \text{with probability one.} \tag{9.14}$$

As for $ArchiveUpdateP_{Q,\epsilon}$, we can also expect that this archiver is generally unbounded problems, i.e., that we obtain

$$\lim_{i \to \infty} A_i = \infty. \tag{9.15}$$

We will hence consider a discretization strategy in the next section.

9.3 $ArchiveUpdateM_{Q,\epsilon}D_x$

As discussed above, it is not wise to merely consider the objective values in order to obtain a discretization of the set of interest in the context of scalar optimization. The general idea of $ArchiveUpdateM_{Q,\epsilon}D_x$ is to keep locally the best found solution within a certain range (using the vector $\Delta \in \mathbb{R}^n$ as discretization parameter in decision variable space) and to discard inferior points in the neighborhood of such point. Algorithm 12 shows the pseudo code of $ArchiveUpdateM_{Q,\epsilon}D_x$, where one candidate solution is used for the update of the archive A_0. The candidate p is rejected (and hence, the new archive A is set to the current archive A_0) if p is either not an ϵ-approximate solution of A_0 (i.e., if $f(x_b) + \epsilon < f(p)$, where x_b is the best found solution so far), or if there exists an archive entry $a \in A_0$ in the neighborhood $B^{\infty}_{\Delta}(p)$ which is at least as good as p (i.e., $f(a) \leq f(p)$), see line 6 of Algorithm 12. If none of the two conditions are true, it follows that (i) p is an approximate solution of the current archive A_0, and (ii) it is the best solution in its neighborhood (measured by Δ in decision variable space). In this case, the new archive will consist of p as well as all other elements of A_0 that are ϵ-approximate solutions, and which are not located in the Δ-neighborhood of p (lines 10 to 14).

Note that in the definition of $ArchiveUpdateM_{Q,\epsilon}D_x$ we have added the best found solution both in the input and in the output. This has just been done for convenience, and the archiver still fits into the framework of the generic stochastic search algorithm. The update of an entire population is straightforward and can be found in Algorithm 3.

Algorithm 12 $\{A, x_b\} := ArchiveUpdateM_{Q,\epsilon}D_x\ (p, A_0, x_{b,0})$

Require: p: candidate solution, A_0: current archive, $x_{b,0}$: current best found solution, $\epsilon \in \mathbb{R}_+$, $\Delta \in \mathbb{R}_+$
Ensure: updated archive A, updated best found solution x_b
if $f(p) < f(x_{b,0})$ **then**
  $x_b := p$
else
  $x_b := x_{b,0}$
end if
if $f(x_b) + \epsilon < f(p)$ or $\{\exists a \in A_0 : p \in B_\Delta^\infty(a)$ and $f(a) \leq f(p)\}$ **then**
  $A := A_0$ ▷ discard p
  return A
end if
$A := \{p\}$
for all $a \in A_0$ **do**
  if $f(a) \leq f(x_b) + \epsilon$ and $a \notin B_\Delta^\infty(p)$ **then**
    $A := A \cup \{a\}$
  end if
end for
return $\{A, x_b\}$

Algorithm 13 $\{A, x_b\} := ArchiveUpdateM_{Q,\epsilon}D_x\ (P, A_0, x_{b,0})$

Require: $P = \{p_1, \ldots, p_s\}$: population, A_0: current archive, $x_{b,0}$: current best found solution, $\epsilon \in \mathbb{R}_+$, $\Delta \in \mathbb{R}_+$
Ensure: updated archive A, updated best found solution x_b
$\{A_1, x_{b,1}\} = ArchiveUpdateM_{Q,\epsilon}D_x\ (p_1, A_0, x_{b,0})$
for $i = 1, \ldots, s-1$ **do**
  $\{A_{i+1}, x_{b,i+1}\} = ArchiveUpdateM_{Q,\epsilon}D_x\ (p_i, A_i, x_{b,i})$
end for
$A := A_s$
$x_b := x_{b,s}$
return $\{A, x_b\}$

The analysis of the final archive A_l after finitely many iterations l is not as straightforward as for the previous archives. It is easy to verify that the element $x_{b,l}$ is indeed the (or one of the) best elements out of $C_l = \bigcup_{i=0}^{l} A_i$, and that the final archive contains only of ϵ-approximate solutions, i.e.,

$$\begin{aligned} x_{b,l} &\in M_{C_l} \quad \text{and} \\ A_l &\subset M_{C_l,\epsilon}. \end{aligned} \tag{9.16}$$

Further approximation qualities, however, cannot be given since the final archive A_l depends on the order of the elements that are considered by the archiver. To see this, consider the one-dimensional problem that is given by the following objective:

$$\begin{aligned} f &: [0, 10] \to \mathbb{R} \\ f(x) &= x^2. \end{aligned} \tag{9.17}$$

For $\epsilon = 1$ it is $M_{Q,\epsilon} = [0, 1]$. Now let A_0 be the empty set, the discretization parameter be given by $\Delta = 0.1$, and the population be

$$P_1 = \{0, 0.05, 0.1, \ldots, 1\}, \tag{9.18}$$

and that all elements are considered by the archiver in this order. Then, an application of $ArchiveUpdateM_{Q,\epsilon}D_x$ leads to the updated archive

$$A_1 = \{0, 0.1, 0.2, \ldots, 1\} \tag{9.19}$$

However, if the entries of P_1 are considered in reversed order the final archive would be given by

$$\tilde{A}_1 = \{0\}. \tag{9.20}$$

Note, however, that all regions of the domain are re-visited after finitely many steps by the generation process during the run of the algorithm. Hence, for this problem one will always obtain archive $\tilde{A}_1$ in the limit.
The following theorem specifies more generally what to expect in the limit.

Theorem 9.1 *Let (SOP) be given and $\epsilon \in \mathbb{R}_+$. Let $x^* \in M_{Q,\epsilon}$ be the unique minimum of f within $Q \cap B^\infty_{\tilde{\Delta}}$, where $\tilde{\Delta} \in \mathbb{R}^n$ with $\tilde{\Delta}_i > \Delta_i$, $i = 1, \ldots, n$. Then, under the assumption (4.1), an application of Algorithm 1, where $ArchiveUpdateM_{Q,\epsilon}D_x$ (Algorithm 13) is used to update the archive, leads to a sequence of archives A_l, $l \in \mathbb{N}$, such that with probability one*

(a) $\exists a_l \in A_l \, : \, a_l \to x^* \;\; \text{for} \;\; l \to \infty$
(b) $A_l \cap B^\infty_\Delta(x^*) = \{a_l\}$

Proof *Ad (a)* By assumption (4.1) on the generation process there exists with probability one a sequence of candidate solutions $p_{l_i} \in Q$ such that $p_{l_i} \in M_{Q,\epsilon}$ and $p_{l_i} \to x^*$. By construction of $ArchiveUpdateM_{Q,\epsilon}D_x$ each candidate p_{l_i} is either discarded if there already exists an archive entry $a_{l_i} \in A_{l_i}$ with $f(a_{l_i}) \leq f(p_{l_i})$ (line 6 of Algorithm 12), or $a_{l_i} := p_{l_i}$ is added to the archive (line 10 of Algorithm 12). Since entries $a \in A_l$ are only replaced if there is a better solution in the Δ-neighborhood of a, there exists with probability for all $l \in \mathbb{N}$ an archive entry $a_l \in A_l \cap B^\infty_\Delta(x^*)$ such that for this sequence it holds $f(a_l) \to f(x^*)$ for $l \to \infty$. Since x^* is the unique solution in $Q \cap B^{\infty}_{\Delta}$, it follows that $a_l \to x^*$ for $l \to \infty$ with probability one, and the claim follows.

Ad (b) Follows by part (a) and the construction of the archiver $ArchiveUpdateM_{Q,\epsilon}D_x$. □

We stress that this result holds in principle for all local isolated minima within $M_{Q,\epsilon}$, but that the limit archive (if existing) heavily depends on the choice of the discretization parameter Δ which makes an in-depth discussion of the archive sizes impossible. In general, the archive will be one if Δ is chosen sufficiently large. If Δ is too small, the archive sizes will be tremendous. If Δ is chosen 'in between', the archive sizes will converge toward the number of local optimal within $M_{Q,\epsilon}$ (and containing approximations of each of them). The choice of Δ is of course highly problem dependent, and no value can be given in general. As a rule of thumb in case, the SOP is derived from a real-world problem, one can choose Δ such that two candidate solutions x_1 and x_2 with $\|x_1 - x_2\| \approx \Delta$ are considered to be (significantly) different for the decision maker.

Finally, note that in the above discussion we have used $\Delta \in \mathbb{R}_+$, i.e., a scalar value for the discretization of the decision variable space. Similar to the previous chapter, we can of course use the vector $\Delta \in \mathbb{R}^n_+$ if necessary. Doing so, one can use the box

$$B^{\infty}_{\Delta}(y) = \{x \in \mathbb{R}^n \ : \ |x_i - y_i| \leq \Delta_i\} \tag{9.21}$$

for the realization of $ArchiveUpdateM_{Q,\epsilon}D_x$.

We will investigate the choice of the discretization parameter and its impact on the final archiver empirically in the next section.

9.4 Numerical Results

Here be briefly demonstrate the effects of the two archivers dealing with the approximation of $M_{Q,\epsilon}$ on three low-dimensional test problems. For an application of the archivers in the context of space mission design, we refer to Schütze et al. (2011b).

9.4.1 Sphere Function

As first example, we consider here is the two-dimensional sphere function which is defined as

$$\begin{aligned} f &: [-2, 2]^2 \to \mathbb{R} \\ f(x) &= x_1^2 + x_2^2. \end{aligned} \tag{9.22}$$

The problem contains one minimizer, $M_Q = \{0\}$, and for $\epsilon = 2$ the set of approximate solutions is the circle with center $(0, 0)^T$ and radius $\sqrt{2}$, $M_{Q,\epsilon} = \{(x, y) : x^2 + y^2 \leq 2\}$.

Figure 9.3 shows some numerical results obtained by $ArchiveUpdateM_{Q,\epsilon}$ for different numbers N of candidate solutions that were drawn uniformly at random from the domain $Q = [-2, 2]^2$. As it can be seen, the archiver eventually captures the entire solution set, which is – as for $ArchiveUpdateP_Q$ and $ArchiveUpdateP_{Q,\epsilon}$ that do not utilize any discretization of the set of interest – only relevant if relatively few candidate solutions are being considered in total, and/or if the time for the function evaluation is relatively costly.

Figure 9.4 shows some final archives of $ArchiveUpdateM_{Q,\epsilon}D_x$ on the same problem. In all cases, $N = 100{,}000$ candidate solutions have been used to feed the archiver. As anticipated, the archive sizes go down with incresing values of Δ. For $\Delta = 2$ – and in fact, for all values larger than $\sqrt{2}$ – the final archive consists only of one point that is located in the vicinity of M_Q.

As discussed above, one can also choose vectors $\Delta \in \mathbb{R}^n_+$ if the discretization has to be done with different values according to each coordinate direction. For this, the box $B^\infty_\Delta(y)$ as defined in (9.21) can be chosen for the realization of $ArchiveUpdateM_{Q,\epsilon}D_x$. We show here exemplary the effect of the resulting archiver on two examples in Fig. 9.5. Figure 9.5a–c show results of the modified $ArchiveUpdateM_{Q,\epsilon}D_x$ for $N_1 = 100{,}000$, $N_2 = 1{,}000{,}000$ and $N_3 = 10{,}000{,}000$ candidate solutions and for $\Delta = (0.2, 2)^T$. Note that Δ_2, the second entry of Δ, is larger than $\sqrt{2}$. Hence, the archiver will have for this problem from one iteration on not have two entries $a_1, a_2 \in A$ with same same or similar x_1 value that have different x_2 values. More precisely, the x_1 values of a_1 and a_2 have to differ at least by Δ_1. Consequently, the archiver will aim for a uniform spread set of solutions along the horizontal cut of $M_{Q,\epsilon}$ (i.e., the x_1-axis intersected with $M_{Q,\epsilon}$). Figure 9.5d shows an analog result for N_3, but this time using $\Delta = (2, 0.2)^T$ resulting in an approximation of the vertical cut of $M_{Q,\epsilon}$.

9.4.2 Himmelblau's Function

Next, we consider the following SOP (Himmelblau 1972):

$$\begin{aligned} f &: [-6, 6]^2 \to \mathbb{R} \\ f(x, y) &= (x^2 + y - 11)^2 + (x + y^2 - 7)^2. \end{aligned} \tag{9.23}$$

This function has four isolated global minimizer within the domain $Q = [-6, 6]^2$. It is $M_Q = \{m_1, m_2, m_3, m_4\}$, where $m_1 = (3, 2)$, $m_1 \approx (-2.8051, 3.1313)$, $m_3 \approx (-3.7793, 3.2832)$, and $m_4 \approx (3.5844, -1.8481)$.

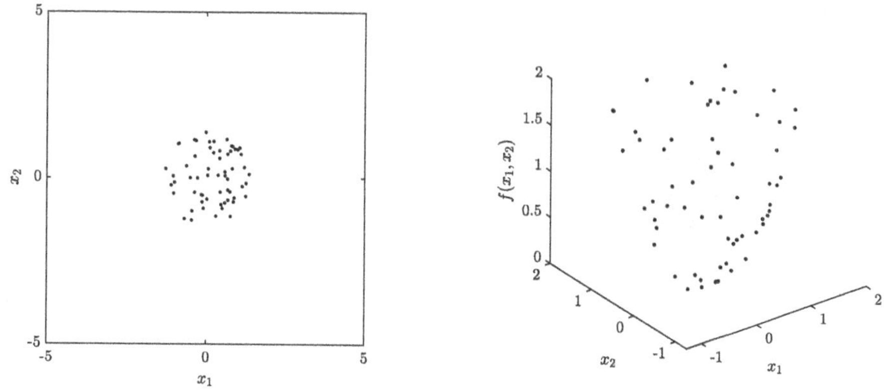

(a) $N = 1,000$, $|A_f| = 61$

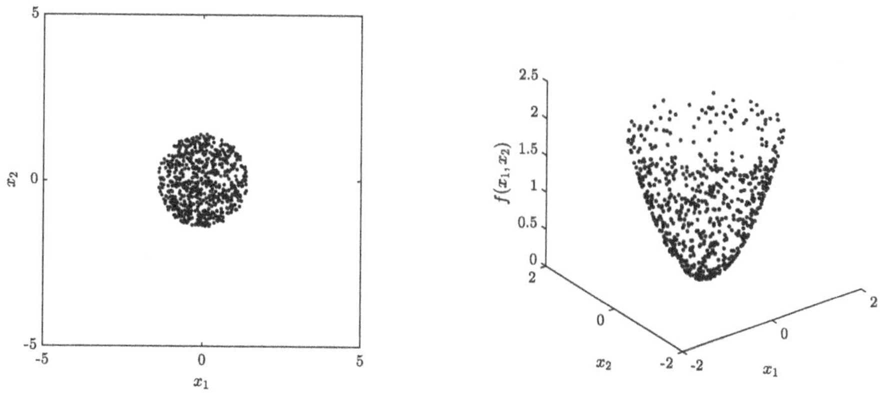

(b) $N = 10,000$, $|A_f| = 635$

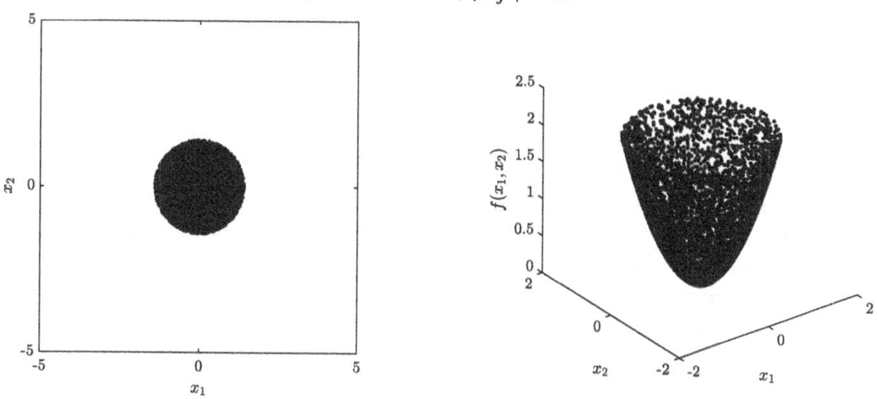

(c) $N = 100,000$, $|A_f| = 6346$

Fig. 9.3 Numerical results of $ArchiveUpdateM_{Q,\epsilon}$ on the sphere function for different values N of candidate solutions. Shown are the final approximations in decision variable space (left) and together with the objective value (right)

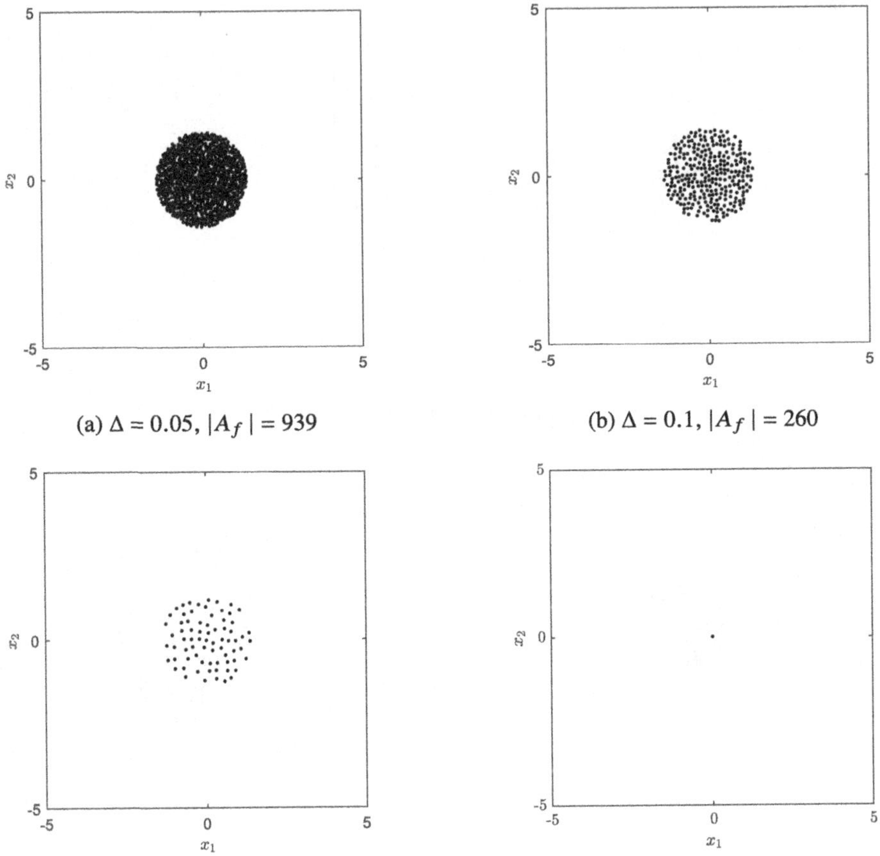

(a) $\Delta = 0.05$, $|A_f| = 939$

(b) $\Delta = 0.1$, $|A_f| = 260$

(c) $\Delta = 0.2$, $|A_f| = 67$

(d) $\Delta = (0.05, 2)^T$, $|A_f| = 1$

Fig. 9.4 Numerical results of $ArchiveUpdateM_{Q,\epsilon}D_x$ on the sphere function for difference values of Δ

Figure 9.6 shows the some numerical results for this problem using $ArchiveUpdateM_{Q,\epsilon}D_x$, $\epsilon = 2$, and different values of Δ. The figures show that also for this problem the archiver is capable of detecting suitable approximations of all approximate solutions. For $\Delta = 1$, the final archiver consists of four elements that tightly approximate M_Q. If Δ would be chosen larger, the final archive would not contain all four minimizer any more. For intance, choose $\Delta = 10$ (or larger). Then, for every point a near enough to any element of M_Q the box $B_\Delta^\infty(a)$ will contain

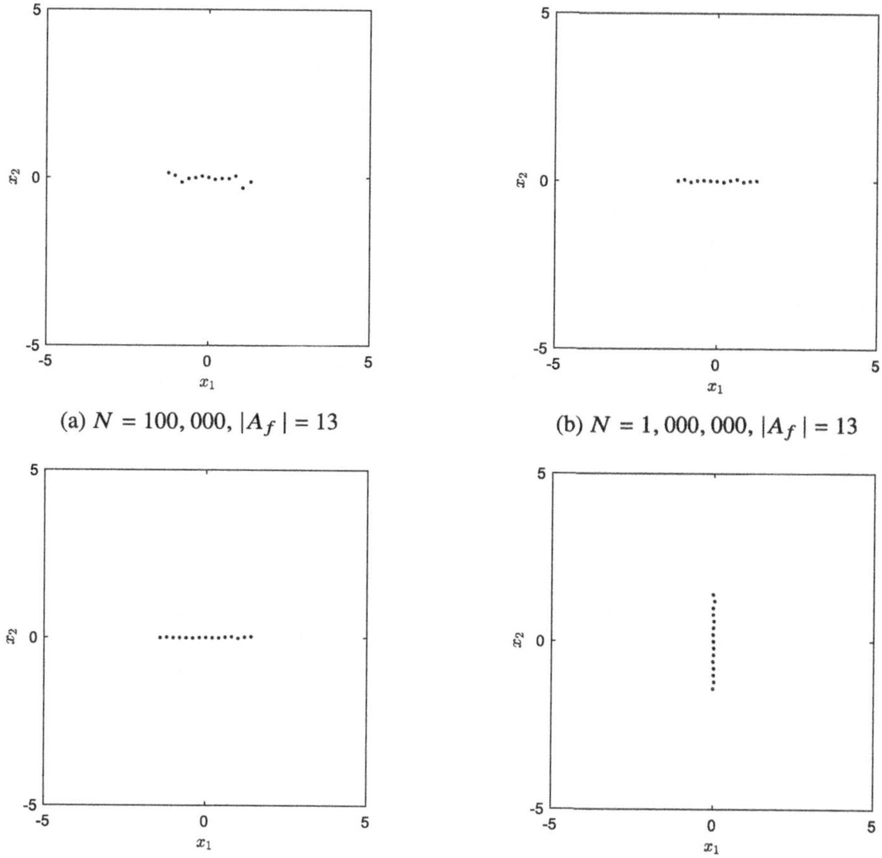

(a) $N = 100,000$, $|A_f| = 13$

(b) $N = 1,000,000$, $|A_f| = 13$

(c) $N = 10,000,000$, $|A_f| = 15$

(d) $N = 10,000,000$, $|A_f| = 1$

Fig. 9.5 Numerical results of $ArchiveUpdateM_{Q,\epsilon}D_x$ on the sphere function for difference values of Δ. In (**a**)–(**c**), $\Delta = (0.2, 2)^T$ has been used, and $\Delta = (2, 0.2)^T$ in (**d**)

the set M_Q as well as all points near to it. The archive will hence eventually only contain *one* element a_j in step j, and a_j is near to a solution $m_i \in M_Q$. Further, the proximity to the element of M_Q may change over the iteration, similar as in the situation discussed above around Fig. 5.1.

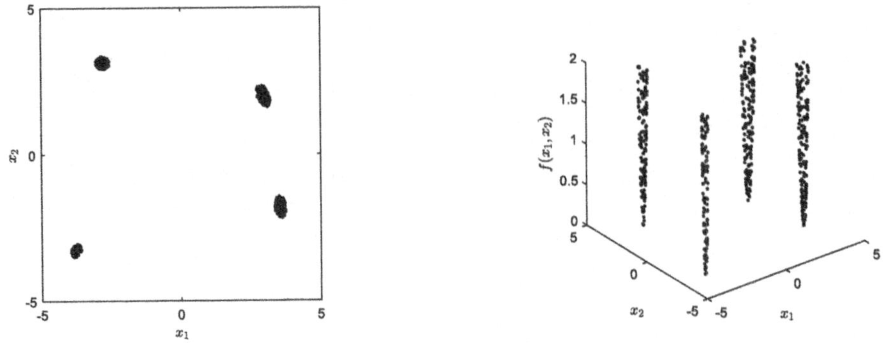

(a) $\Delta = 0.001$, $|A_f| = 5731$

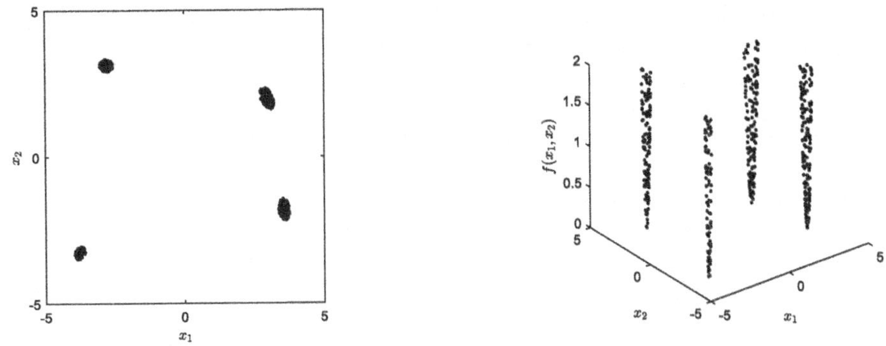

(b) $\Delta = 0.01$, $|A_f| = 439$

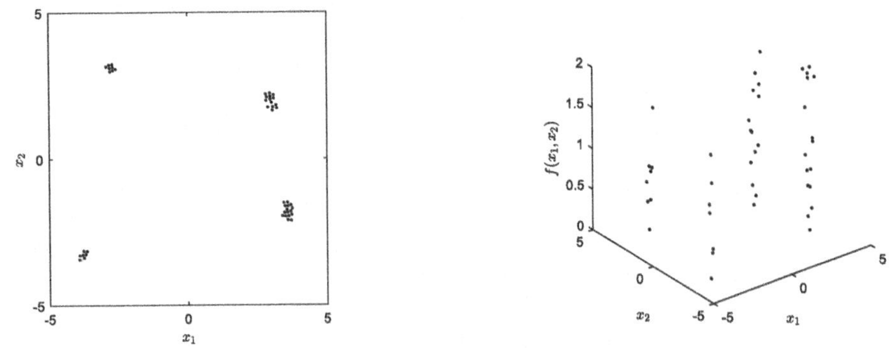

(c) $\Delta = 0.1$, $|A_f| = 45$

Fig. 9.6 Numerical results of $ArchiveUpdateM_{Q,\epsilon}D_x$ on the Himmelblau function

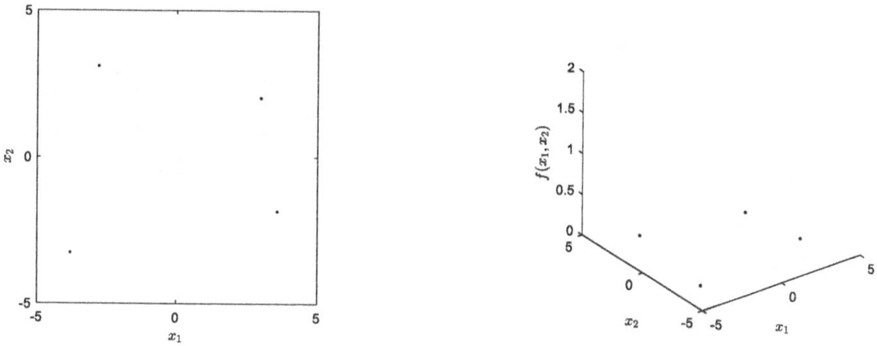

(d) $\Delta = 1$, $|A_f| = 4$

Fig. 9.6 (continued)

9.4.3 Eggholder Function

Finally, we consider the multi-modal SOP that is also known as the eggholder function:

$$f : [-600, 600]^2 \to \mathbb{R}$$
$$f(x) = -(x_2 + 47)\sin\sqrt{\left|x_2 + \frac{x_1}{2} + 47\right|} - x_1 \sin\sqrt{|x_1 - (x_2 + 47)|}. \tag{9.24}$$

Figure 9.7 shows some results of $ArchiveUpdateM_{Q,\epsilon}D_x$ on this problem for $\epsilon = 500$ and $N = 100{,}000$ candidate solutions. For small values of Δ, a total of 30 connected components of $M_{Q,\epsilon}$ can be identified. If the value of Δ is increased, the components are represented by fewer elements, and eventually, several local optimal are grouped together as described in the previous example.

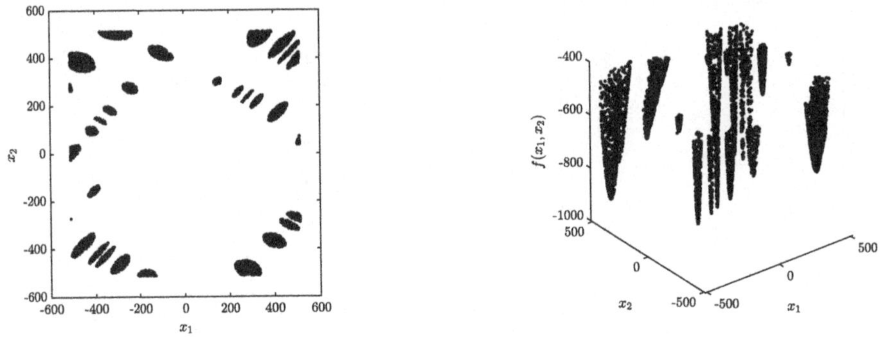

(a) $\Delta = 1$, $|A_f| = 5580$

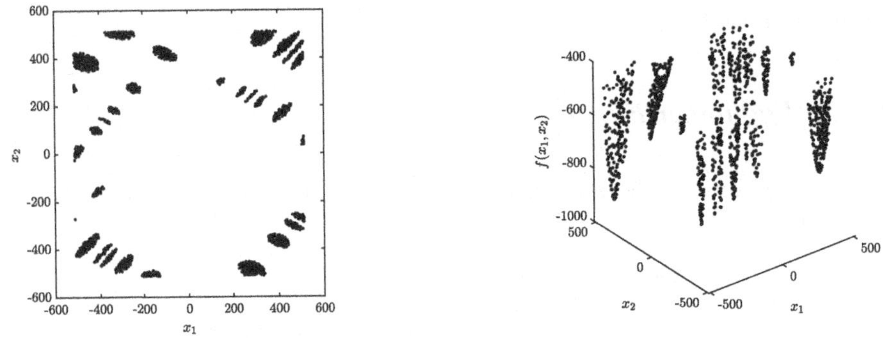

(b) $\Delta = 5$, $|A_f| = 1142$

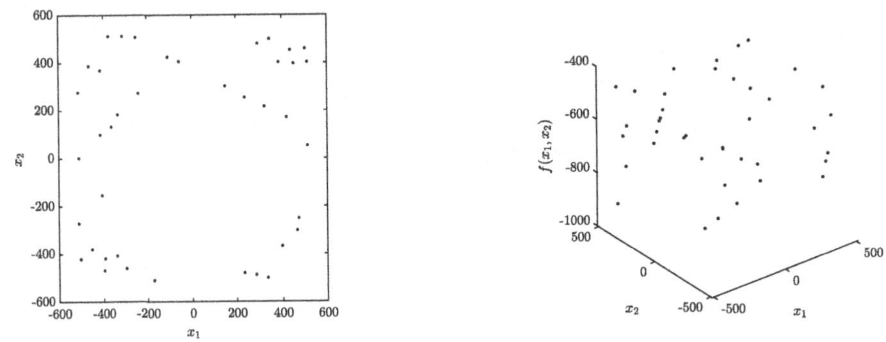

(c) $\Delta = 50$, $|A_f| = 42$

Fig. 9.7 Numerical results of $ArchiveUpdateM_{Q,\epsilon}D_x$ on the Eggholder function

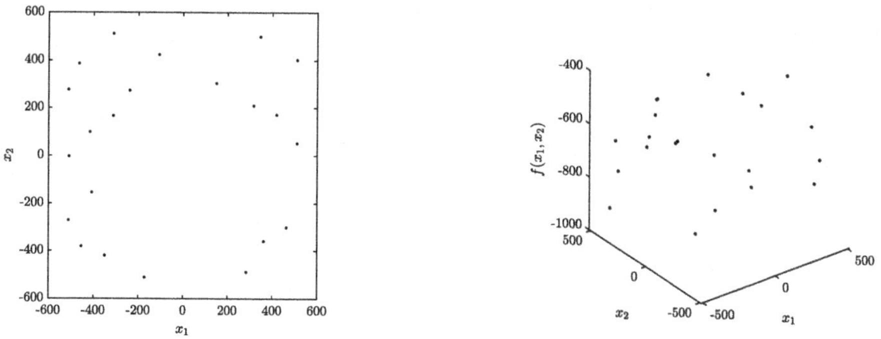

(d) $\Delta = 100$, $|A_f| = 23$

Fig. 9.7 (continued)

Chapter 10
Using Archivers Within MOEAs

As discussed above, all of the strategies presented in this book can be used jointly with in principle any set-based multi-objective search algorithm, either as an external archive to the base algorithm, or directly as part of the selection process. In order to maintain the convergence properties of the chosen archiving strategy, it is important to assure that assumption (4.1) on the process to generate new solutions is satisfied. We will in this chapter discuss four evolutionary algorithms that are coupled with archivers. The first three algorithms, NSGA-II-A, MOEA/D-A, and SMS-EMOA-A, aim for Pareto front approximations and are straightforward couplings of the base MOEAs with external archives. The numerical results we present indicate the strength of such an extension. The fourth algorithm, NϵSGA, has been designed to compute suitable finite size approximations of the set of approximate solutions of a given MOP. In this method, the generational operators and the selection mechanism have already been designed to feed the archiver.

10.1 NSGA-II-A, MOEA/D-A, and SMS-EMOA-A

Here we show exemplary on three algorithms that the strategies we have proposed in the previous chapters can easily be integrated into any given MOEA. For this, we have decided to take one dominance based, one decomposition based, and one indicator based algorithm as base MOEA: NSGA-II, MOEA/D, and SMS-EMOA, respectively.

O. Schütze and C. Hernández, *Archiving Strategies for Evolutionary Multi-objective Optimization Algorithms*, Studies in Computational Intelligence 938,
https://doi.org/10.1007/978-3-030-63773-6_10

Algorithm 14 shows the pseudo code of NSGA-II-A, the archive equipped version of NSGA-II. Compared to its base algorithm, NSGA-II-A only has two modifications: (i) the archiver has been added (lines 4 and 13), and (ii) the algorithm returns the union of the final population P_t and the final archive A_t (since both sets are evolved over time during the run of the algorithm). $ArchiveUpdate$ stands for any possible (external) archive including the ones discussed in this book. It is important to note that the set R_t (i.e., the result of the mutation) is used to update the archive. One possible alternative would be to feed the archive only with the set P_t, i.e., *after* NSGA-II has done the selection. On the one hand, P_t contains fewer elements than R_t and contains the promising candidates (from the viewpoint of NSGA-II). The use of P_t instead of R_t would hence reduce the computational overhead induced by the external archive. On the other hand, it is questionable if in that case the condition (4.1) on the generation process would be satisfied (regardless of the choice of the mutation operator). Consequently, it is questionable if the resulting algorithm—base MOEA plus external archive—still possesses the desired convergence properties. In general, we recommend feeding the external archiver with candidate solutions *before* the selection has been done. We will see the benefits (albeit for another algorithm) in the numerical results section.

Algorithm 15 shows the pseudo code of MOEA/D-A. As for the previous algorithm, there are only two changes compared to its base algorithm MOEA/D: the additional archiving strategy, and the return of $P_t \cup A_t$ instead of only the final archive A_t. Other variants of MOEA/D that use external archives are proposed in Cai et al. (2015), Wang et al. (2018), Tanabe and Ishibuchi (2018).

Finally, we present the archive equipped version of SMS-EMOA in Algorithm 16. Also for this method, the only two changes compared to SMS-EMOA are the additional use of the archiver (lines 3 and 12), and the return of the extended set $A_t \cup P_t$.

Similarly, in principle, any MOEA—and any other set based multi-objective search algorithm—can be enhanced with an external archive. We will demonstrate in the sequel on some numerical examples the benefit of such a combination.

10.2 NϵSGA

The next algorithm is the Non-ϵ-dominated Sorting Genetic Algorithm (NϵSGA, Hernández et al. 2020). NϵSGA aims for the computation of the set of approximate solutions of a MOP and makes use of the archivers presented in Chap. 8. Algorithm 17 shows the pseudo code of the algorithm, where $ArchiveP_{Q,\epsilon}D_{xy}$ is used (while it can easily be modified to use any of the other archivers from Chapter 8).

Algorithm 14 NSGA-II-A

Require: population size $npop$
Ensure: final approximation $P_t \cup A_t$
$P_0 = InitializePopulation(npop)$
$F_0 = FastNonDominatedSort(P_0)$
$C_0 = CrawdingDistance(P_0)$
$A_0 = ArchiveUpdate(P_0, \emptyset)$
$t = 1$
while StoppingCriteria() **do**
  $Q_t = CrossOver(P_t)$
  $R_t = Mutation(Q_t)$
  $S_t = P_{t-1} \cup R_t$
  $\{F_t^1, \ldots, F_t^v\} = FastNonDominatedSort(S_t)$
  $C_t = CrawdingDistance(F_t^1, \ldots, F_t^v)$
  $P_t = Selection(F_t^1, \ldots, F_t^v, C_t)$
  $A_t = ArchiveUpdate(R_t, A_{t-1})$
  $t = t + 1$
end while
return $P_t \cup A_t$

Algorithm 15 MOEA/D-A

Require: $npop$: population size, N: the number of subproblems, $\lambda^1, \ldots, \lambda^N$: weight vectors, T: number of weight vectors in the neighborhood of each weight vector
Ensure: final approximation $P_t \cup A_t$
$D = \{D \in d^{i,j}\}$ where $d^{i,j} = \|\lambda^i - \lambda^j\|$ for $i, j = 1, \ldots, T$
$B^i = \{i^1, \ldots, i^T\}$ where $\lambda^{i1}, \ldots, \lambda^{iT}$ are the T closest vectors to λ^i
$P_0 = InitializePopulation(npop)$
$FV_0 = EvaluatePopulation(P_0)$
$z = InitializeZ()$
$A_0 = ArchiveUpdate(P_0, \emptyset)$
$t = 1$
while StoppingCriteria() **do**
  for $i = 1, \ldots, N$ **do**
    $p_t = CrossOver(B^i)$
    $q_t = Mutation(p_t)$
    $r_t = Improve(q_t)$
    $z = UpdateZ()$
    $Updateneighboursolutions$
    $UpdateEP$
  end for
  $A_t = ArchiveUpdate(\{q_t\}, A_{t-1})$
  $t = t + 1$
end while
return $P_t \cup A_t$

Algorithm 16 SMS-EMOA-A

Require: population size $npop$
Ensure: final approximation $A_t \cup A_t$
1: $P_0 = InitializePopulation(npop)$
2: $F_0 = FastNonDominatedSort(P_0)$
3: $A_0 = ArchiveUpdate(P_0, \emptyset)$
4: $t = 1$
5: **while** StoppingCriteria() **do**
6: $\quad p_t = CrossOver(P_t)$
7: $\quad q_t = Mutation(p_t)$
8: $\quad S_t = P_{t-1} \cup \{q_t\}$
9: $\quad \{F_t^1, \ldots, F_t^v\} = FastNonDominatedSort(S_t)$
10: $\quad r_t = \arg\min_{s \in R_v} \mathcal{S}(F_t^v) - \mathcal{S}(F_t^v\{s\})$
11: $\quad P_t = S_t \backslash \{r_t\}$
12: $\quad A_t = ArchiveUpdate(\{q_t\}, A_{t-1})$
13: $\quad t = t + 1$
14: **end while**
15: **return** $P_t \cup A_t$

In the first step, N ϵ SGA initializes two sub-populations, P_1 and R_1, at random. The aim of the sequence P_t, $t \in \mathbb{N}$, is to obtain a well-spread set of solutions along the Pareto front, and the aim of the sequence R_t is to find well-spread approximate solutions (ideally of both $P_{Q,\epsilon}$ and $F(P_{Q,\epsilon})$ since $ArchiveP_{Q,\epsilon}D_{xy}$ is used). P_t is evolved using the same principles as for the classical NSGA-II algorithm. In each iteration step, the individuals are ranked as in NSGA-II, however, with the modification that $-\epsilon$-dominance is used instead of the classical dominance relation. Next, NϵSGA applies in line 13 a density estimator based on the nearest neighbors of each solution both in decision variable and objective space. The algorithm then selects 50 percent of the new population based on the obtained values in objective space, and the remaining 50 percent based on the values in decision variable space. This is done in order to increase the diversity in the populations. Next, $ArchiveP_{Q,\epsilon}D_{xy}$ is updated with the evolved sets O_i and S_i. Finally, if a specified number of generations is reached, NϵSGA performs a migration between the two populations which allows them to keep nearly optimal solutions that are similar in objective space but differ in decision space. For more details on the parts of the algorithm and further discussion, we refer to Hernández et al. (2020).

Algorithm 17 Non ϵ-Dominated Sorting EMOA

Require: $ngen$: number of generations, $npop$: population size, $\epsilon \in \mathbb{R}^n_+$, $\delta_x \in \mathbb{R}_+$, $\delta_y \in \mathbb{R}_+$, $nf \in [1, \ldots, ngen]$, $nr \in [1, \ldots, npop]$, $ratio \in [0, 1]$

Ensure: final approximation A_{ngen}

Generate initial population P_1 with size $npop$

Generate initial population R_1 with size $\lfloor npop * ratio \rfloor$

$A \leftarrow ArchivePQ_{,\epsilon}D_{xy}(P_0, \emptyset, \epsilon, \delta_x, \delta_y)$

for $i = 1 \ldots ngen - 1$ **do**

  Select $\lceil \frac{npop}{2} \rceil$ Parents Q_i with Tournament

  $\tilde{O}_i \leftarrow SBXCrossover(Q_i)$

  $O_i \leftarrow PolynomialMutation(\tilde{O}_i)$

  $\tilde{P}_i \leftarrow P_i \cup O_i$

  $F \leftarrow$ Non ϵ-DominatedSorting($\tilde{P}_i$)

  $P_{i+1} = \emptyset$

  $j = 1$

  while $|P_{i+1}| + |F_j| \leq npop$ **do**

    $NearestNeighborDistance(F_j)$ ▷ both in parameter and objective space

    $P_{i+1} = P_{i+1} \cup F_j$

    $j \leftarrow j + 1$

  end while

  $Sort(F_i, \prec_n)$

  $P_{i+1} \leftarrow P_{i+1} \cup F_j[1 : npop - |P_{i+1}|]$

  $S_i \leftarrow$ Evolve R_i as in NSGA-II

  $R_{i+1} \leftarrow$ Select from S_i using usual non-dominated ranking and crawding distance

  $A_{i+1} \leftarrow ArchivePQ_{,\epsilon}D_{xy}(O_i \cup S_i, A_i, \epsilon, \delta_x, \delta_y)$

  if $mod(i, nf) == 0$ **then**

    Exchange nr individuals at random between P_{i+1} and R_{i+1}.

  end if

end for

return A_{ngen}

10.3 Numerical Results

Here we evaluate the performance of the above described algorithms. For NSGA-II, MOEA/D and SMS-EMOA we have used the publicly available implementations in PlatEMO (Tian et al. 2017). For MOEA/D we have used the Tchebycheff decomposition approach.

We first evaluate the performance of NSGA-II-A, MOEA/D-A, and SMS-EMOA-A compared to their base algorithms. The comparisons are divided into three parts: in the first part, we assume that the budget of functions evaluations to be spent in total is relatively small (which is the case if the function evaluation is expensive). In that case, it makes sense to store all obtained non-dominated solutions, and hence, to use $ArchiveUpdateP_Q$ as an external archiver. This will not be suitable for larger budgets of function evaluations, and we will hence in the second part use $ArchiveUpdateTight2$ (i.e., an archiver that applies a discretization strategy) as external archiver using a larger budget of function values. All the tests will be performed on the test problems we have used so far to demonstrate the effect of the archivers. Since all of these problems are low-dimensional and represent no challenge for state-of-the-art solvers, we will in the third part repeat the comparisons that use $ArchiveUpdateTight2$, but now on the entire DTLZ benchmark suite with higher dimensional search spaces.

We will first consider the scenario that the overall function budget to be spent is relatively small. For this, we will consider the bi-objective problems with a budget of $N = 20{,}000$ function evaluations, and the three objective problem CONV3 with $N = 50{,}000$. Since in this case it is still possible to store all obtained non-dominated solutions, we will use $ArchiveUpdateP_Q$ as external archiver. The archive sizes for all MOEAs and problems are set to 200. Table 10.1 shows the obtained approximation qualities $\Delta_2(F(O_f), F(P_Q))$ of the final outcome set O_f and standard deviations for the three standalone MOEAs together with their three variants that use external archives. Hereby, the outcome set of each MOEA is the final population ($O_f = P_f$),

Table 10.1 Approximation qualities $\Delta_2(F(O_f), F(P_Q))$ and standard deviations for the different problems and algorithms, based on 30 independent runs. For CONV2, DENT, and SSW we have used $N = 20,000$ function evaluations, and $N = 50,000$ for CONV3

Problem	NSGA-II	NSGA-II-A	MOEA/D	MOEA/D-A	SMS-EMOA	SMS-EMOA-A
CONV2	0.50878 (0.0082825)	0.093894 (0.00073922)	0.69432 (0.0015411)	0.18744 (0.016183)	0.4569 (0.0017684)	0.089897 (0.00054048)
DENT	0.36743 (0.11757)	0.38945 (0.18474)	1.6131 (0.13595)	1.2085 (0.23954)	0.56513 (0.12705)	0.55552 (0.15526)
SSW	0.45696 (0.010337)	0.2119 (0.0043064)	4.7928 (0.43339)	1.0234 (0.088509)	0.42423 (0.0036422)	0.19244 (0.0021341)
CONV3	0.57175 (0.010754)	0.2359 (0.0021724)	0.61818 (0.0038244)	0.24788 (0.0021788)	0.47882 (0.003138)	0.20492 (0.001052)

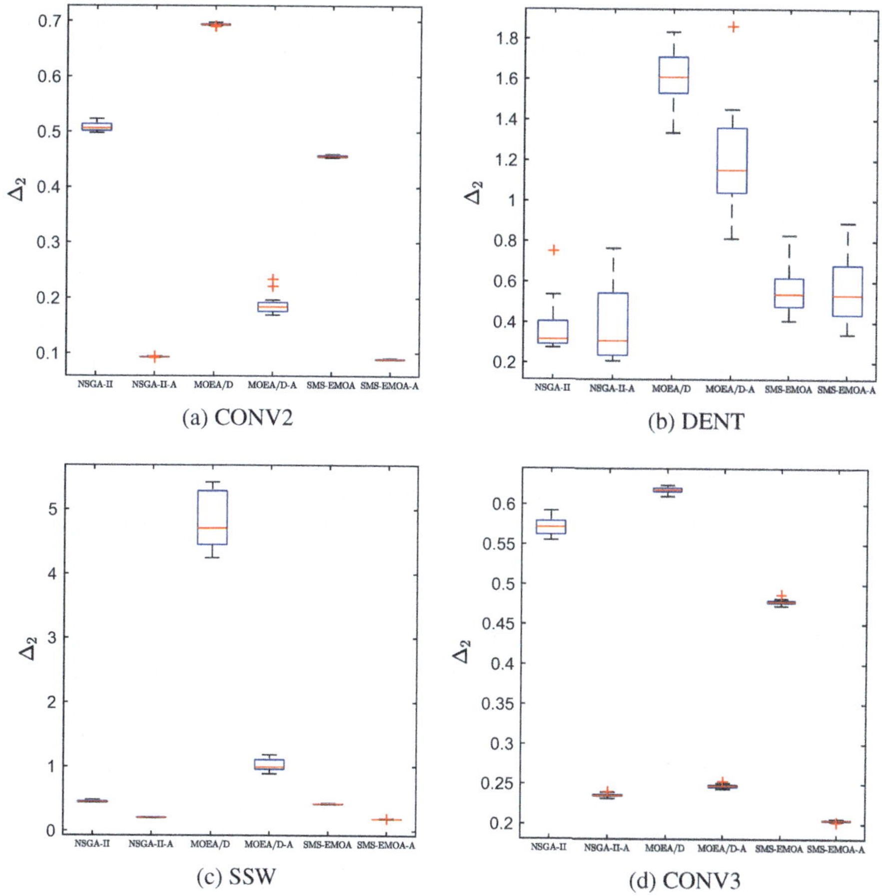

Fig. 10.1 Box plots for the approximation qualities $\Delta_2(F(A_f), F(P_Q))$ for the different problems and algorithms, based on 30 independent runs

and the outcome of the archive equipped MOEA is the union of the final population and the final archive ($O_f = P_f \cup A_f$). Note that since $ArchiveUpdateP_Q$ stores all obtained non-dominated solutions, it will hence also contain all the elements of the final population (i.e., $P_f \subset A_f$). It is hence in this case sufficient to only present the final archive (which will change once we use a discretization strategy). Figure 10.1 shows the resulting box plots of the approximation qualities. Similarly, Table 10.2 and Fig. 10.2 show the final archive/population sizes, and Table 10.3 and Fig. 10.3 the respective running times. Figures 10.4, 10.5, 10.6, 10.7 show the median plots out of the 30 independent runs, and Fig. 10.8 shows the convergence plots.

As it can be seen—and as anticipated—the approximation qualities are (much) better for all archive equipped MOEAs while in turn archive sizes and running times are much larger for these algorithms due to the overhead generated by the external

Table 10.2 Archive sizes and standard deviations of the final archives/populations for the different problems and algorithms, based on 30 independent runs. For CONV2, DENT, and SSW we have used $N = 20,000$ function evaluations, and $N = 50,000$ for CONV3

Problem	NSGA-II	NSGA-II-A	MOEA/D	MOEA/D-A	SMS-EMOA	SMS-EMOA-A
CONV2	200 (0)	9215.8 (118.8314)	200 (0)	9705.2 (248.5799)	200 (0)	11125.9 (131.1479)
DENT	200 (0)	979.6 (53.6032)	200 (0)	2051.45 (73.8843)	200 (0)	1541.35 (41.1496)
SSW	200 (0)	2891.45 (68.467)	200 (0)	1470.5 (480.378)	200 (0)	4043.15 (61.4485)
CONV3	200 (0)	6786.35 (159.8714)	190 (0)	9757.45 (179.7576)	200 (0)	12058.75 (126.2607)

Table 10.3 Running and standard deviations for the different problems and algorithms, based on 30 independent runs. For CONV2, DENT, and SSW we have used $N = 20,000$ function evaluations, and $N = 50,000$ for CONV3

Problem	NSGA-II	NSGA-II-A	MOEA/D	MOEA/D-A	SMS-EMOA	SMS-EMOA-A
CONV2	0.77514 (0.33505)	57.3954 (8.3585)	8.7394 (1.2462)	127.8908 (78.7487)	269.0248 (475.181)	511.1748 (649.8752)
DENT	0.76508 (0.18695)	3.606 (1.5334)	9.1341 (2.4982)	27.9084 (4.7512)	49.1256 (9.33)	65.286 (8.9558)
SSW	1.1626 (0.62517)	12.1651 (3.0311)	11.5529 (1.9675)	26.0932 (4.95)	109.331 (19.6423)	194.3387 (110.6781)
CONV3	4.1079 (4.8168)	41.4112 (7.6692)	17.1164 (3.0152)	283.9647 (397.5685)	1046.6213 (415.8236)	1453.231 (601.6486)

archive. While it can be argued that the additional use of the archiver is not needed for both NSGA-II and SMS-EMOA to solve these rather easy test problems, the situation changes for MOEA/D. Note that for the problems CONV2 and DENT the archiver helps to maintain some solutions along the flat regions of the Pareto fronts. As discussed above, in some scenarios it is desired to obtain gap free approximations and hence also to cover these regions. The advantage of the external archiver becomes much clearer for SSW (see Fig. 10.6). While the standalone MOEA/D has not yet approached a suitable approximation of the Pareto front, the situation is completely different for the equipped MOEA. This example already shows that if we would have fed MOEA/D-A only with candidate solutions generated by MOEA/D *after* the selection, then the approximation of the final archive would not be satisfying. It is hence advisable to feed the archiver with all candidate solutions x for which the objective values $F(x)$ are already computed by the base algorithm.

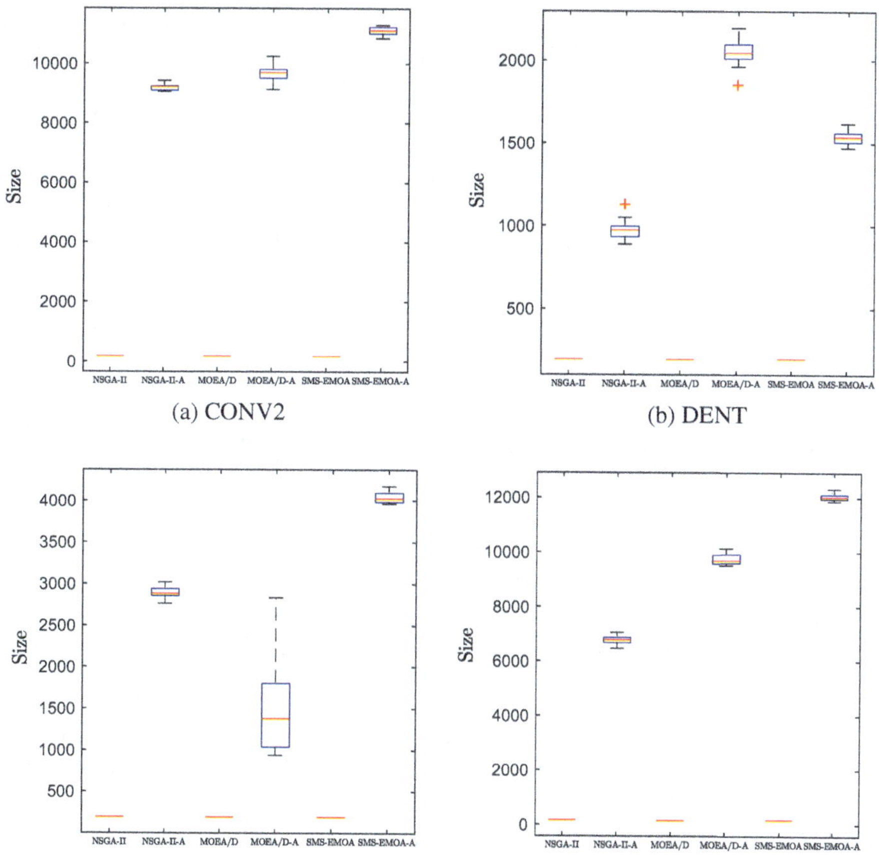

(a) CONV2 (b) DENT

(c) SSW (d) CONV3

Fig. 10.2 Box plots for the final archive/population sizes for the different problems and algorithms, based on 30 independent runs

These examples, even if only easy test problems are used, already show that it might be beneficial to store solutions externally to the used population. The additional storage comes of course with a higher computational time, however, the number of function evaluations spent is the same for both algorithms. If the overall budget of function calls is relatively low and/or each function evaluation is expensive, then it makes sense to store all non-dominated solutions obtained during the search. If the number of function evaluations exceeds a certain limit, however, it is not beneficial any more to use $ArchiveUpdateP_Q$ as external archiver. In that case, archive sizes and running times would increase significantly without offering much better approximation qualities compared to the base algorithms. In this case, it is rather advisable to take an archiver that uses a discretization strategy, which we will do in the following.

Table 10.4 Approximation qualities $\Delta_2(F(A_f), F(P_Q))$ and standard deviations for the different problems and algorithms, based on 30 independent runs. For CONV2, DENT, and SSW we have used $N = 50,000$ function evaluations, and $N = 100,000$ for CONV3

Algorithm	CONV2	DENT	SSW	CONV3
NSGA-II	0.72406 (0.016331)	0.36832 (0.010014)	0.66786 (0.017276)	0.69017 (0.019099)
A_f	1.2936 (0.039002)	0.56931 (0.017497)	0.8077 (0.022433)	0.51973 (0.0093841)
NSGA-II-A	0.67111 (0.014804)	0.33414 (0.0081383)	0.55795 (0.012896)	0.47492 (0.0068743)
MOEA/D	0.94013 (0.00023927)	1.5463 (0.11511)	5.5977 (0.20535)	0.73343 (0.001417)
A_f	1.3026 (0.034847)	1.3128 (0.2409)	1.0701 (0.11656)	0.52625 (0.0064451)
MOEA/D-A	0.70801 (0.01496)	1.2494 (0.24558)	1.0655 (0.11108)	0.49261 (0.0063459)
SMS-EMOA	0.64295 (0.0039832)	0.61059 (0.17215)	0.59957 (0.0059084)	0.58027 (0.027501)
A_f	1.3098 (0.034469)	0.74369 (0.13246)	0.80687 (0.017715)	0.52456 (0.0097769)
SMS-EMOA-A	0.60955 (0.0038728)	0.57827 (0.16996)	0.52493 (0.0085326)	0.45921 (0.0052403)

Table 10.5 Archive sizes and standard deviations of the final archives/populations for the different problems and algorithms, based on 30 independent runs. For CONV2, DENT, and SSW we have used $N = 50,000$ function evaluations, and $N = 100,000$ for CONV3

Algorithm	CONV2	DENT	SSW	CONV3
NSGA-II	100 (0)	100 (0)	100 (0)	250 (0)
A_f	26.8 (1.7045)	32.9 (1.7137)	51.45 (2.7237)	146.45 (7.1339)
NSGA-II-A	126.8 (1.7045)	132.9 (1.7137)	151.45 (2.7237)	246.45 (7.1339)
MOEA/D	100 (0)	100 (0)	100 (0)	231 (0)
A_f	26.7 (1.9494)	15.95 (2.5849)	36.55 (4.2855)	143.05 (6.194)
MOEA/D-A	126.7 (1.9494)	115.95 (2.5849)	136.55 (4.2855)	234.05 (6.194)
SMS-EMOA	100 (0)	100 (0)	100 (0)	250 (0)
A_f	26.3 (1.3416)	27.8 (2.9842)	51.4 (2.4794)	144.9 (5.4763)
SMS-EMOA-A	126.3 (1.3416)	127.8 (2.9842)	151.4 (2.4794)	244.9 (5.4763)

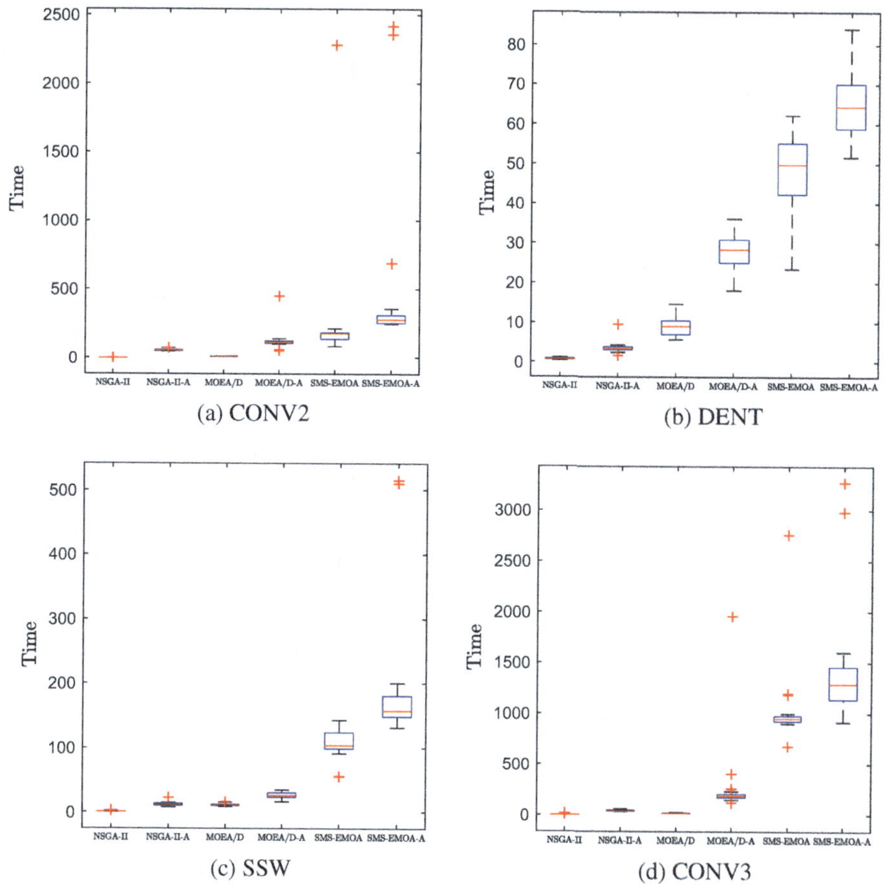

Fig. 10.3 Box plots for the running times for the different problems and algorithms, based on 30 independent runs

Table 10.6 Running and standard deviations for the different problems and algorithms, based on 30 independent runs. For CONV2, DENT, and SSW we have used $N = 50,000$ function evaluations, and $N = 100,000$ for CONV3

Algorithm	CONV2	DENT	SSW	CONV3
NSGA-II	1.6828 (0.099292)	1.7372 (0.11637)	1.9334 (0.15062)	4.8396 (0.38173)
NSGA-II-A	3.1979 (0.20199)	3.1373 (0.22673)	3.6922 (0.27315)	9.9611 (0.79593)
MOEA/D	14.9341 (1.1489)	16.277 (2.1305)	17.3881 (2.0143)	31.7241 (4.1322)
MOEA/D-A	17.3751 (1.3349)	18.4923 (2.4262)	19.8586 (2.3081)	38.5891 (4.9937)
SMS-EMOA	82.7374 (6.0482)	40.52 (2.7589)	69.2963 (7.9408)	357.953 (34.36)
SMS-EMOA-A	85.4054 (6.2512)	42.9576 (2.9262)	72.1014 (8.2645)	365.3557 (35.1287)

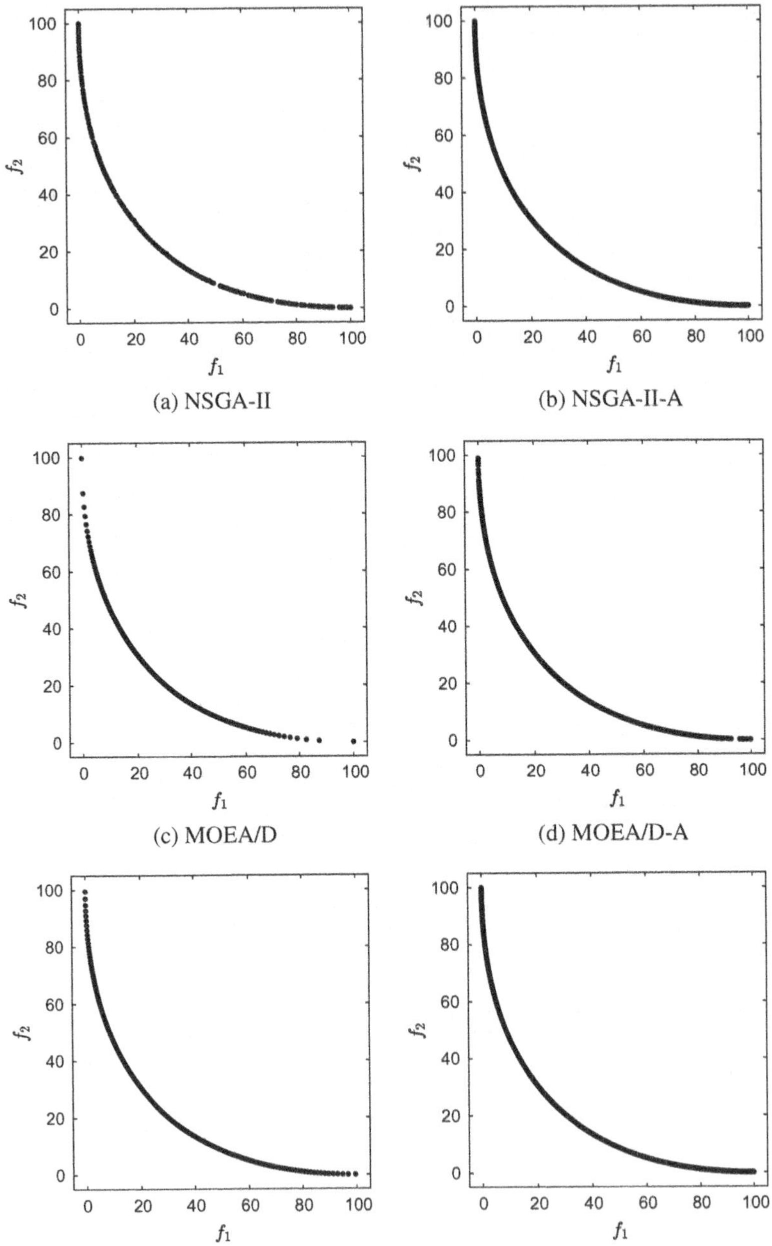

(a) NSGA-II (b) NSGA-II-A

(c) MOEA/D (d) MOEA/D-A

(e) SMS-EMOA (f) SMS-EMOA-A

Fig. 10.4 Numerical results of the three base MOEAs and their archive equipped variants on CONV2 using $N = 20,000$ function evaluations

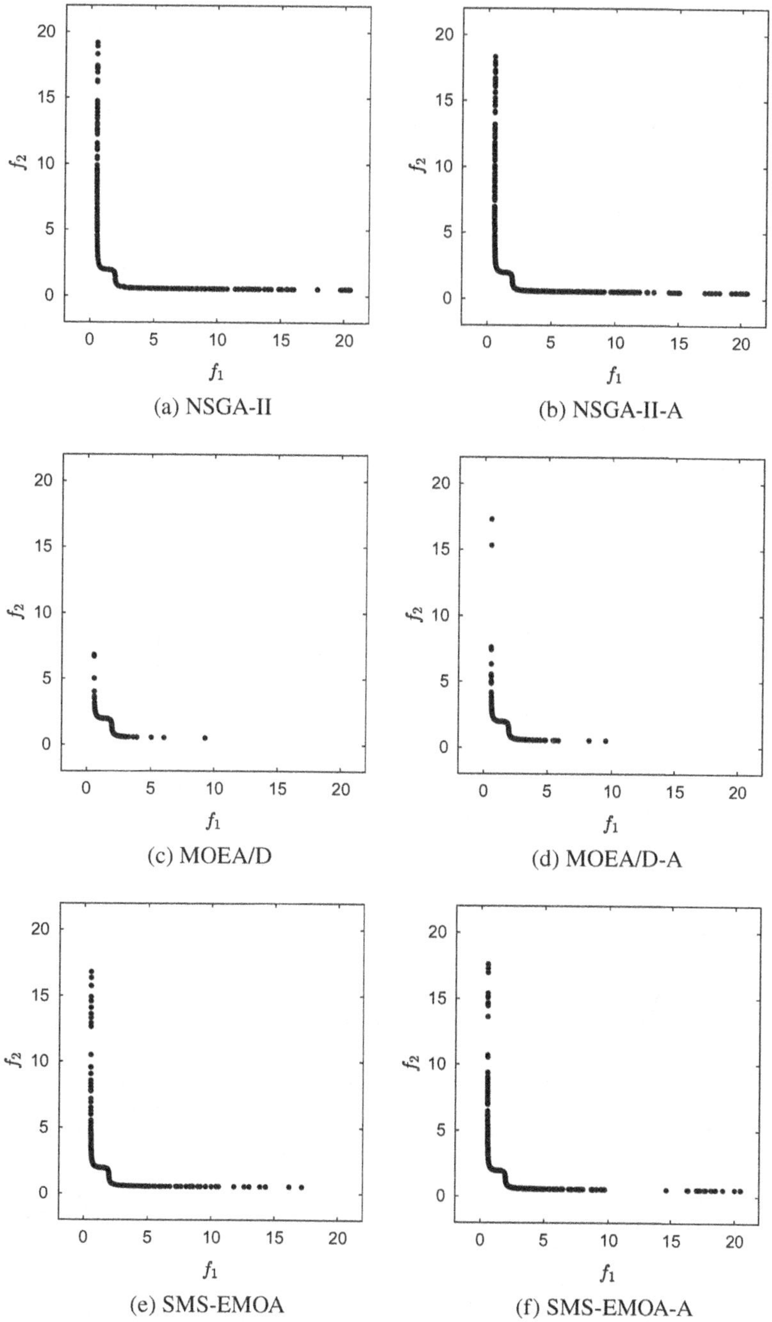

(a) NSGA-II (b) NSGA-II-A

(c) MOEA/D (d) MOEA/D-A

(e) SMS-EMOA (f) SMS-EMOA-A

Fig. 10.5 Numerical results of the three base MOEAs and their archive equipped variants on DENT using $N = 20,000$ function evaluations

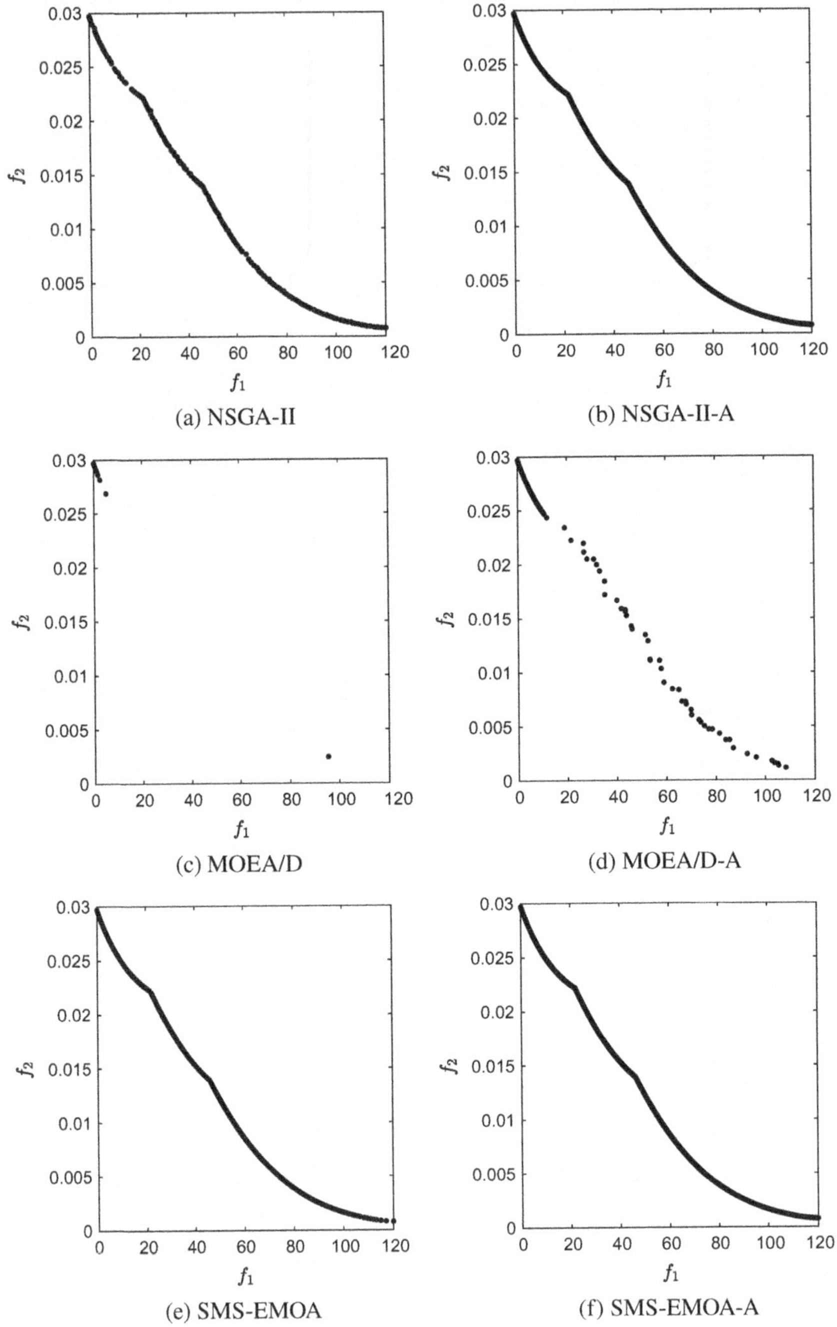

Fig. 10.6 Numerical results of the three base MOEAs and their archive equipped variants on SSW using $N = 20,000$ function evaluations

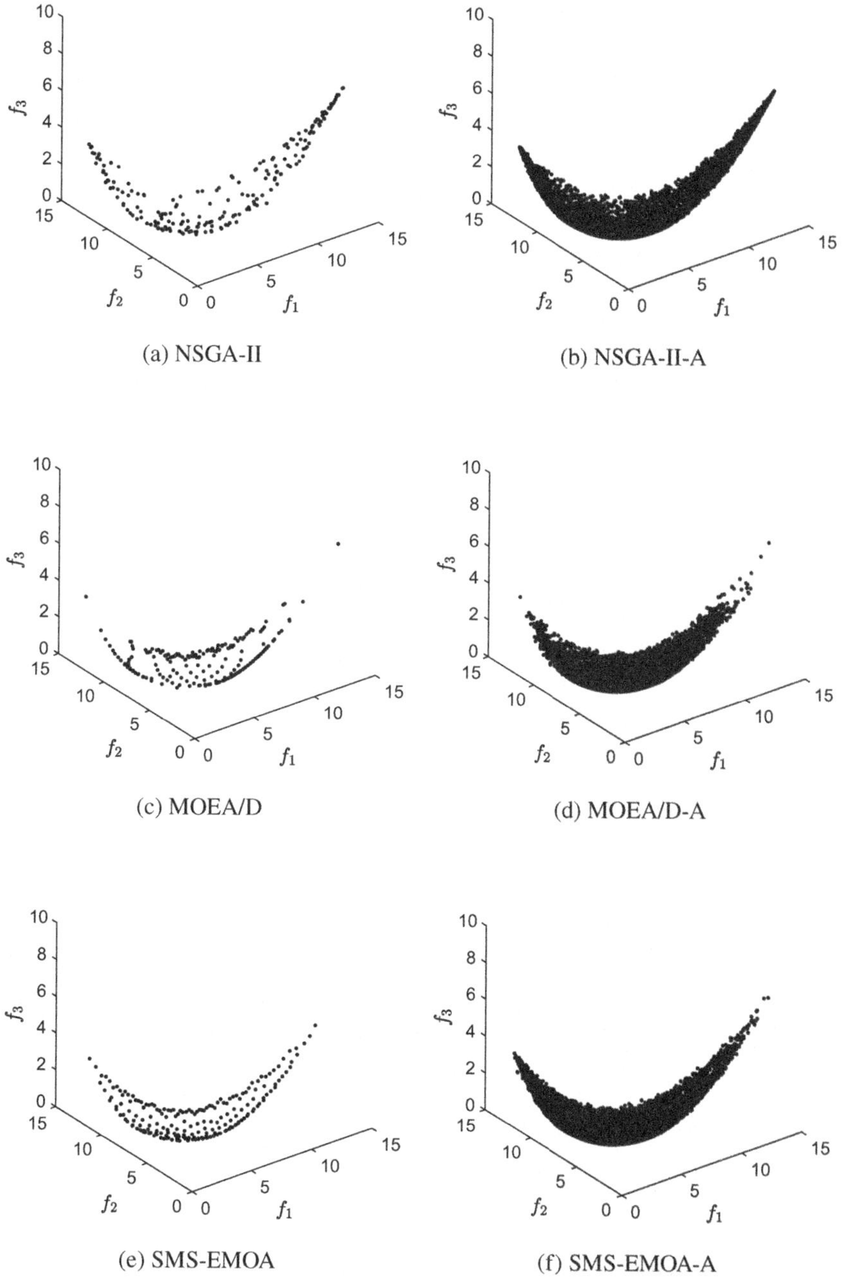

(a) NSGA-II (b) NSGA-II-A

(c) MOEA/D (d) MOEA/D-A

(e) SMS-EMOA (f) SMS-EMOA-A

Fig. 10.7 Numerical results of the three base MOEAs and their archive equipped variants on CONV3 using $N = 50,000$ function evaluations

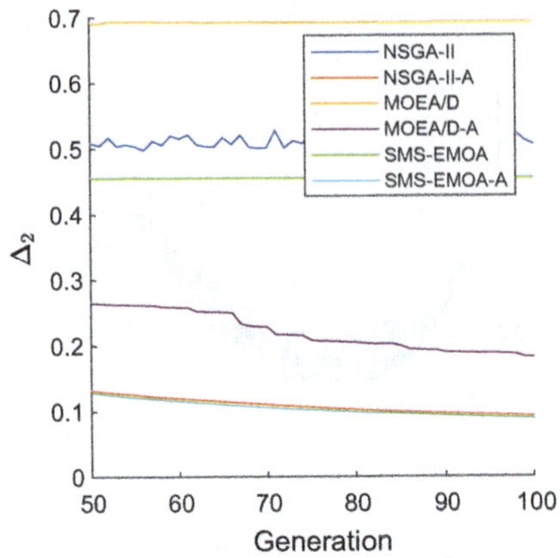

(a) CONV2

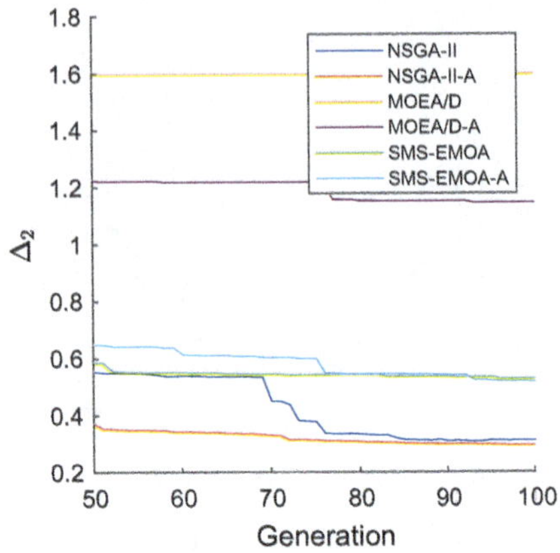

(b) DENT

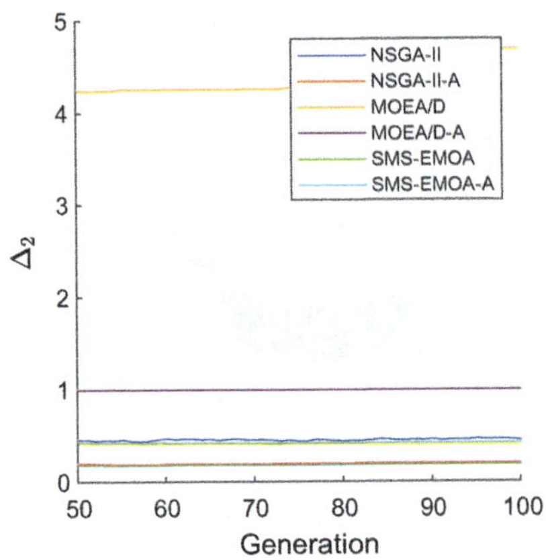

(c) SSW

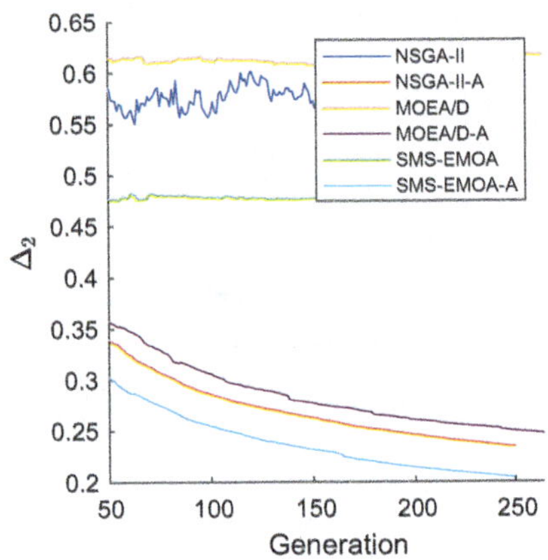

(d) CONV3

Fig. 10.8 Convergence plots

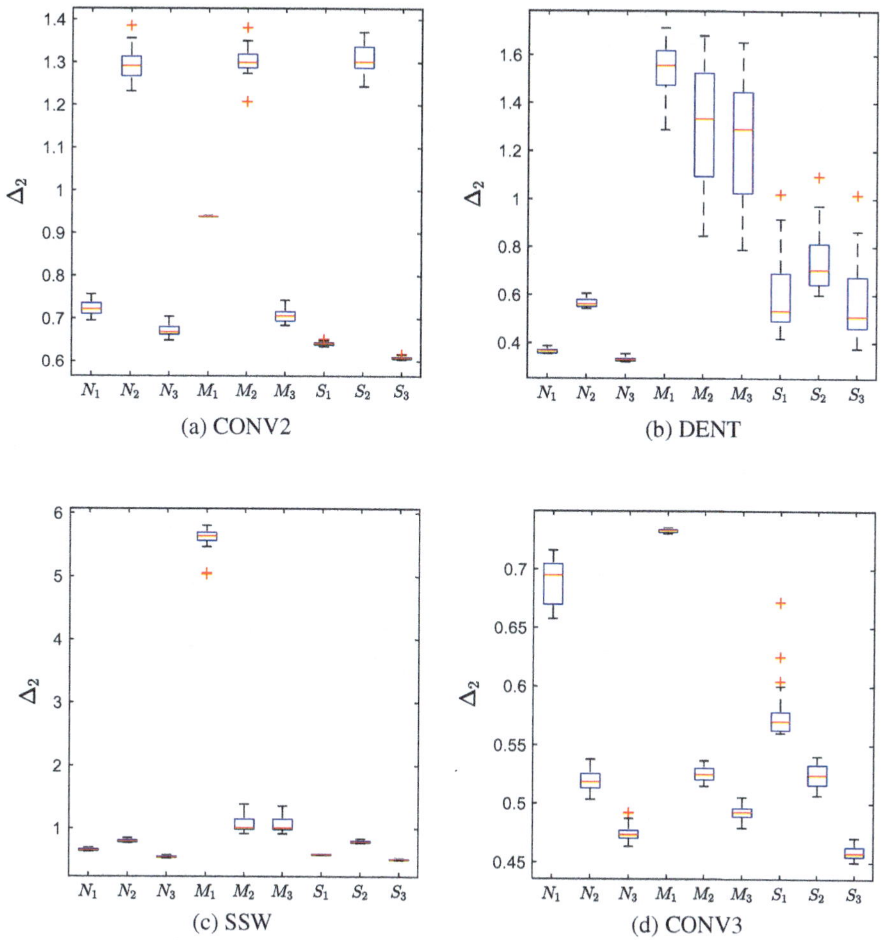

Fig. 10.9 Box plots for the approximation qualities $\Delta_2(F(A_f), F(P_Q))$ for the different problems and algorithms, based on 30 independent runs

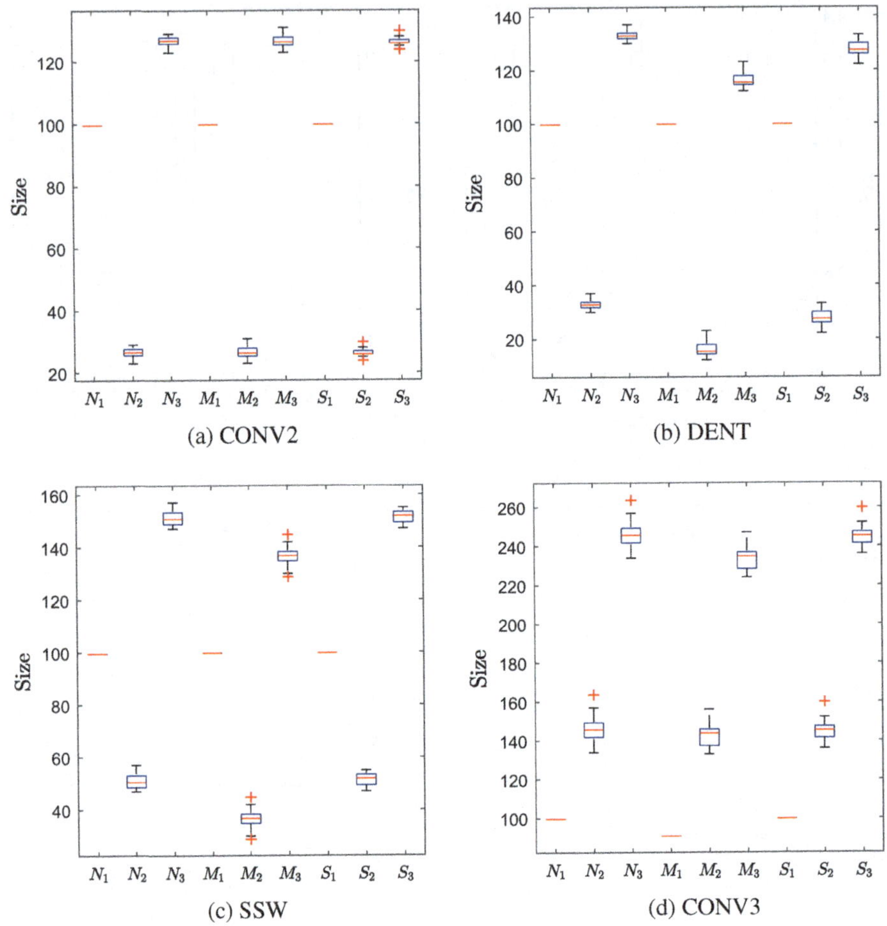

Fig. 10.10 Box plots for the final archive/population sizes for the different problems and algorithms, based on 30 independent runs

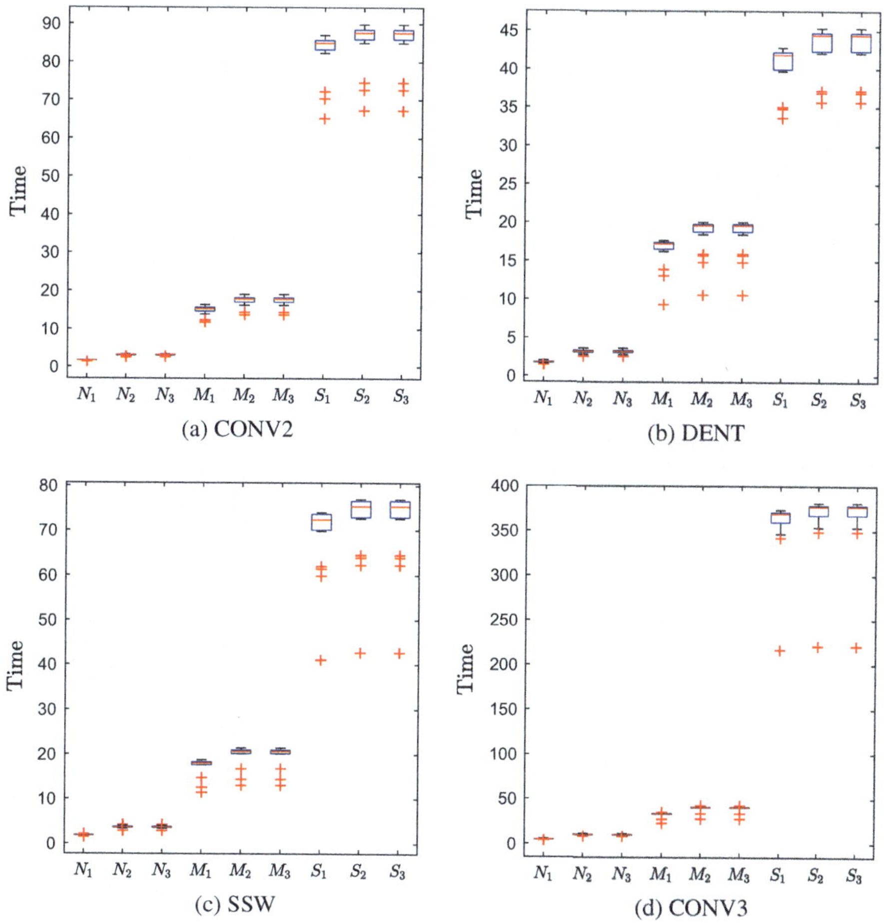

Fig. 10.11 Box plots for the running times for the different problems and algorithms, based on 30 independent runs

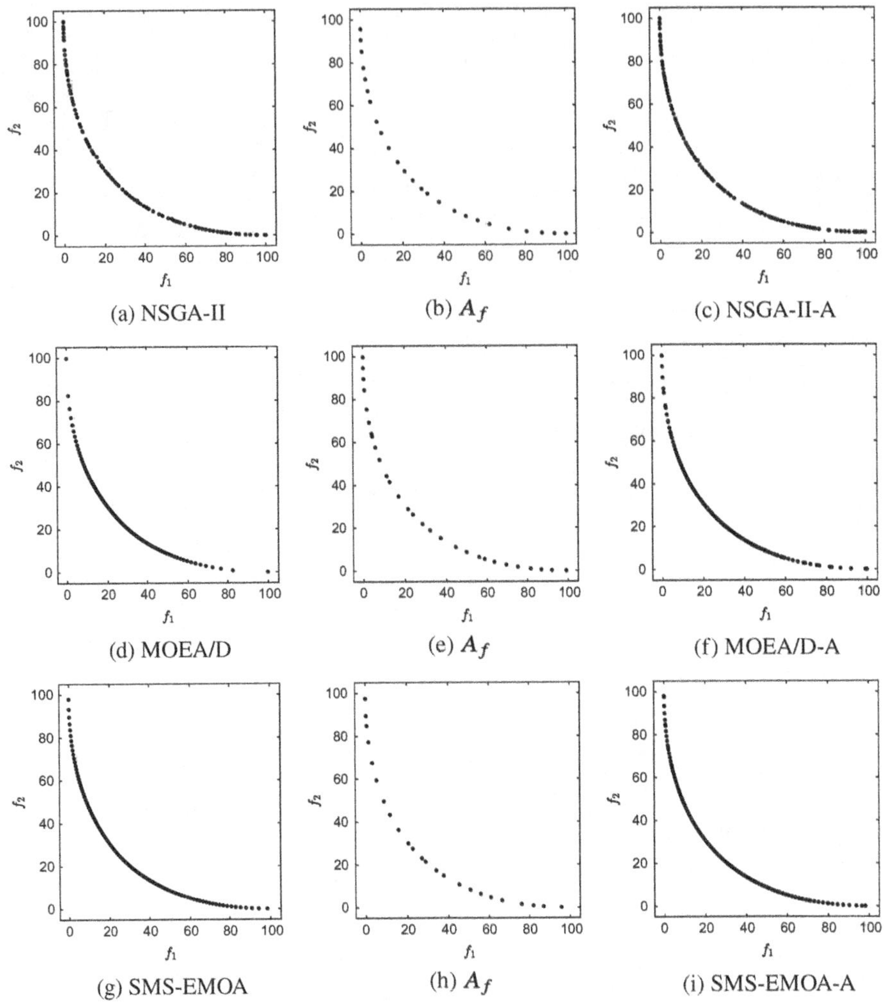

Fig. 10.12 Numerical results of the three base MOEAs and their archive equipped variants on CONV2 using $N = 50,000$ function evaluations. The middle column shows the final exernal archive A_f for each problem, and the right column the union of A_f and the final population

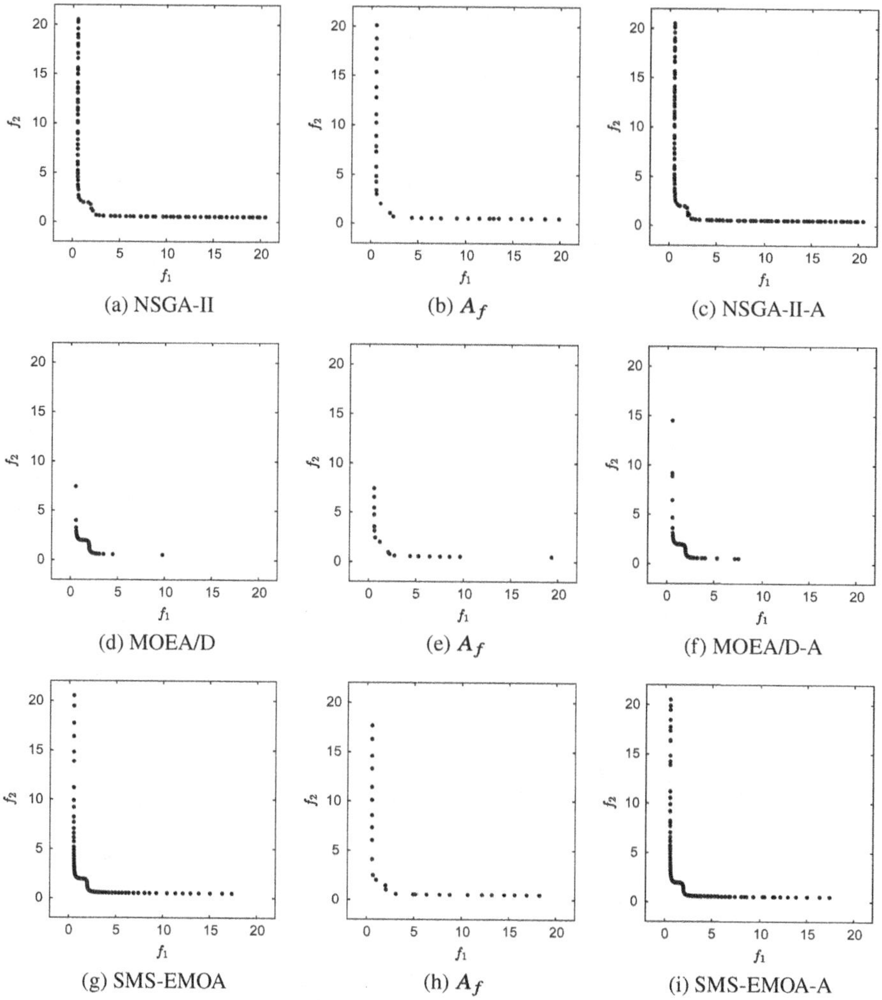

(a) NSGA-II (b) $\boldsymbol{A_f}$ (c) NSGA-II-A

(d) MOEA/D (e) $\boldsymbol{A_f}$ (f) MOEA/D-A

(g) SMS-EMOA (h) $\boldsymbol{A_f}$ (i) SMS-EMOA-A

Fig. 10.13 Numerical results of the three base MOEAs and their archive equipped variants on DENT using $N = 50,000$ function evaluations. The middle column shows the final exernal archive A_f for each problem, and the right column the union of A_f and the final population

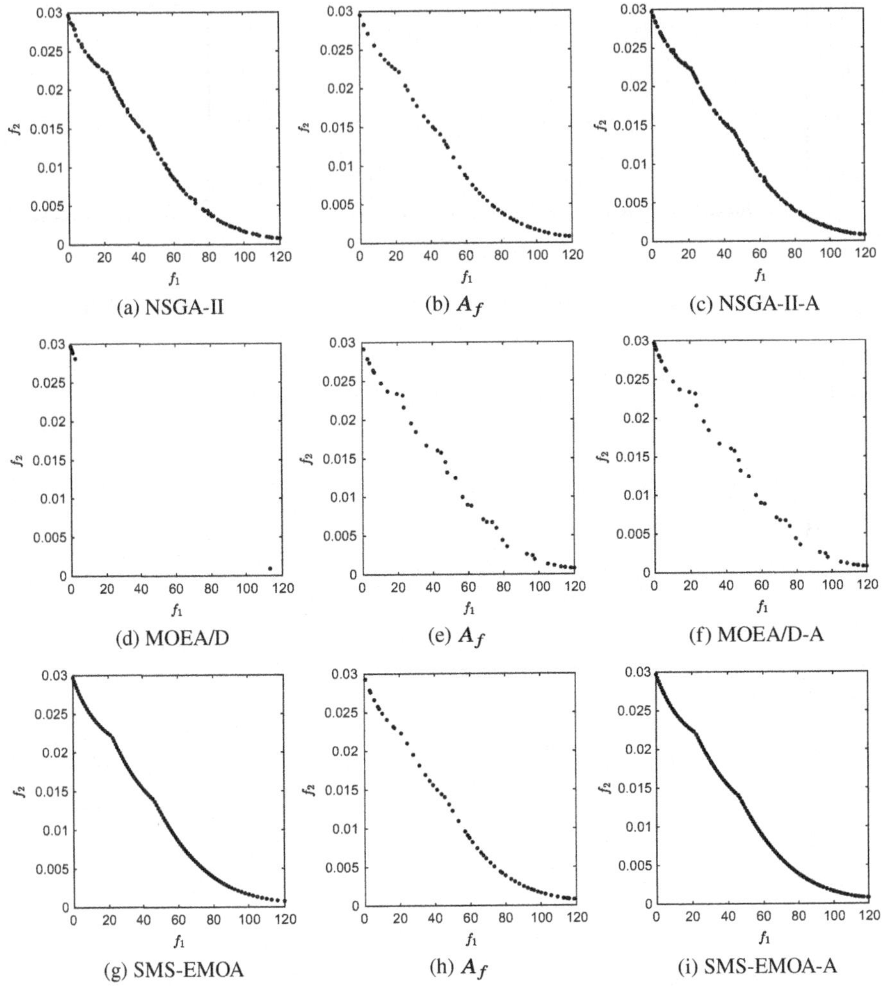

Fig. 10.14 Numerical results of the three base MOEAs and their archive equipped variants on SSW using $N = 50,000$ function evaluations. The middle column shows the final exernal archive A_f for each problem, and the right column the union of A_f and the final population

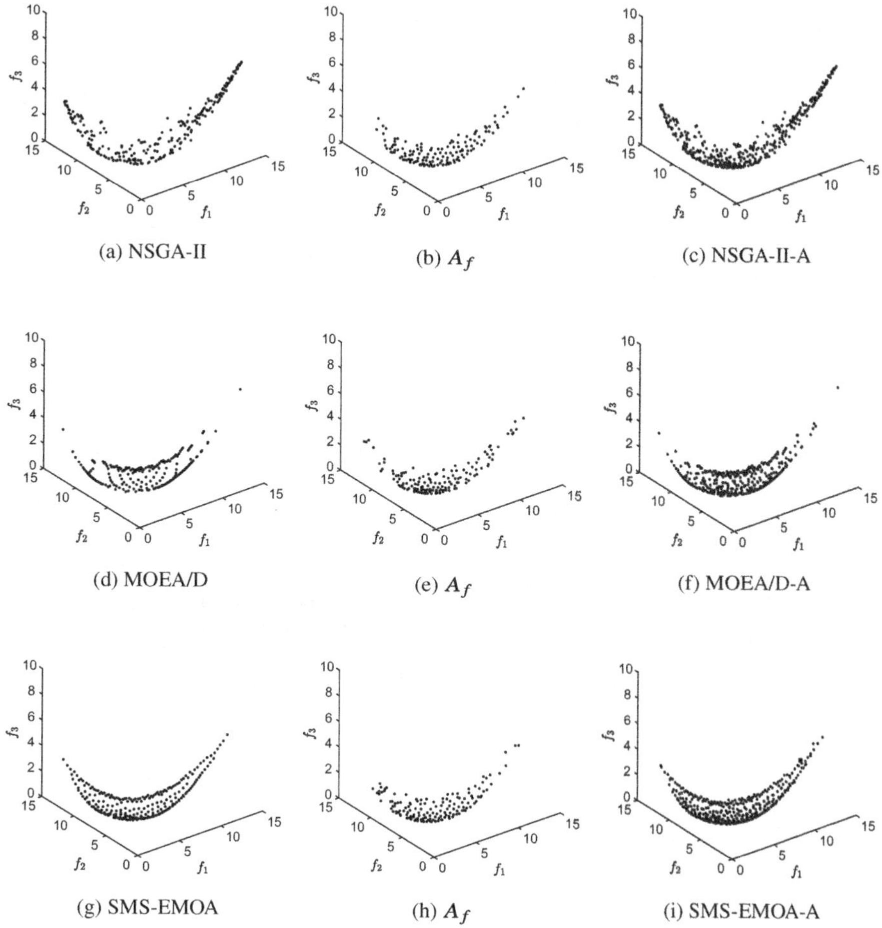

Fig. 10.15 Numerical results of the three base MOEAs and their archive equipped variants on CONV3 using $N = 100,000$ function evaluations. The middle column shows the final exernal archive A_f for each problem, and the right column the union of A_f and the final population

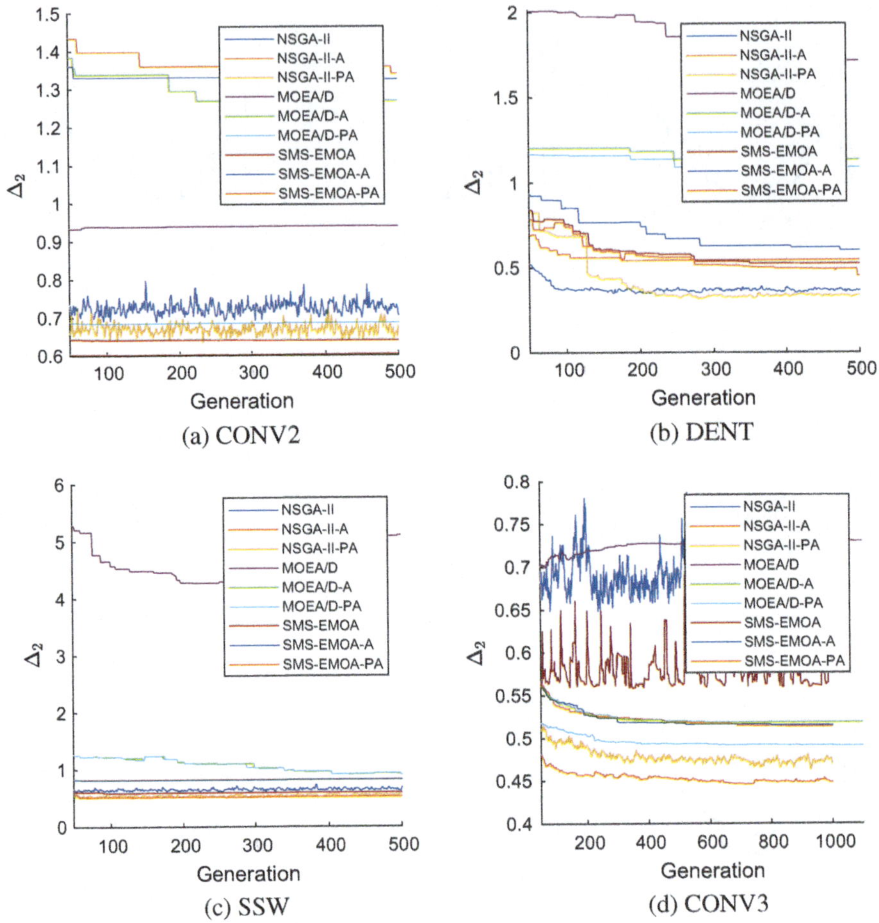

Fig. 10.16 Convergence plots part 2

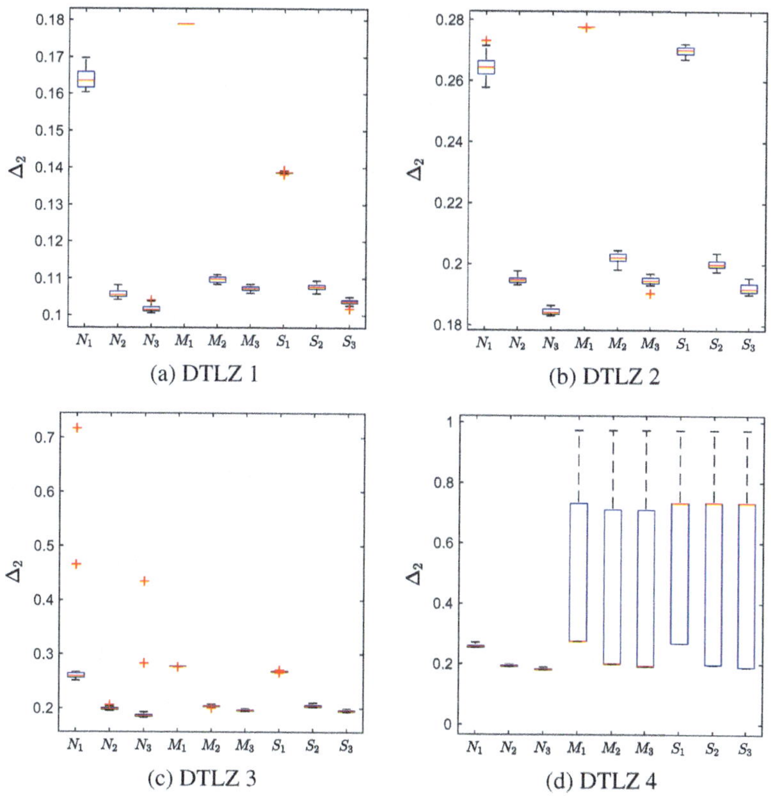

Fig. 10.17 Box plots for the approximation qualities $\Delta_2(F(A_f), F(P_Q))$ for the different algorithms on the DTLZ problems, based on 30 independent runs

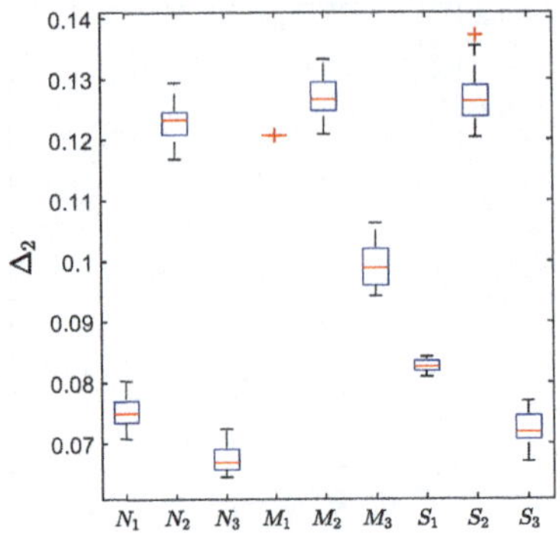

(e) DTLZ 5

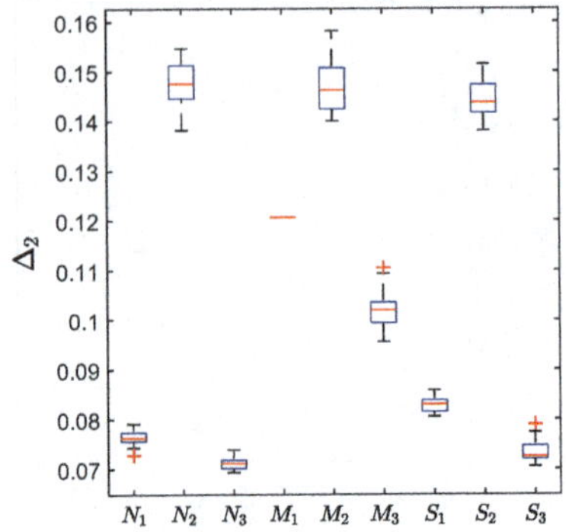

(f) DTLZ 6

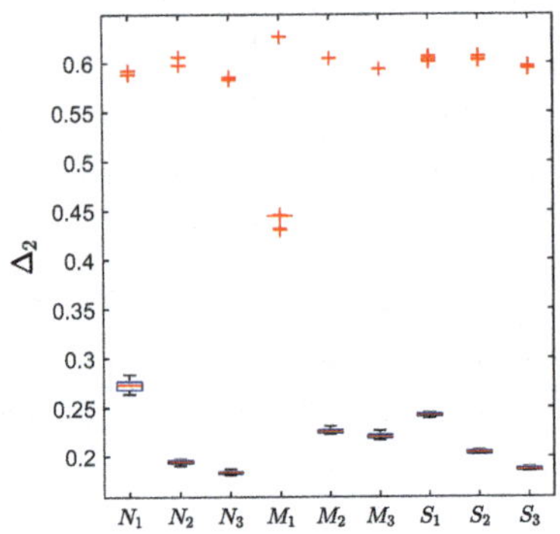

(g) DTLZ 7

Fig. 10.17 (continued)

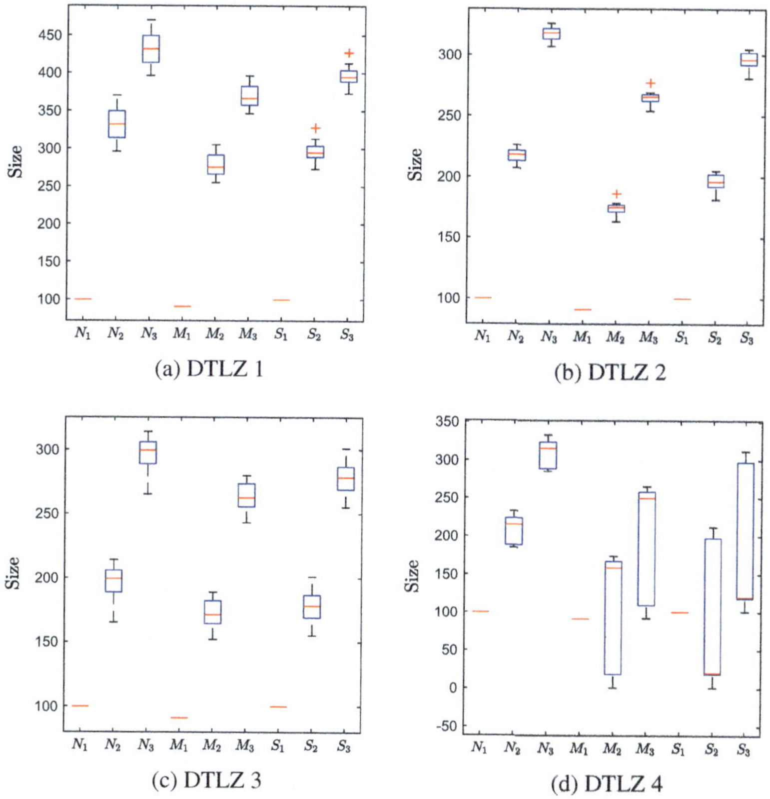

Fig. 10.18 Box plots for the final archive/population sizes for the different algorithms on the DTLZ problems, based on 30 independent runs

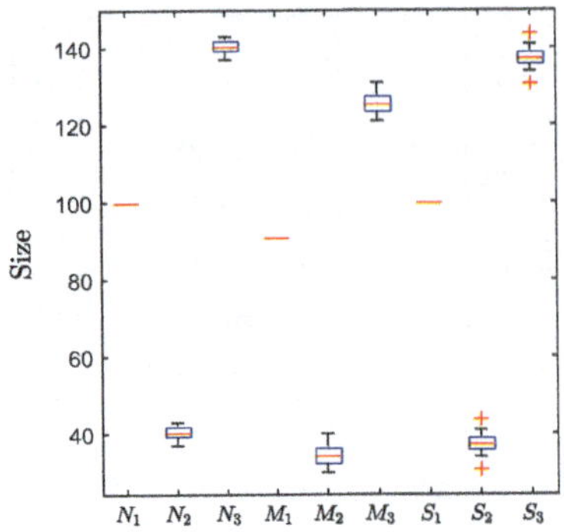

(e) DTLZ 5

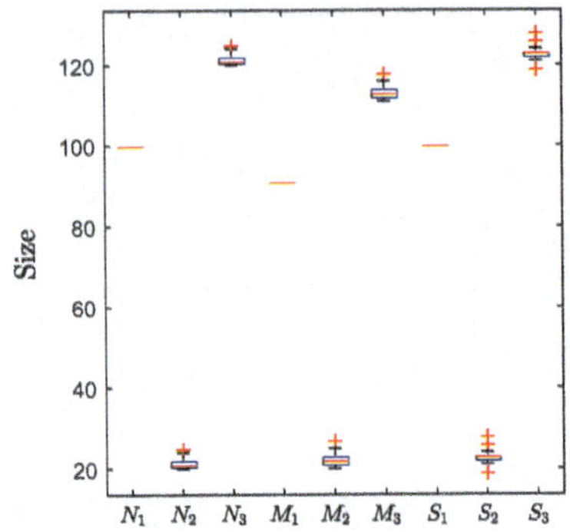

(f) DTLZ 6

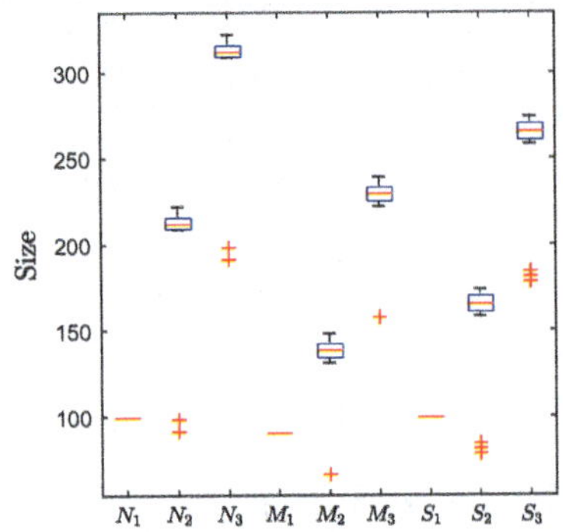

(g) DTLZ 7

Fig. 10.18 (continued)

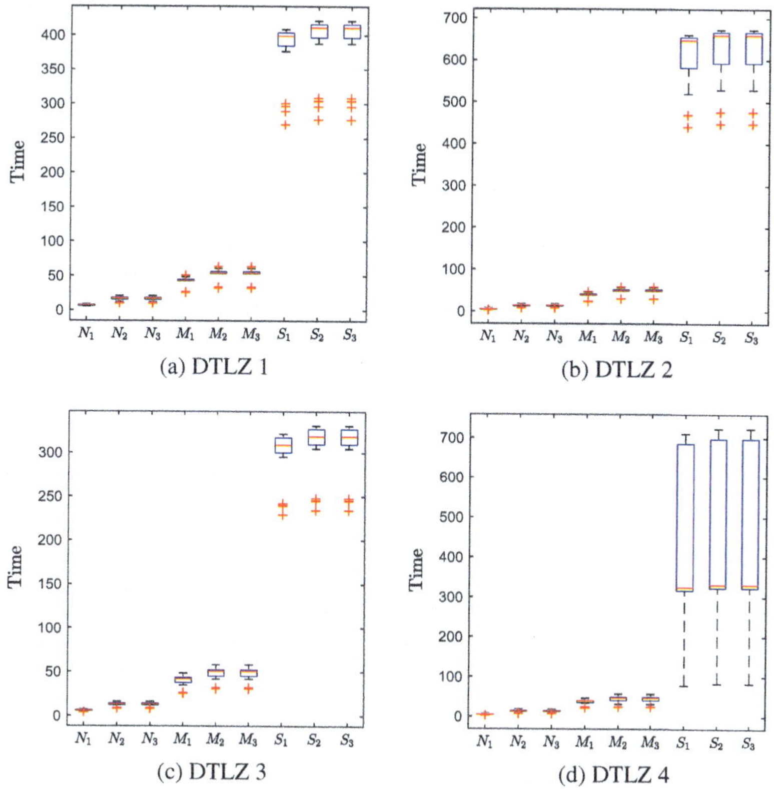

Fig. 10.19 Box plots for the running times for the different problems on the DTLZ problems, based on 30 independent runs

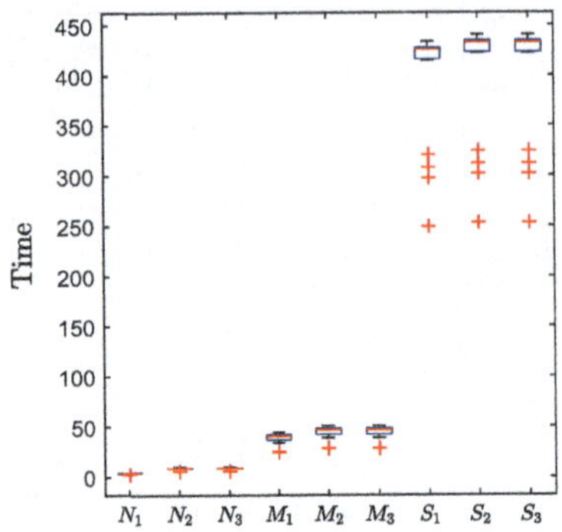

(e) DTLZ 5

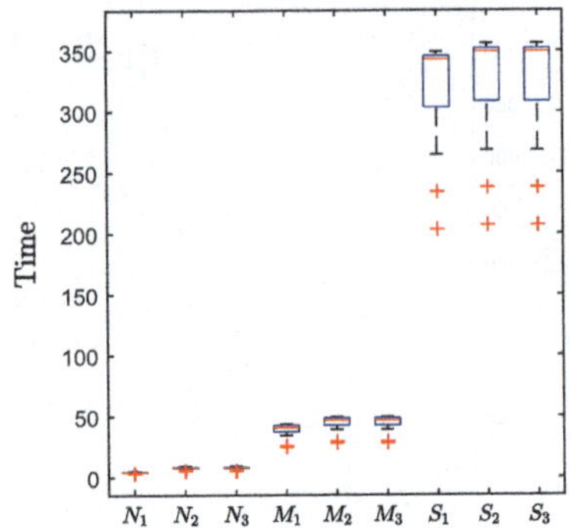

(f) DTLZ 6

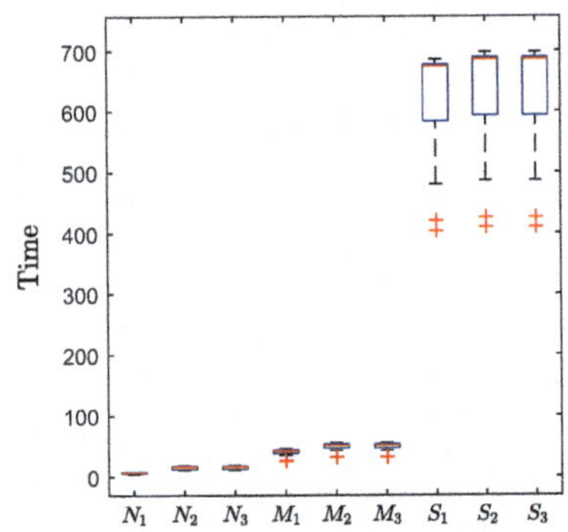

(g) DTLZ 7

Fig. 10.19 (continued)

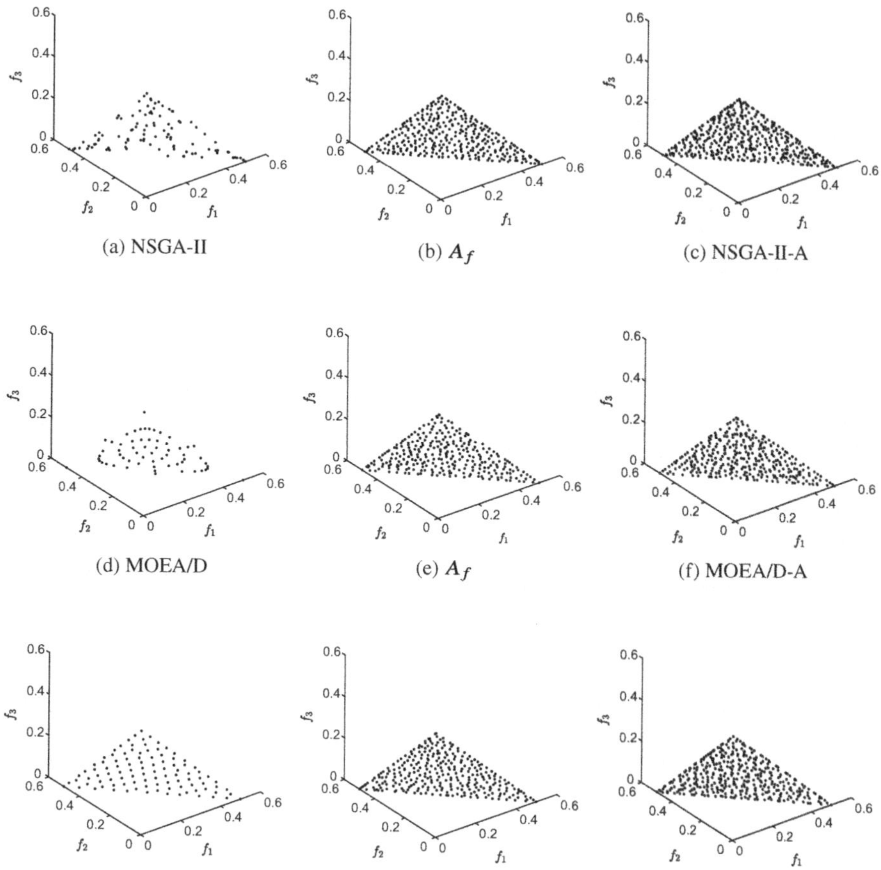

(a) NSGA-II (b) A_f (c) NSGA-II-A

(d) MOEA/D (e) A_f (f) MOEA/D-A

(g) SMS-EMOA (h) A_f (i) SMS-EMOA-A

Fig. 10.20 Numerical results of the three base MOEAs and their archive equipped variants on DTLZ1 using 125, 000 function evaluations. The middle column shows the final exernal archive A_f for each problem, and the right column the union of A_f and the final population

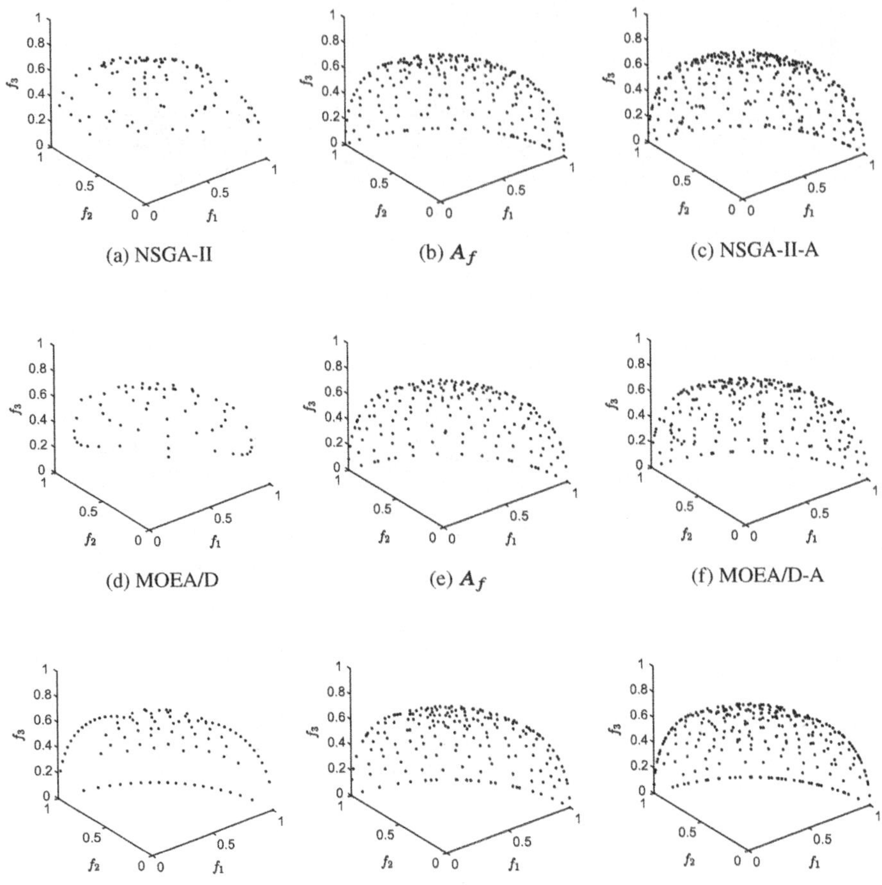

(a) NSGA-II (b) $\boldsymbol{A_f}$ (c) NSGA-II-A

(d) MOEA/D (e) $\boldsymbol{A_f}$ (f) MOEA/D-A

(g) SMS-EMOA (h) $\boldsymbol{A_f}$ (i) SMS-EMOA-A

Fig. 10.21 Numerical results of the three base MOEAs and their archive equipped variants on DTLZ2 using $125,000$ function evaluations. The middle column shows the final exernal archive A_f for each problem, and the right column the union of A_f and the final population

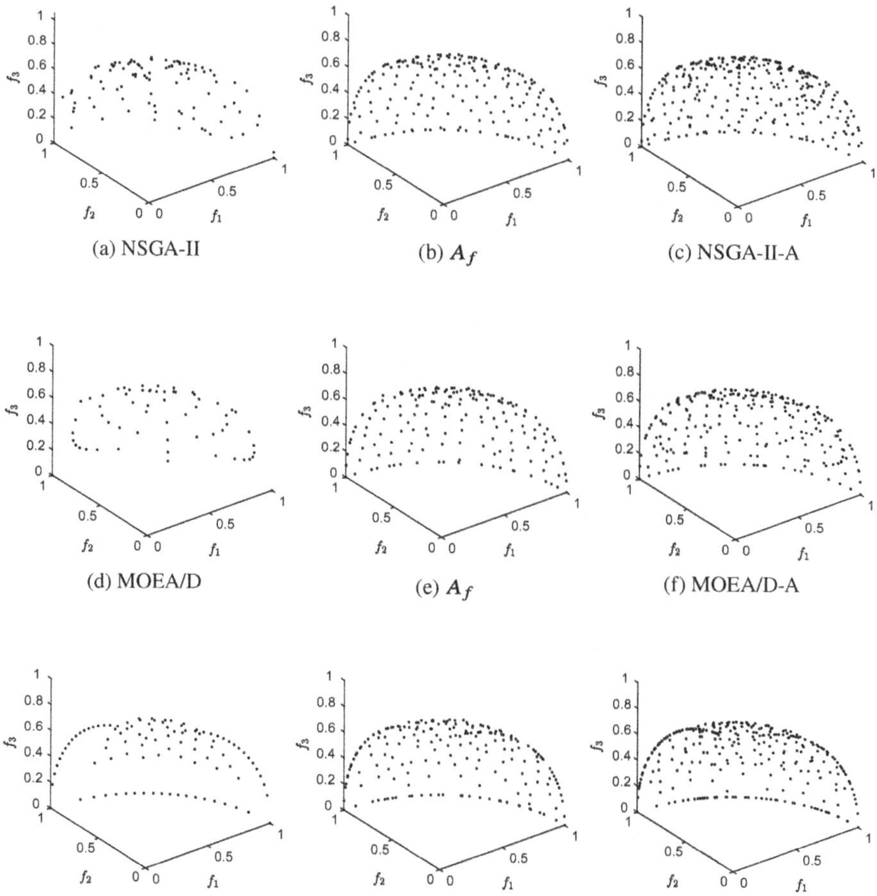

Fig. 10.22 Numerical results of the three base MOEAs and their archive equipped variants on DTLZ3 using 125, 000 function evaluations. The middle column shows the final exernal archive A_f for each problem, and the right column the union of A_f and the final population

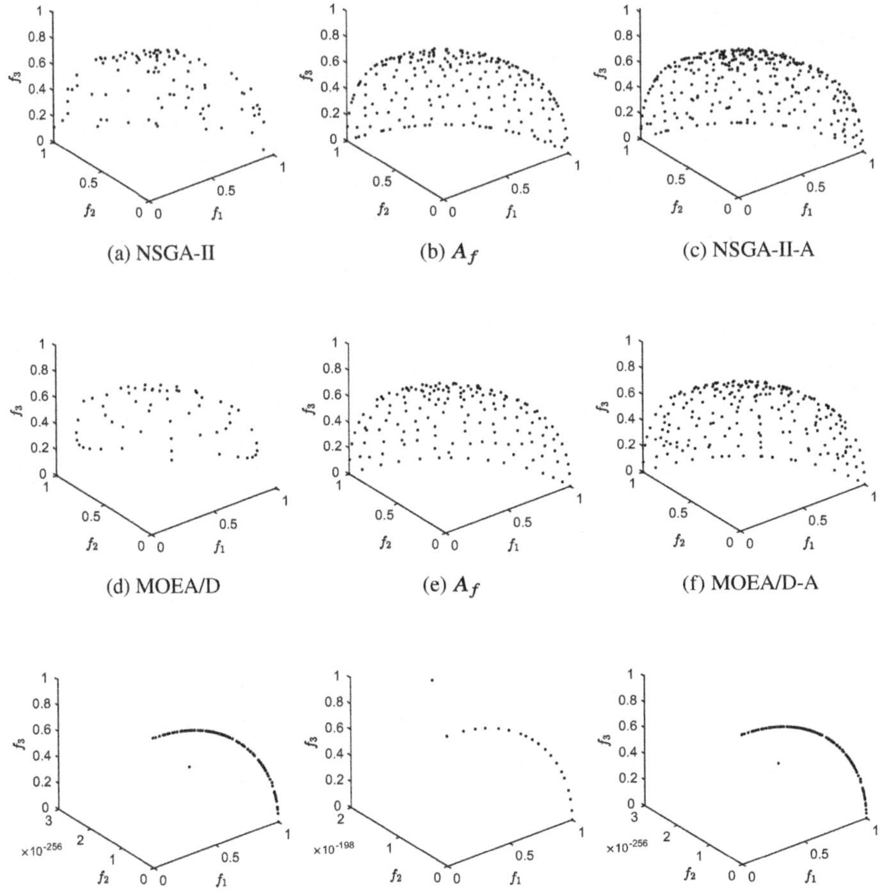

(a) NSGA-II (b) A_f (c) NSGA-II-A

(d) MOEA/D (e) A_f (f) MOEA/D-A

(g) SMS-EMOA (h) A_f (i) SMS-EMOA-A

Fig. 10.23 Numerical results of the three base MOEAs and their archive equipped variants on DTLZ4 using 125, 000 function evaluations. The middle column shows the final exernal archive A_f for each problem, and the right column the union of A_f and the final population

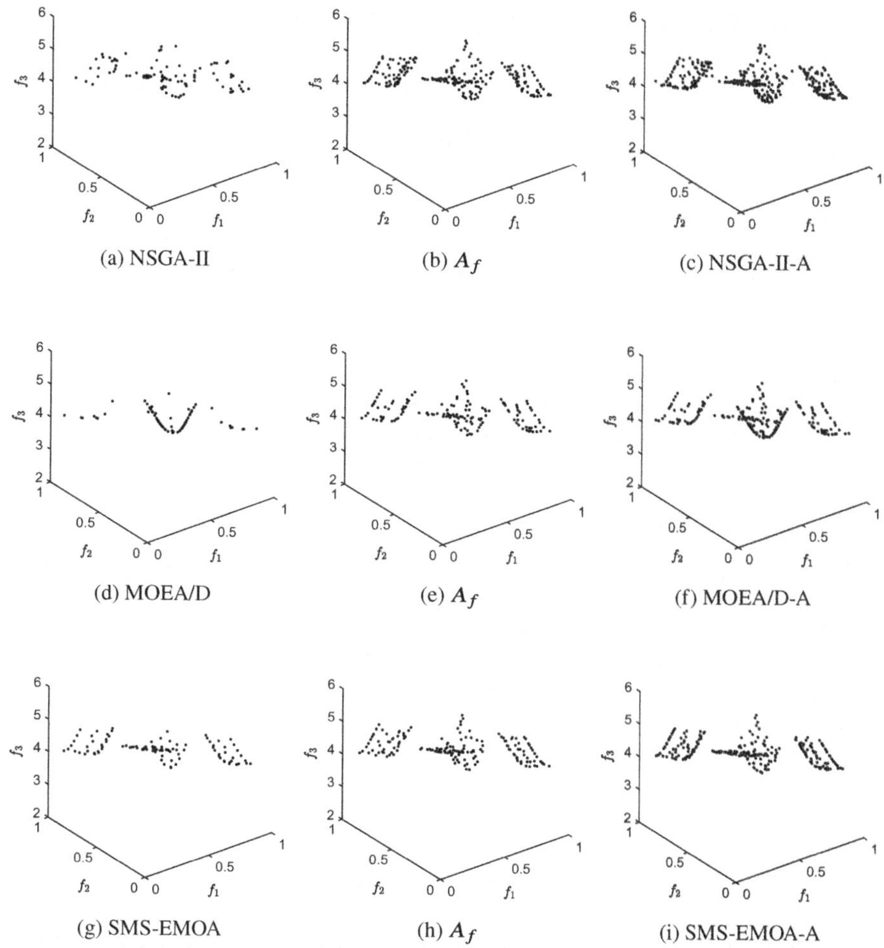

Fig. 10.24 Numerical results of the three base MOEAs and their archive equipped variants on DTLZ7 using 125, 000 function evaluations. The middle column shows the final exernal archive A_f for each problem, and the right column the union of A_f and the final population

We will in the following consider the same test problems as above, but double the budget of function evaluations to $N = 50,000$ for CONV2, DENT, and SSW, and to $N = 100,000$ for CONV3 (Fig. 10.5). The population sizes for the MOEAs have been set to 100 for the bi-objective problems, and to 250 for the three objective problem. As external archiver we have chosen $ArchiveUpdateTight2$ where we have used $\epsilon = (5,5)^T$ and $\Delta = 5$ for CONV2, $\epsilon = (2,2)^T$ and $\Delta = 1$ for DENT, $\epsilon = (5, 0.005)^T$ and $\Delta = 2$ for SSW, and $\epsilon = (0.25, 0.25, 0.25)^T$ and $\Delta = 0.5$ for CONV3.

Table 10.4 and Fig. 10.9 show as above the approximation qualities in terms of Δ_2, Table 10.5 and Fig. 10.10 the obtained archive sizes, and Table 10.6 and Fig. 10.11 the respective running times. Figures 10.12 to 10.15 show the median plots, and Fig. 10.16 the convergence plots. For sake of comparison and to see the effect of the external archiver, we show separately the final archive of a combined MOEA (denoted by A_f) as well as the result of the algorithm (i.e. $A_f \cup P_f$). In the box plots, N_1 denotes the result of the standalone NSGA-II, N_2 the result of the final archive, and N_3 the result of the combined MOEA. Analogous for the other algorithms (M_1 to M_3 and S_1 to S_3).
From the figures and tables we see in principle similar results as above—better approximation qualities for a higher cost—but to a much lesser extent in particular regarding archive sizes and running times. The approximation qualities of the combined MOEAs yield better values in all cases. Note that this is would not be the case if only the final archive A_f would be considered. This, however, entirely depends on the choice of the design parameters (which are ϵ and Δ for $ArchiveUpdateTight2$), and that in most cases $F(A_f)$ already yields a (nearly) perfect Pareto front approximation—according to the scope of the archiver and, again, to the choices of the design parameters. As in the previous part, archive sizes and running times increase compared to the base MOEAs, but the differences are not that significant any more, at least for MOEA/D (10 to 20% increase of running time) and for SMS-EMOA (2 to 6% increase of running time). The running time of NSGA-II-A is roughly the double compared to NSGA-II.

So far, we only have considered low dimensional problems that are relatively easy to solve for any set based multi-objective solver. We will hence in the third and last part investigate the treatment of higher-dimensional problems. For this, we have chosen to use the DTLZ test suite (Deb et al. 2002b). These functions are widely used in the EMO community, and all of the three base MOEAs have already shown their ability to solve them (Figs. 10.13, 10.14, 10.14, 10.16).

Figures 10.17, 10.18, and 10.19 show the box plots for the approximation qualities, sizes of the final archives/populations, and the running times, respectively. Figures 10.20 to 10.24 show the medians for the problems DTLZ 1-4 and DTLZ 7. For the computations, we have used $n = 12$ decision variables for DTLZ 1-6, and $n = 24$ for DTLZ7, and $k = 3$ objectives. The population sizes have been set to 100 for NSGA-II and SMS-EMOA, and to 91 for MOEA/D. The budget of function evaluations has been set to $N = 100,000$ for all problems and algorithms.

Table 10.7 Chosen design parameters for the application of NϵSGA

Design parameter	Value
Crossover probability	0.9
Mutation probability	$1/n$
Population size (npop)	100
No. of generations (ngen)	100
η_c	20
η_m	20
nr	3
nf	10

The results confirm the observations made in the previous part: the approximation qualities improve in all cases, and in some cases even significantly. In turn, the running times of the composed algorithms increase, but again, the difference is not so significant any more at least for MOEA/D and SMS-EMOA: an increase of 15 to 25 percent for MOEA/D and 1.5 to 3 percent for SMS-EMOA. For NSGA-II, there is still an increase of 100 to 150 percent. However, we stress that for all of the chosen academic functions the time for each function evaluations is in the range of milliseconds. For increasing function evaluation time the additional overhead generated by the external archiver will reduce significantly while yielding the same benefit in terms of the approximation quality.

Though all considered test problems have been relatively easy to solve by all algorithms, the potential benefit of using an external archive could already be demonstrated. The benefit will even increase for the treatment of more complex and higher dimensional problems which we will not address here.

Finally, we present some numerical results of NϵSGA. Since this is the only solver that explicitly aims to obtain approximations of the entire set of ϵ-efficient solutions, we omit a comparison to other algorithms here (and refer for more discussion on this to Hernández et al. 2020).

For the execution of NϵSGA we have used the design parameters that are shown in Table 10.7. For the archiver $ArchiveUpdateP_{Q,\epsilon}D_{xy}$ we have used the same values of ϵ, Δ_x and Δ_y for the discretization as specified in Sect. 8.6. Figures 10.25 and 10.26 show the median results (out of 30 independent runs) for the problems DEB99, TWO-TO-ONE, SYM-PART, SSW, OMNI-TEST, and LSS. For sake of completeness we also show the approximation qualities in Figs. 10.27 and 10.28, the running times in Figure 10.29 and the sizes of the final archives in Fig. 10.30. For all problems, the obtained approximations are essentially perfect according to the values of ϵ, Δ_x and Δ_y. To obtain these results, we have used only 10 percent of the budget used Section 8.6, where $N = 100,000$ randomly chosen points have been used to update the archive. For NϵSGA, we have set both the population size and the number of generations to 100 leading to a maximal number of 10,000 function calls spent per run of the algorithm. Further, the "bloating effect" of the approximation of the solution set is reduced which can be seen most significantly on DEB99 (for this, compare the results in Figs. 8.14 and 10.25) (Figs. 10.21, 10.22 and 10.23).

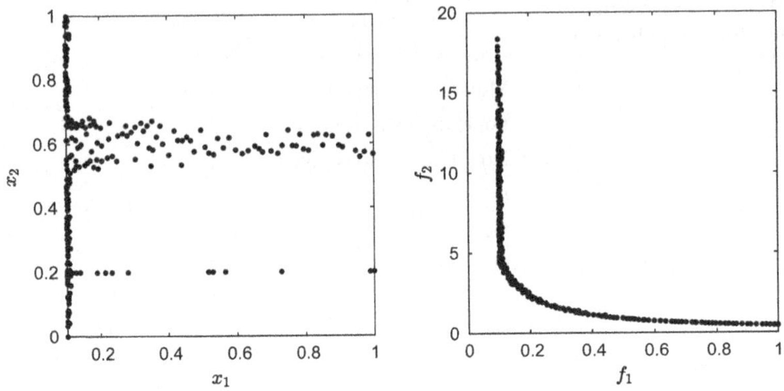

(a) DEB99

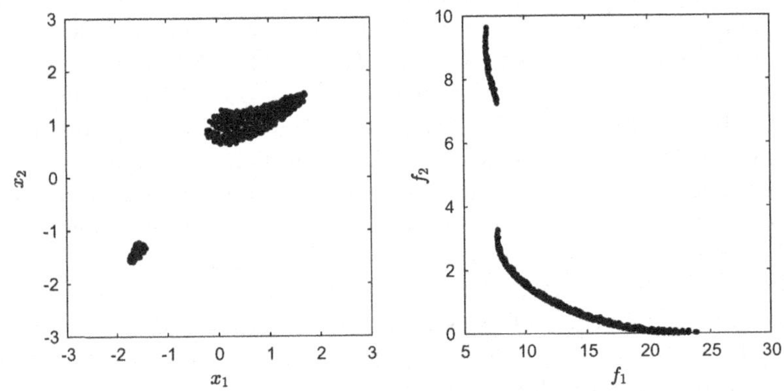

(b) TWO-To-ONE

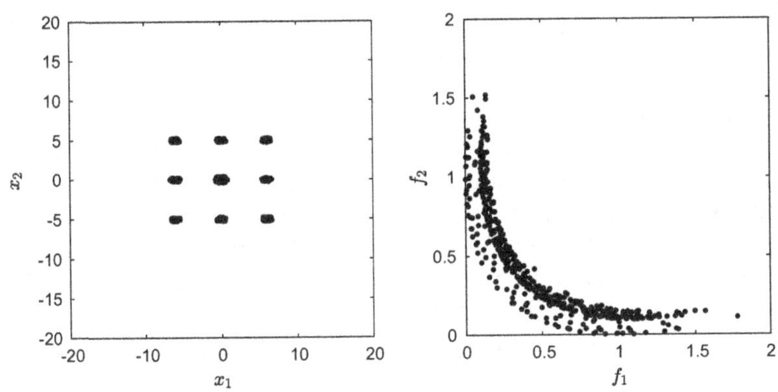

(c) SYM-PART

Fig. 10.25 Numerical results of NϵSGA for the problems DEB99, TWO-TO-ONE, and SYM-PART

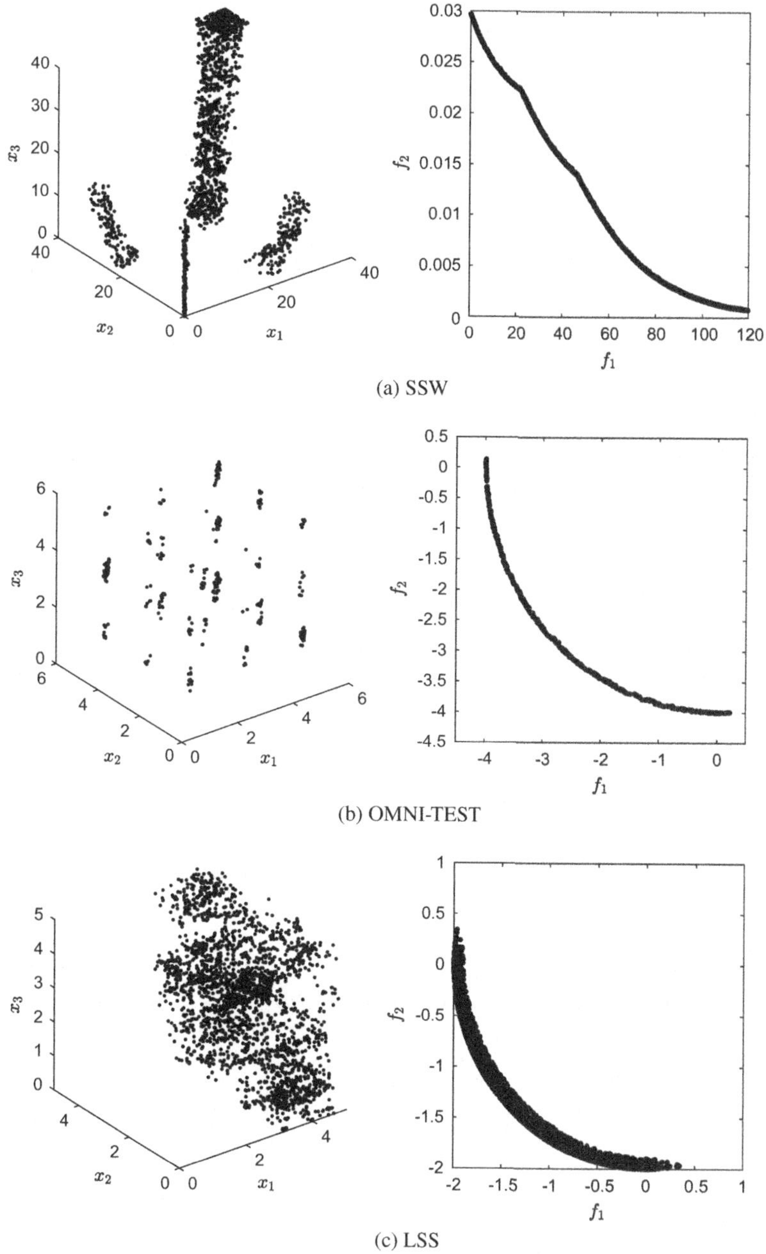

Fig. 10.26 Numerical results of NϵSGA for the problems SSW, OMNI-TEST, and LSS

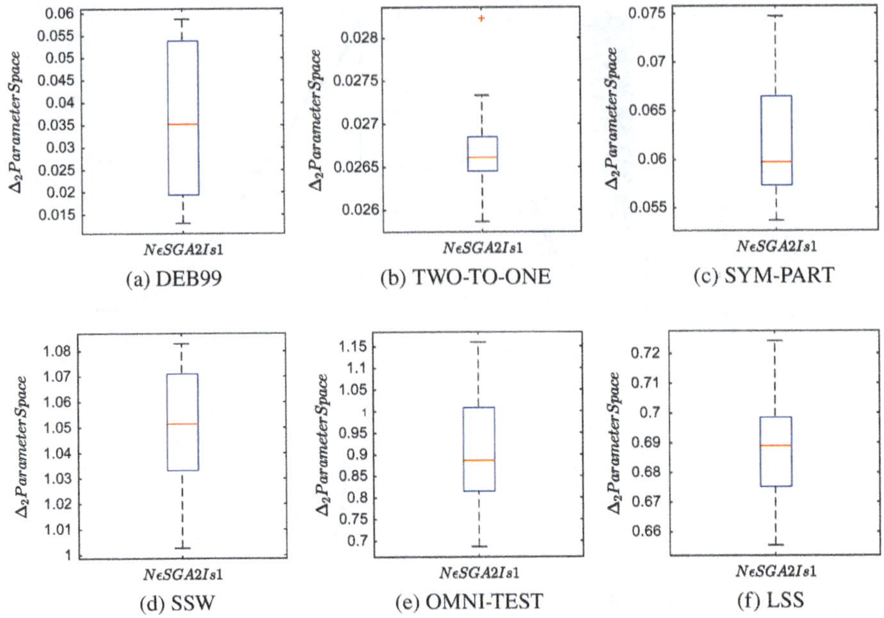

Fig. 10.27 Box plots for the approximation qualities $\Delta_2(A_f, P_{Q,\epsilon})$ obtained by NϵSGA, based on 30 independent runs

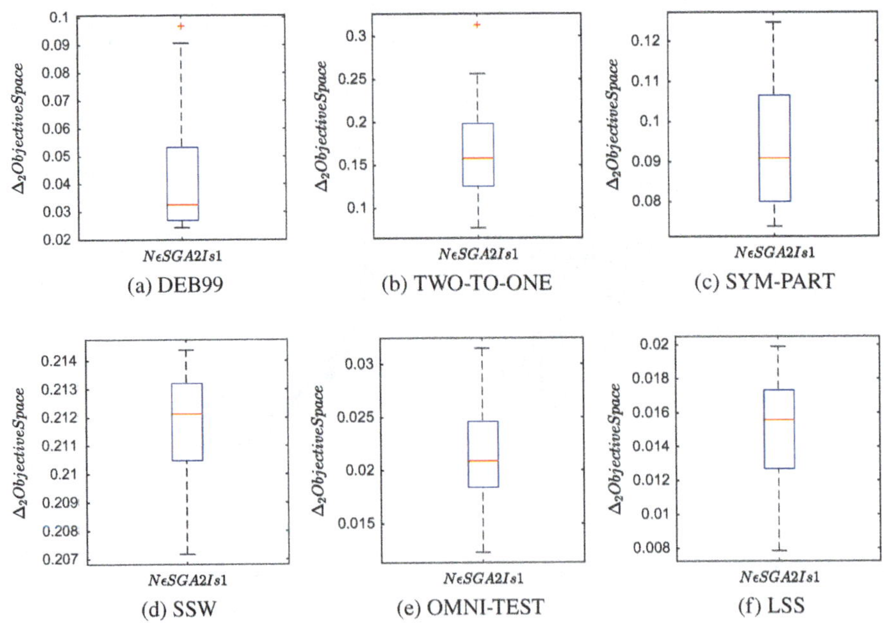

Fig. 10.28 Box plots for the approximation qualities $\Delta_2(F(A_f), F(P_{Q,\epsilon}))$ obtained by NϵSGA, based on 30 independent runs

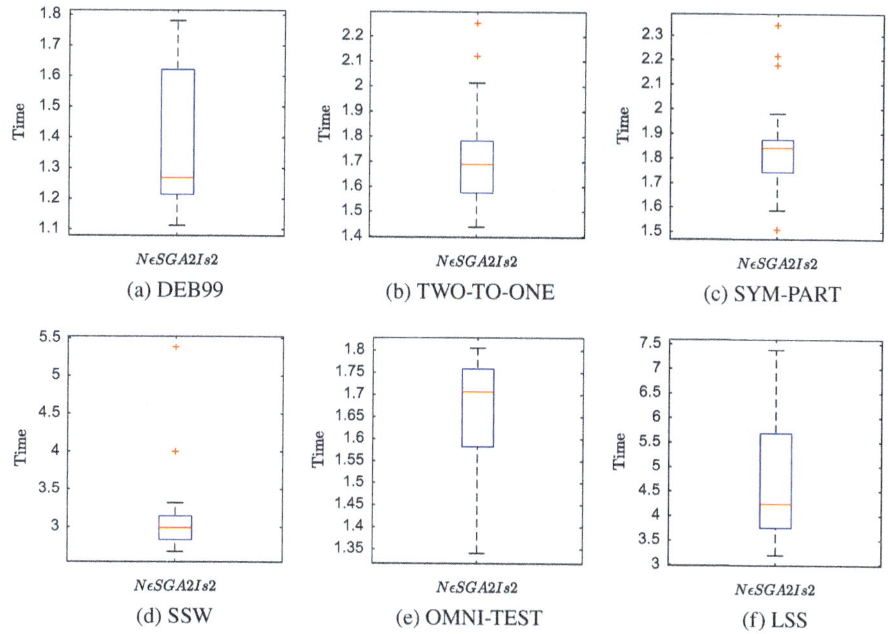

Fig. 10.29 Box plots for the running times of NϵSGA for the different problems, based on 30 independent runs

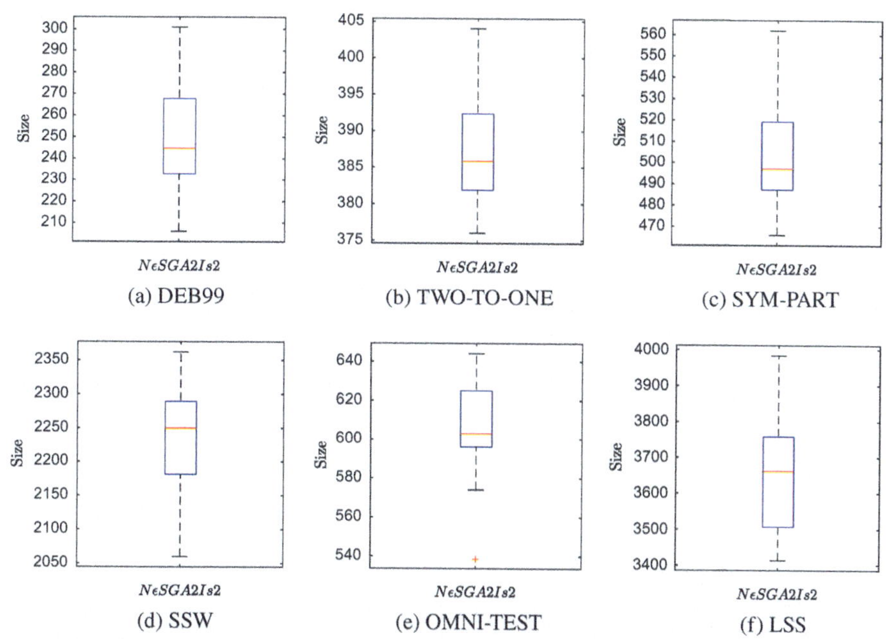

Fig. 10.30 Box plots for the final archive sizes obtained by NϵSGA for the different problems, based on 30 independent runs

Appendix A
Test Problems

Here we briefly present the test problems we have used throughout this book as demonstrators. For more information we refer to the publications where these function have originally been proposed (Deb 2001; Deb et al. 2002b; Schaeffler et al. 2002; Okabe et al. 2004; Schütze 2004; Preuss et al. 2006; Rudolph et al. 2007; Emmerich and Deutz 2007; Deb and Tiwari 2008; Witting 2012) as well as on the works (Collette and Siarry 2004; Huband et al. 2006; Köppen and Yoshida 2007; Schütze et al. 2011a), where some more discussion can be found on these and other test functions. We show the sets of approximate solutions $P_{Q,\epsilon}$ as well as their images $F(P_{Q,\epsilon})$ for one particular value of ϵ only if they are used in this book.

O. Schütze and C. Hernández, *Archiving Strategies for Evolutionary Multi-objective Optimization Algorithms*, Studies in Computational Intelligence 938,
https://doi.org/10.1007/978-3-030-63773-6

A.1 CONV2

Formulation:

$$F : Q \subset \mathbb{R}^2 \to \mathbb{R}^2$$

$$F(x) = \begin{pmatrix} x_1^2 + x_2^2 \\ (x_1 - 10)^2 + x_2^2 \end{pmatrix} \tag{A.1}$$

where

$$Q = [0, 10] \times [-5, 5]$$

Characteristics: linear Pareto set ($P_Q = [0, 10] \times \{0\}$), convex Pareto front, unimodal (convex) problem (Fig. A.1).

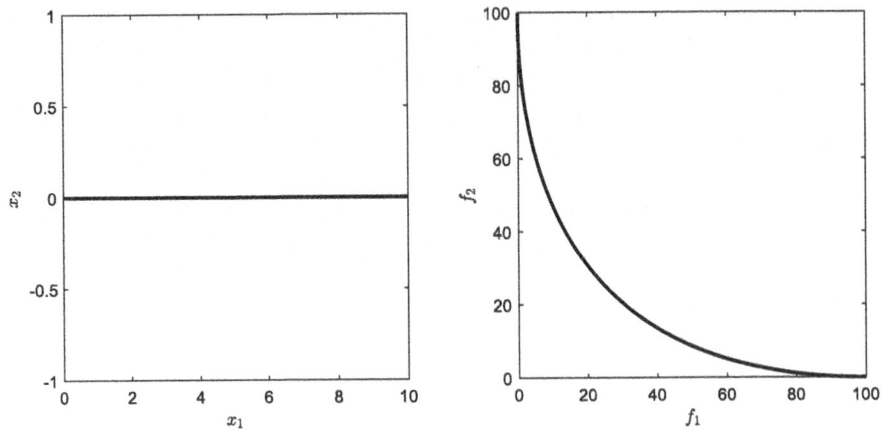

(a) P_Q (b) $F(P_Q)$

Fig. A.1 Pareto set and front of CONV2

A.2 DENT

Formulation (Witting 2012):

$$F : Q \subset \mathbb{R}^2 \to \mathbb{R}^2$$

$$F(x) = \begin{pmatrix} \frac{1}{2}(\sqrt{1+(x_1+x_2)^2} + \sqrt{1+(x_1-x_2)^2} + x_1 - x_2) + \alpha e^{-(x_1-x_2)^2} \\ \frac{1}{2}(\sqrt{1+(x_1+x_2)^2} + \sqrt{1+(x_1-x_2)^2} - x_1 + x_2) + \alpha e^{-(x_1-x_2)^2} \end{pmatrix}$$

where:

$$\alpha = 0.85,$$

$$Q = [-10, 10] \times [-10, 10] \tag{A.2}$$

Characteristics: linear Pareto set (P_Q is the line segment that connects $(-10, -10)^T$ and $(10, 10)^T$), convex-concave Pareto front, unimodal problem (Fig. A.2).

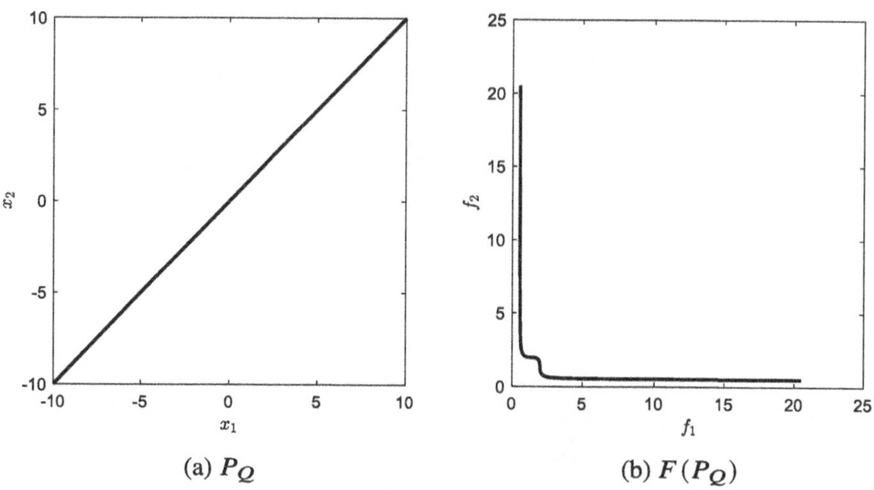

Fig. A.2 Pareto set and front of DENT

A.3 SSW

Formulation (Schaeffler et al. 2002):

$$
\begin{aligned}
F &: Q \subset \mathbb{R}^3 \to \mathbb{R}^3 \\
F(x) &= (f_1(x), f_2(x))^T \\
&\text{where:} \\
f_1(x) &= \sum_{j=1}^{n} x_j, \qquad f_2(x) = 1 - \prod_{j=1}^{n} (1 - w_j(x_j)), \qquad \text{(A.3)} \\
w_j(z) &= \begin{cases} 0.01 \cdot \exp(-(\frac{z}{20})^{2.5}) & \text{for } \; j = 1, 2 \\ 0.01 \cdot \exp(-\frac{z}{15}) & \text{for } \; j > 2 \end{cases}, \\
Q &= [0, 40]^n \quad \text{(we have used } n = 3\text{)}
\end{aligned}
$$

Characteristics: nonlinear and disconnected Pareto set, convex-concave Pareto front, unimodal problem (Fig. A.3).

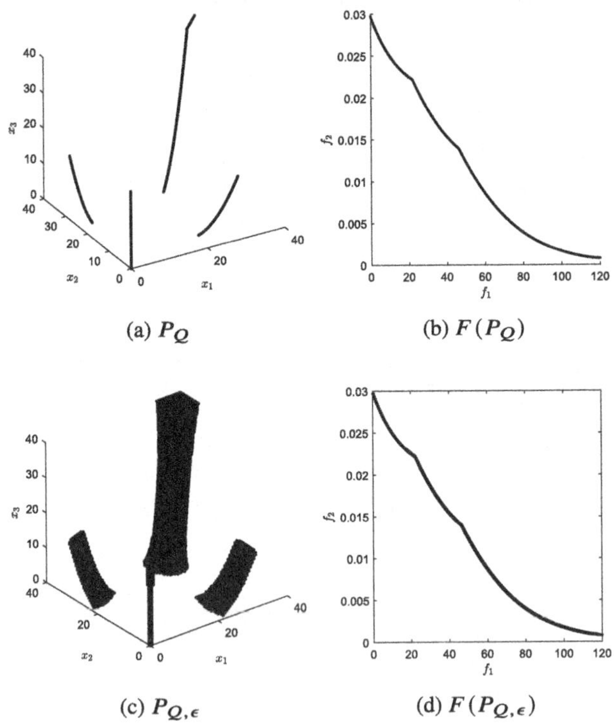

(a) P_Q (b) $F(P_Q)$

(c) $P_{Q,\epsilon}$ (d) $F(P_{Q,\epsilon})$

Fig. A.3 Pareto set/front of SSW and approximate solutions for $\epsilon = (0.001, 0.000001)$

A.4 CONV3

Formulation (Schütze 2004):

$$
\begin{aligned}
F &: Q \subset \mathbb{R}^3 \to \mathbb{R}^3 \\
F(x) &= (f_1(x), f_2(x), f_3(x))^T \\
&\text{where:} \\
f_i(x) &= \|x - a^i\|_2^2 \\
a^1 &= (-1, -1, \ldots, -1)^T \\
a^2 &= (1, 1, \ldots, 1)^T \\
a^3 &= (-1, 1, \ldots, -1, 1)^T \\
Q &= [-3, 3]^n \quad \text{(we have used } n_1 = 3 \text{ and } n_2 = 30) \qquad \text{(A.4)}
\end{aligned}
$$

Characteristics: linear Pareto set (the 2-simplex that is defined by the vertices a^1, a^2 and a^3), convex Pareto front, unimodal (convex) problem (Fig. A.4).

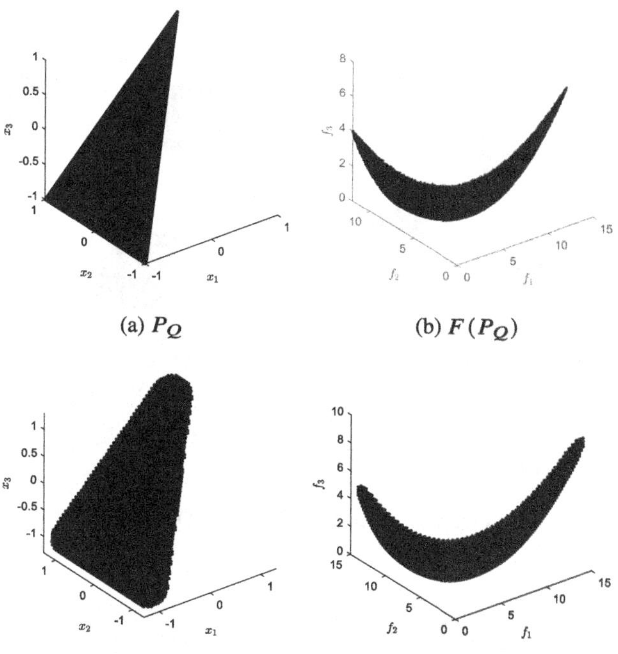

(a) P_Q (b) $F(P_Q)$

(c) $P_{Q,\epsilon}$ (d) $F(P_{Q,\epsilon})$

Fig. A.4 Pareto set/front of CONV3 and for $n_1 = 3$ and approximate solutions for $\epsilon = (0.1\ 0.1\ 0.1)$

A.5 DTLZ 7

Formulation (Deb et al. 2002b):

$$F : Q \subset \mathbb{R}^n \to \mathbb{R}^k$$

$$F(x) = \begin{pmatrix} x_1 \\ x_2 \\ \vdots \\ x_{k-1} \\ 1 + g(x)h(x, g(x)) \end{pmatrix}$$

where:

$$g(x) = 1 + \frac{9}{l} \sum_{i=n-l}^{n} x_i, \qquad h(x, g(x)) = k - \sum_{i=1}^{k-1} \frac{x_i}{1 + g(x)}(1 + \sin(3\pi x_i)),$$

$$Q = [0, 1]^n \qquad \text{(we have used } n = 3 \text{ and } k = 3) \tag{A.5}$$

Characteristics: linear and disconnected Pareto set, non-linear Pareto front, multi-modal problem (Fig. A.5).

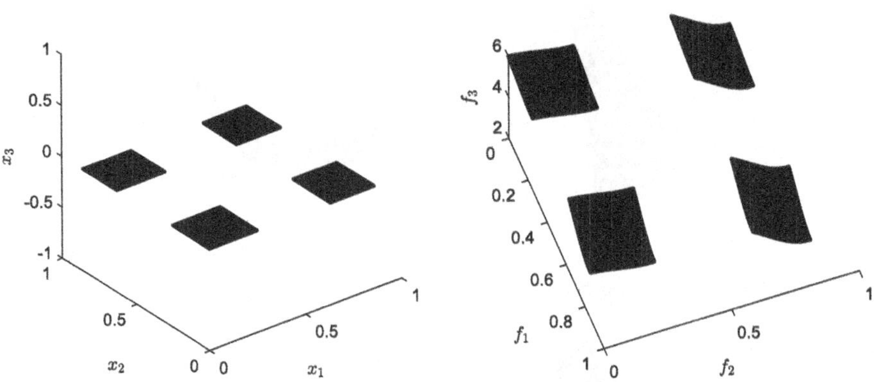

(a) P_Q (b) $F(P_Q)$

Fig. A.5 Pareto set and front of DTLZ7

A.6 DEB99

Formulation (Deb 2001):

$$
\begin{aligned}
&F : Q \subset \mathbb{R}^2 \to \mathbb{R}^2, \\
&F(x) = \begin{pmatrix} x_1 \\ \frac{g(x)}{x_1} \end{pmatrix} \\
&\text{where:} \\
&g(x) = 2 - \exp\left(-\left(\frac{x_2 - 0.2}{0.004}\right)^2\right) - 1.56\exp\left(-\left(\frac{x_2 - 0.6}{0.4}\right)^2\right), \\
&Q = [0.1, 1] \times [0, 1]
\end{aligned}
\tag{A.6}
$$

Characteristics: linear Pareto set ($P_Q = [0.1, 1] \times \{0.2\}$), convex Pareto front, unimodal function (Fig. A.6).

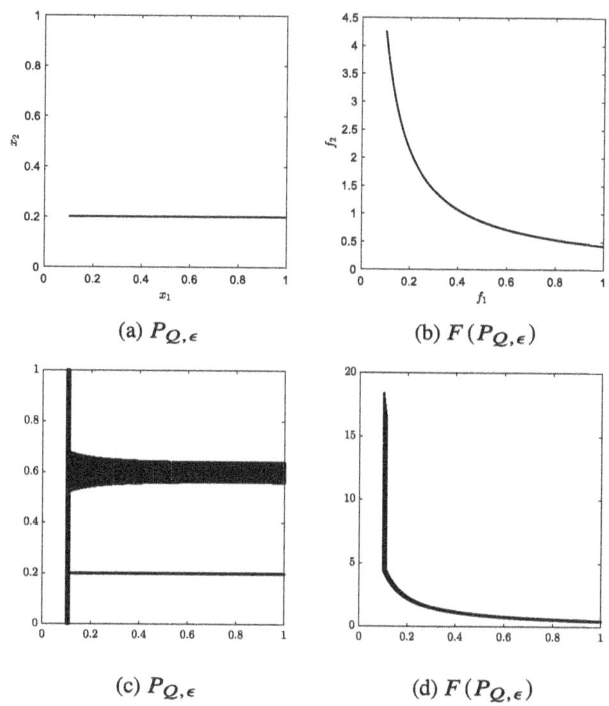

Fig. A.6 Pareto set/front of DEB99 and approximate solutions for $\epsilon = (0.0110.011)$

A.7 TWO-ON-ONE

Formulation (Deb and Tiwari 2008):

$$
\begin{aligned}
& F : Q \subset \mathbb{R}^2 \to \mathbb{R}^2, \\
& F(x) = \begin{pmatrix} x_1^4 + x_2^4 - x_1^2 + x_2^2 - c * x_1 * v(2) + d * x_1 + 20 \\ (x_1 - k)^2 + (x_2 - l)^2 \end{pmatrix} \\
& \quad \text{where:} \\
& k = 0, \quad l = 1, \quad c = 10, \quad d = 0.25 \\
& Q = [-3, 3]^2
\end{aligned} \tag{A.7}
$$

Characteristics: disconnected (2 components) and nonlinear Pareto set, unimodal function (Fig. A.7).

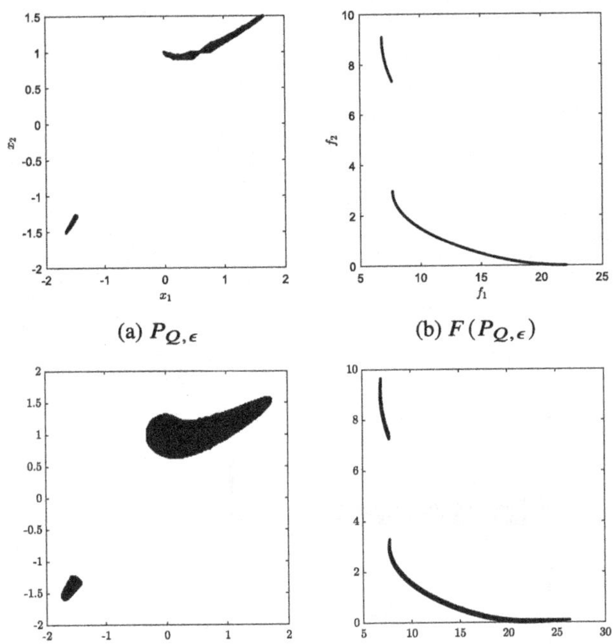

(a) $P_{Q,\epsilon}$ (b) $F(P_{Q,\epsilon})$ (c) $P_{Q,\epsilon}$ (d) $F(P_{Q,\epsilon})$

Fig. A.7 Pareto set/front of TWO-ON-ONE and approximate solutions for $\epsilon = (0.1\,0.1)$

A.8 SYM-PART

Formulation (Rudolph et al. 2007):

$$
\begin{aligned}
&F : Q \subset \mathbb{R}^2 \to \mathbb{R}^2\\
&F(x) = (f_1(x), f_2(x))^T\\
&\text{where:}\\
&f_1(x) = (x_1 - t_1(c + 2a) + a)^2 + (x_2 - t_2 b)^2,\\
&f_2(x) = (x_1 - t_1(c + 2a) - a)^2 + (x_2 - t_2 b)^2,\\
&t1 = sign(x_1)\min(ceil((|x_1| - a - (c/2))/(2a + c)), 1),\\
&t2 = sign(x_2)\min(ceil((|x_2| - b/2)/b), 1),\\
&a = 0.5, \quad b = 5, \quad c = 5,\\
&Q = [-20, 20]^2
\end{aligned}
\tag{A.8}
$$

Characteristics: disconnected (9 components) and linear (on each sub-domain) Pareto set, convex Pareto front, unimodal on each sub-domain (Fig. A.8).

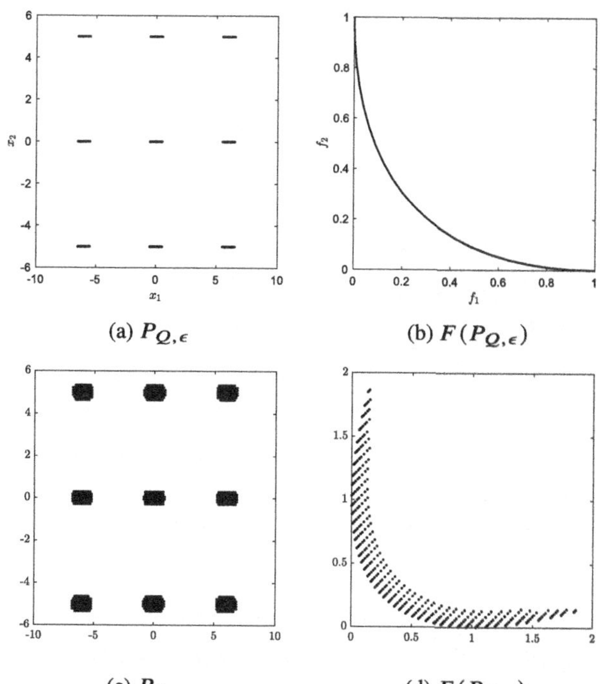

Fig. A.8 Pareto set/front of SYM-PART and approximate solutions for $\epsilon = (0.15\, 0.15)$

A.9 OMNI-TEST

Formulation (Deb and Tiwari 2008):

$$
\begin{aligned}
F &: Q \subset \mathbb{R}^n \to \mathbb{R}^2, \\
F(x) &= \begin{pmatrix} \sum_{i=1}^{n} (\sin(\pi x_i)) \\ \sum_{i=1}^{n} (\cos(\pi x_i)) \end{pmatrix} \\
&\text{where:} \\
Q &= [0, 6]^n \quad \text{(we have used } n = 4\text{)}
\end{aligned}
\tag{A.9}
$$

Characteristics: disconnected (27 components) and linear (on each sub-domain) Pareto set, convex Pareto front (Fig. A.9).

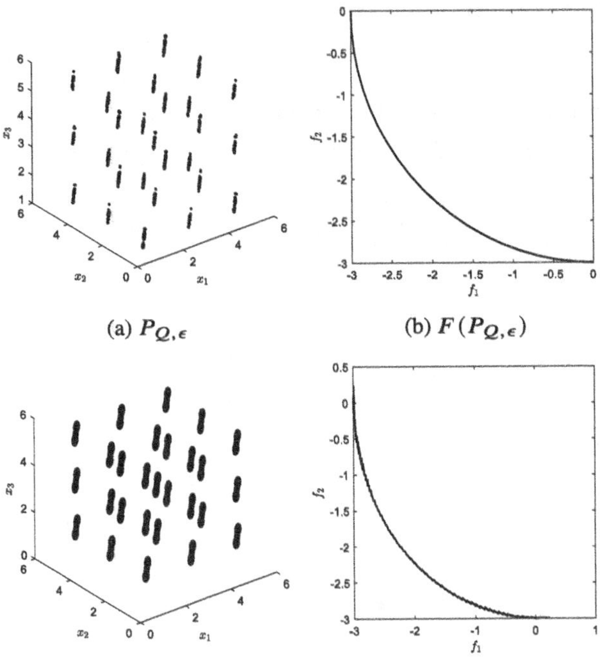

(a) $P_{Q,\epsilon}$ (b) $F(P_{Q,\epsilon})$

(c) $P_{Q,\epsilon}$ (d) $F(P_{Q,\epsilon})$

Fig. A.9 Pareto set/front of OMNI-TEST and approximate solutions for $\epsilon = (0.01\ 0.01)$

A.10 LSS

Formulation (Emmerich and Deutz 2007):

$$
\begin{aligned}
F :& Q \subset \mathbb{R}^5 \to \mathbb{R}^2, \\
F(x) &= \begin{pmatrix} (1 + r(x)) \cos(x_1) \\ (1 + r(x)) \sin(x_1) \end{pmatrix} \\
& \text{where:} \\
g(x) &= \frac{1}{|x| - 1} \sum_{i=2}^{n} x, \\
r(x) &= \sin(\pi * g(x))^2, \\
Q &= [0, 5]^n \quad \text{(we have used } n = 3\text{)}
\end{aligned}
\tag{A.10}
$$

Characteristics: disconnected (5 components) and linear (on each sub-domain) Pareto set, convex Pareto front (Fig. A.10).

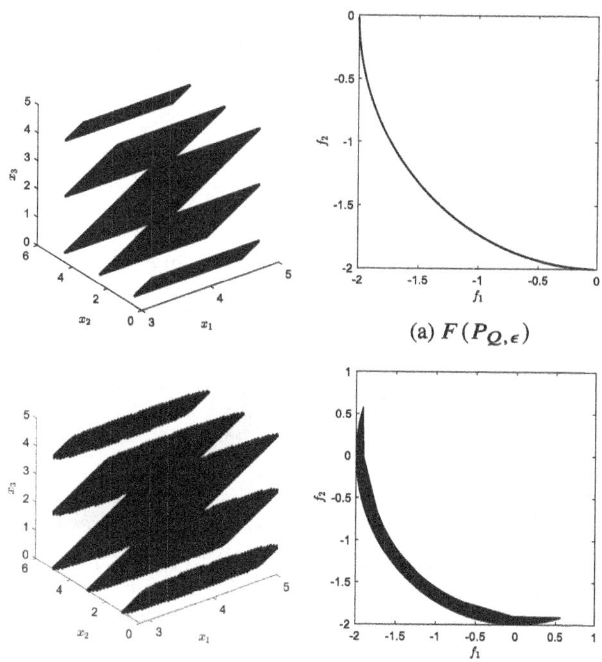

(a) $F(P_{Q,\epsilon})$

(b) $P_{Q,\epsilon}$ (c) $F(P_{Q,\epsilon})$

Fig. A.10 Pareto set/front of LSS and approximate solutions for $\epsilon = (0.1\ 0.1)$

A.11 OKA1

Formulation (Okabe et al. 2004):

$$F : Q \subset \mathbb{R}^2 \to \mathbb{R}^2,$$

$$F(x) = \begin{pmatrix} xp_1(x) \\ \sqrt{2\pi} - \sqrt{abs(xp_1(x))} + 2\sqrt{abs(xp_2(x)) - 3\cos xp_1(x) - 3} \end{pmatrix}$$

where:

$$xp_1(x) = \cos(\pi/12)x_1 - \sin(\pi/12)x_2,$$
$$xp_2(x) = \sin(\pi/12)x_1 + \cos(\pi/12)x_2,$$
$$Q = [6\sin(\pi/12), 6\sin(\pi/12) + 2\pi\cos(\pi/12)] \times [-2\pi\sin(\pi/12), 6\cos(\pi/12)] \tag{A.11}$$

Characteristics: connected and nonlinear Pareto set, convex Pareto front (Fig. A.11).

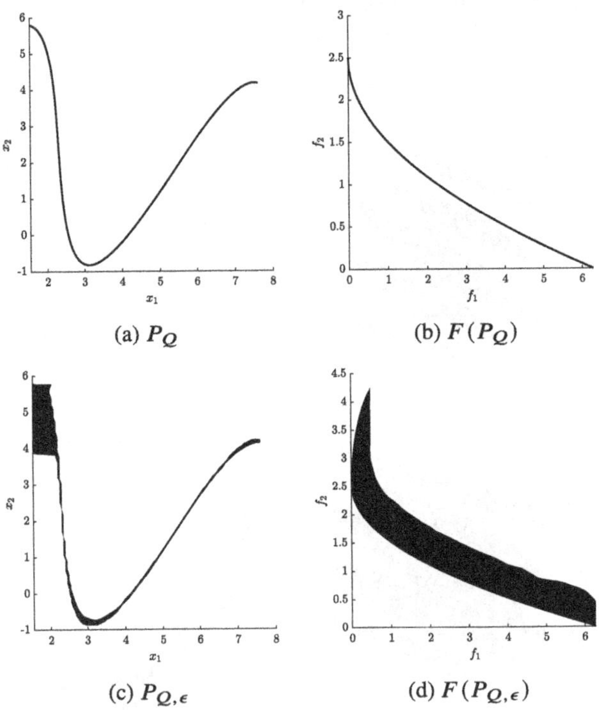

(a) P_Q (b) $F(P_Q)$ (c) $P_{Q,\epsilon}$ (d) $F(P_{Q,\epsilon})$

Fig. A.11 Pareto set/front of OKA1 and approximate solutions for $\epsilon = (0.5\,0.5)$

Appendix B
Archivers

Here we give a brief overview of all archivers that have been discussed in this book. For all archivers for multi-objective problems we assume to deal with the minimization of a function $F : Q \subset \mathbb{R}^n \to \mathbb{R}^k$ (i.e., we are given n decision variables and k objective functions). For the archivers for scalar optimization problems we assume to deal with the minimization of a function $f : Q \subset \mathbb{R}^n \to \mathbb{R}$. In all cases, a given finite archive $A_0 \subset \mathbb{R}^n$ is updated by a finite set (or population) P, and the archiver returns the updated archive A. The design parameters for the discretization of the set of interest are given by ϵ, Δ_x and Δ_y. $\Theta \in (0, 1)$ as well as $\tilde{\Delta}$, $\Delta^* \in (0, \Delta)$ are safety factors that are needed for theoretical purposes, and can be set to $\Theta = 1$, $\tilde{\Delta} = \Delta$ and $\Delta^* = 1$, respectively. We will provide the modifications of the archivers in case Δ_x or Δ_y is a vector.

Implementations of all the archivers in MATLAB together with the test examples used in this book can be found in the GitHup software development platform under https://github.com/NumericalEvolutionaryOptimization/archivers.

O. Schütze and C. Hernández, *Archiving Strategies for Evolutionary Multi-objective Optimization Algorithms*, Studies in Computational Intelligence 938,
https://doi.org/10.1007/978-3-030-63773-6

B.1 *ArchiveUpdateP*$_Q$

Algorithm $A := ArchiveUpdateP_Q(P, A_0)$

Require: population P, archive A_0
Ensure: updated archive A

$A := A_0$
for all $p \in P$ **do**
  if $\nexists a \in A : a \prec p$ **then**
    $A := A \cup \{p\}$
  end if
  for all $a \in A$ **do**
    if $p \prec a$ **then**
      $A := A \backslash \{a\}$
    end if
  end for
end for
return A

Brief Description:

Aims for the approximation of the entire Pareto set/front of a given MOP, stores all non-dominated points out of the considered candidate solutions, see Sect. 5.2 for details.

Main Results:

Approximation quality: Theorems 5.1–5.3.

Upper bound on archive size: (unbounded).

B.2 *ArchiveUpdateEps1*

Algorithm $A := ArchiveUpdateEps1\ (P, A_0)$

Require: P: current population, A_0: current archive, $\epsilon \in \mathbb{R}^k_+$, $\Theta \in (0, 1)$
Ensure: A: updated archive
1: $A := A_0$
2: **for all** $p \in P$ **do**
3:     **if** $\exists a \in A : a \prec_{\Theta\epsilon} p$ **then**
4:         CONTINUE ▷ *do not execute lines* 6 – 11
5:     **end if**
6:     **for all** $a \in A$ **do**
7:         **if** $p \prec a$ **then**
8:             $A := A \backslash \{a\}$
9:         **end if**
10:    **end for**
11:    $A := A \cup \{p\}$
12: **end for**
13: **return** A

Brief Description:

Aims to maintain a ϵ-approximate Pareto front of a given MOP, see Sect. 6.2 for details.

Main Results:

Approximation quality: Lemma 6.2, Theorem 6.1.

Upper bound on archive size: Theorem 6.2.

B.3 *ArchiveUpdateEps2*

Algorithm $A := ArchiveUpdateEps2\ (P, A_0)$

Require: P: current population, A_0: current archive, $\epsilon \in \mathbb{R}^k_+$, $\Theta \in (0, 1)$
Ensure: A: updated archive

$A := A_0$
for all $p \in P$ **do**
  if $\nexists a \in A : a \prec_{\Theta\epsilon} p$ **then**
    $A := A \cup \{p\}$
  end if
  for all $a \in A$ **do**
    if $p \prec a$ **then**
      $A := A \cup \{p\} \backslash \{a\}$
    end if
  end for
end for
return A

Brief Description:

Aims to maintain a ϵ-Pareto front of a given MOP, see Sect. 6.3 for details.

Main Results:

Approximation quality: Lemma 6.4, Theorem 6.3.

Upper bound on archive size: Theorem 6.4.

B.4 *ArchiveUpdateTight1*

Algorithm $A := ArchiveUpdateTight1_{\epsilon,\tilde{\Delta}}\ (P, A_0)$

Require: P: current population, A_0: current archive, $\epsilon \in \mathbb{R}^k_+$, $\Theta \in (0, 1)$, $0 < \tilde{\Delta} < \Delta$
Ensure: A: updated archive
1: $A := A_0$
2: **for all** $p \in P$ **do**
3:    **if** $(\exists a \in A : a \prec p)$ or $(\exists a_1 \in A :\ a_1 \prec_{\Theta\epsilon} p$ and $\exists a_2 \in A : d_\infty(F(a_2), F(p)) \leq \tilde{\Delta})$ **then**
4:       CONTINUE ▷ *do not execute lines* 6 – 11
5:    **end if**
6:    **for all** $a \in A$ **do**
7:       **if** $p \prec a$ **then**
8:          $A := A \backslash \{a\}$
9:       **end if**
10:    **end for**
11:    $A := A \cup \{p\}$
12: **end for**
13: **return** A

Brief Description:

Aims to maintain a $(\Delta, \Theta\epsilon_m)$-tight ϵ-approximate Pareto front of a given MOP, see Sect. 7.2 for details.

Main Results:

Approximation quality: Theorems 7.1 and 7.2.

Upper bound on archive size: as for *ArchiveUpdateEps1* (see Theorem 6.2).

B.5 *ArchiveUpdateTight2*

Algorithm $A := ArchiveUpdateTight2_{\epsilon,\tilde{\Delta}}\ (P, A_0)$

Require: P: current population, A_0: current archive, $\epsilon \in \mathbb{R}^k_+$, $\Theta \in (0, 1)$, $0 < \tilde{\Delta} < \Delta$
Ensure: A: updated archive

$A := A_0$
for all $p \in P$ **do**
  if $(\nexists a_1 \in A : a_1 \prec_{\Theta\epsilon} p)$ or $(\nexists a_2 \in A :\ a_2 \prec p$ and $\forall a \in A : d_\infty(F(a), F(p)) > \tilde{\Delta})$ **then**
    $A := A \cup \{p\}$
  end if
  for all $a \in A$ **do**
    if $p \prec a$ **then**
      $A := A \cup \{p\} \backslash \{a\}$
    end if
  end for
end for
return A

Brief Description:

Aims to maintain a Δ-tight ϵ-Pareto front of a given MOP, see Sect. 7.3 for details.

Main Results:

Approximation quality: Theorems 7.3 and 7.4.

Upper bound on archive size: as for $ArchiveUpdateEps2$ (see Theorem 6.4).

B.6 $ArchiveUpdateP_{Q,\epsilon}$

Algorithm $A := ArchiveUpdateP_{Q,\epsilon}\ (P, A_0)$

Require: P: population, A_0: current archive, $\epsilon \in \mathbb{R}^k_+$
Ensure: updated archive A

$A := A_0$
for all $p \in P$ **do**
    if $\nexists a \in A : a \prec_{-\epsilon} p$ **then**
        $A := A \cup \{p\}$
    end if
    for all $a \in A$ **do**
        if $p \prec_{-\epsilon} a$ **then**
            $A := A \backslash \{a\}$
        end if
    end for
end for
return A

Brief Description:

Aims for the approximation of the entire set of approximate solutions $P_{Q,\epsilon}$ and its image $F(P_{Q,\epsilon})$ of a given MOP, see Sect. 8.2 for details.

Main Results:

Approximation quality: Theorems 8.4 and 8.5.

Upper bound on archive size: (unbounded).

B.7 *ArchiveUpdate* $P_{Q,\epsilon}D_x$

Algorithm $A := ArchiveUpdateP_{Q,\epsilon}D_x\ (P, A_0, \Delta_x)$

Require: population P, archive A_0, $\epsilon \in \mathbb{R}^k_+$, $\Delta_x \in \mathbb{R}_+$, $\Delta^* \in (0, \Delta_x)$
Ensure: updated archive A
1: $A := A_0$
2: **for all** $p \in P$ **do**
3:    **if** $\nexists a_1 \in A : a \prec_{-\epsilon} p$ and $\nexists a_2 \in A : d_H(a_2, p) \leq \Delta^*$ **then**
4:       $A := A \cup \{p\}$
5:       **for all** $a \in A$ **do**
6:          **if** $p \prec_{-\epsilon} a$ **then**
7:             $A := A \backslash \{a\}$
8:          **end if**
9:       **end for**
10:    **end if**
11: **end for**
12: **return** A

Brief Description:

Aims for a uniform approximation $P_{Q,\epsilon}$ of a given MOP, see Sect. 8.4 for details.

Modification:

If $\Delta_x \in \mathbb{R}^n_+$, and $\Delta^*_x \in \mathbb{R}^n_+$ with $\Delta^*_i \in (0, \Delta_{x,i})$, $i = 1, \ldots, n$, then one can replace the last condition in line 3 by

$$\nexists a_2 \in A : \ |a_{2,i} - p_i| \leq \Delta^*_i, \quad i = 1, \ldots, n$$

B.8 $ArchiveUpdateP_{Q,\epsilon}D_y$

Algorithm $A := ArchiveUpdateP_{Q,\epsilon}D_y\ (P, A_0, \Delta_y)$

Require: population P, archive A_0, $\epsilon \in \mathbb{R}^k_+$, $\Delta_y \in \mathbb{R}_+$, $\Delta^* \in (0, \Delta)$
Ensure: updated archive A
1: $A := A_0$
2: **for all** $p \in P$ **do**
3:   **if** $\nexists a_1 \in A : a \prec_{-\epsilon} p$ and $\nexists a_2 \in A : d_H(F(a_2), F(p)) \leq \Delta^*$ **then**
4:     $A := A \cup \{p\}$
5:     **for all** $a \in A$ **do**
6:       **if** $p \prec_{-(\epsilon+1\Delta)} a$ **then**
7:         $A := A \backslash \{a\}$
8:       **end if**
9:     **end for**
10:   **end if**
11: **end for**
12: **return** A

Brief Description:

Aims for a uniform approximation $F(P_{Q,\epsilon})$ of a given MOP, see Sect. 8.3 for details.

Main Results:

Approximation quality: Theorems 8.6 and 8.7.

Upper bound on archive size: Theorem 8.8.

Modification:

If $\Delta_y \in \mathbb{R}^k_+$, and $\Delta^* \in \mathbb{R}^k_+$ with $\Delta^*_i \in (0, \Delta_{x,i})$, $i = 1, \ldots, k$, then one can replace the last condition in line 3 by

$$\nexists a_2 \in A : |f_i(a_2) - f_i(p)| \leq \Delta^*_i, \quad i = 1, \ldots, k$$

B.9 $ArchiveUpdateP_{Q,\epsilon}D_{xy}$

$A := ArchiveUpdateP_{Q,\epsilon}D_{xy}\ (P, A_0, \epsilon, \Delta_x, \Delta_y)$

Require: population P, archive A_0, $\epsilon \in \mathbb{R}^k_+$, $\Delta_x \in \mathbb{R}_+$, $\Delta_y \in \mathbb{R}_+$, $\Delta_x^* \in (0, \Delta_x)$, $\Delta_y^* \in (0, \Delta_y)$
Ensure: updated archive A
$A := A_0$
for all $p \in P$ **do**
  if $\nexists a_1 \in A : a_1 \prec_{-\epsilon} p$ and $\nexists a_2 \in A : (d_H(F(a_2), F(p)) \leq \Delta_y^*$ and $d_H(a_2, p) \leq \Delta_x^*)$ **then**
    $A \leftarrow A \cup \{p\}$
    $\hat{A} = \{a_1 \in A | \nexists a_2 \in A : a_2 \prec_{-(\epsilon+\Delta_y)} a_1\}$
    for all $a \in A \backslash \hat{A}$ **do**
      if $p \prec_{-(\epsilon+\Delta_y)} a$ and $dist(a, \hat{A}) \geq 2\Delta_x$ **then**
        $A \leftarrow A \backslash \{a\}$
      end if
    end for
  end if
end for
return A

Brief Description:

Aims for a uniform approximation of both $P_{Q,\epsilon}$ and $F(P_{Q,\epsilon})$ of a given MOP, see Sect. 8.5 for details for details.

Main Results:

Approximation quality: Theorem 8.9.

Upper bound on archive size: discussed in Sect. 8.5.

Modifications:

The modifications for $\Delta_x \in \mathbb{R}^n_+$ and $\Delta_y \in \mathbb{R}^k_+$ are as for $ArchiveUpdateP_{Q,\epsilon}D_x$ and $ArchiveUpdateP_{Q,\epsilon}D_y$, respectively.

B.10 $ArchiveUpdateM_{Q,\epsilon}$

$A := ArchiveUpdateM_{Q,\epsilon}\ (P, A_0)$

Require: P: population, A_0: current archive, $\epsilon \in \mathbb{R}_+$
Ensure: updated archive A

```
A := A_0
for all p ∈ P do
    if ∄a ∈ A : f(a) + ε ≤ f(p) then
        A := A ∪ {p}
    end if
    for all a ∈ A do
        if f(p) + ε < f(a) then
            A := A\{a}
        end if
    end for
end for
return A
```

Brief Description:

Aims for the approximation of the entire set of approximate solutions $M_{Q,\epsilon}$ and its image $f(M_{Q,\epsilon})$ of a given SOP, see Sect. 9.2 for details.

Main Results:

Approximation quality: Corollaries 9.2 and 9.3.

Upper bound on archive size: (unbounded).

B.11 $ArchiveUpdateM_{Q,\epsilon}D_x$

$\{A, x_b\} := ArchiveUpdateM_{Q,\epsilon}D_x\ (p, A_0, x_{b,0})$

Require: p: candidate solution, A_0: current archive, $x_{b,0}$: current best found solution, $\epsilon \in \mathbb{R}_+$, $\Delta \in \mathbb{R}_+$

Ensure: updated archive A, updated best found solution x_b

if $f(p) < f(x_{b,0})$ **then**
  $x_b := p$
else
  $x_b := x_{b,0}$
end if
if $f(x_b) + \epsilon < f(p)$ or $\{\exists a \in A_0 : p \in B_\Delta^\infty(a)$ and $f(a) \le f(p)\}$ **then**
  $A := A_0$ ▷ discard p
  return A
end if
$A := \{p\}$
for all $a \in A_0$ **do**
  if $f(a) \le f(x_b) + \epsilon$ and $a \notin B_\Delta^\infty(p)$ **then**
    $A := A \cup \{a\}$
  end if
end for
return $\{A, x_b\}$

Brief Description:

Aims for a uniform approximation of $M_{Q,\epsilon}$ of a given SOP, see Sect. 9.3 for details.

Main Results:

Approximation quality: Theorem 9.1.

Upper bound on archive size: discussed in Sect. 9.3.

Modification:

If $\Delta \in \mathbb{R}^n_+$, one can use

$$B_\Delta^\infty(y) = \{x \in \mathbb{R}^n : |x_i - y_i| \le \Delta_i\}$$

References

Abbass, H. (2004). An inexpensive cognitive approach for bi-objective optimization using bliss points and interaction. In *PPSN '04: Proceedings of the 10th International Conference on Parallel Problem Solving From Nature* (pp. 712–721).

Aguilera-Rueda, V.-J., Cruz-Ramírez, N., & Mezura-Montes, E. (2020). Data-driven Bayesian network learning: A bi-objective approach to address the bias-variance decomposition. *Mathematical and Computational Applications*, *25*(2).

Aguirre, P., Krauskopf, B., & Osinga, H. M. (2014). Global invariant manifolds near a Shilnikov homoclinic bifurcation. *Journal of Computational Dynamics*, *1*(1), 1–38.

Ahn, C. W., Kim, E., Kim, H.-T., Lim, D.-H., & An, J. (2010). A hybrid multiobjective evolutionary algorithm: Striking a balance with local search. *Mathematical and Computer Modelling*, *52*, 2048–2059.

Altwaijry, N., & El Bachir Menai, M. (2012). Data structures in multi-objective evolutionary algorithms. *Journal of Computer Science and Technology*, *27*, 1197–1210.

Batista, L. S., Campelo, F., Guimaraes, F. G., & Ramírez, J. A. (2011). A comparison of dominance criteria in many-objective optimization problems. In *2011 IEEE Congress on Evolutionary Computation (CEC)* (pp. 1–8). IEEE.

Beume, N., Naujoks, B., & Emmerich, M. (2007). SMS-EMOA: Multiobjective selection based on dominated hypervolume. *European Journal of Operational Research*, *181*(3), 1653–1669.

Bezerra, L. C. T., López-Ibáñez, M., & Stützle, T. (2019). Archiver effects on the performance of state-of-the-art multi- and many-objective evolutionary algorithms. In *GECCO '19: Proceedings of the Genetic and Evolutionary Computation Conference* (pp. 620–628).

Bogoya, J. M., Vargas, A., Cuate, O., & Schütze, O. (2018). A (p, q)-averaged Hausdorff distance for arbitrary measurable sets. *Mathematical and Computational Applications*, *23*(3).

Bogoya, J. M., Vargas, A., & Schütze, O. (2019). The averaged Hausdorff distances in multi-objective optimization: A review. *Mathematics*, *7*(10).

Bosman, P. A. N. (2011). On gradients and hybrid evolutionary algorithms for real-valued multi-objective optimization. *IEEE Transactions on Evolutionary Computation*, *16*(1), 51–69.

Bosman, P. A., & De Jong, E. D. (2005). Exploiting gradient information in numerical multi-objective evolutionary optimization. In *Proceedings of the 7th annual conference on Genetic and evolutionary computation* (pp. 755–762). ACM.

O. Schütze and C. Hernández, *Archiving Strategies for Evolutionary Multi-objective Optimization Algorithms*, Studies in Computational Intelligence 938,
https://doi.org/10.1007/978-3-030-63773-6

Botello-Aceves, S., Aguirre, A. H., & Valdez, S. I. (2020). Computation of the improvement directions of the Pareto front and its application to moeas. In *Genetic and Evolutionary Computation Conference (GECCO-2020)* (pp. 480–488).

Bringmann, K., & Friedrich, T. (2014). Convergence of hypervolume-based archiving algorithms. *IEEE Transactions on Evolutionary Computation, 18*(5), 643–657.

Brockhoff, D., Tran, T.-D., & Hansen, N. (2015). Benchmarking numerical multiobjective optimizers revisited. In *Proceedings of the 2015 Annual Conference on Genetic and Evolutionary Computation* (pp. 639–646).

Brown, M., & Smith, R. E. (2005). Directed multi-objective optimization. *International Journal of Computers, Systems, and Signals, 6*(1), 3–17.

Cai, X., Li, Y., Fan, Z., & Zhang, Q. (2015). An external archive guided multiobjective evolutionary algorithm based on decomposition. *IEEE Transactions on Evolutionary Computation, 19*, 508–523.

Chavent, M., & Lechevallier Y. (2002). Dynamical clustering of interval dat: Optimization of an adequacy criterion based on Hausdorff distance. In K. Jajuga, Sokolowski, & H. H. Bock (Eds.), *Classification, clustering, and data analysis* (pp. 53–60). Studies in classification, data analysis, and knowledge organization. Berlin, Heidelberg: Springer.

Coello, C. A. C., & Cortés, N. C. (2005). Solving multiobjective optimization problems using an artificial immune system. *Genetic Programming and Evolvable Machines, 6*(2), 163–190.

Coello Coello, C. A., Lamont, G. B., & Van Veldhuizen, D. A. (2007). *Evolutionary algorithms for solving multi-objective problems* (2nd ed.). New York: Springer.

Collette, Y., & Siarry, P. (2004). *Multiobjective optimization—Principles and case studies*. Springer.

Corne, D. W., & Knowles, J. D. (2003). Some multiobjective optimizers are better than others. In *Proceedings of the IEEE Congress on Evolutionary Computation* (pp. 2506–2512). IEEE Press.

Crepinsek, M., Liu, S.-H., & Mernik, M. (2013). Exploration and exploitation in evolutionary algorithms: A survey. *ACM Computing Surveys, 35*, 1–33.

Cuate, O., & Schütze, O. (2019). Variation rate to maintain diversity in decision space within multi-objective evolutionary algorithms. *Mathematical and Computational Applications, 24*(3).

Cui, Y., Geng, Z., Zhu, Q., & Han, Y. (2017). Review: Multi-objective optimization methods and application in energy saving. *Energy, 125*(15), 681–704.

Cuate, O., Ponsich, A., Uribe, L., Zapotecas, S., Lara, A., & Schütze, O. (2020). A new hybrid evolutionary algorithm for the treatment of equality constrained MOPs. *Mathematics, 8*.

Das, I., & Dennis, J. E. (1998). Normal-boundary intersection: A new method for generating the Pareto surface in nonlinear multicriteria optimization problems. *SIAM Journal on Optimization, 8*(3), 631–657.

de Jong, E. D. (2007). A monotonic archive for Pareto-coevolution. *Evolutionary Computation, 15*, 61–93.

Deb, K. (2001). *Multi-objective optimization using evolutionary algorithms*. Chichester, UK: Wiley. ISBN 0-471-87339-X.

Deb, K. (2015). *Multi-objective evolutionary algorithms* (pp. 995–1015). Berlin Heidelberg: Springer.

Deb, K. (2020). Evolutionary multi-objective optimization: past, present and future. In *GECCO'20: Proceedings of the 22th Annual Conference on Genetic and Evolutionary Computation* (pp. 343–372).

Deb, K., & Agrawal, S. (1999). *A niched-penalty approach for constraint handling in genetic algorithms* (pp. 235–243). Berlin Heidelberg: Springer.

Deb, K., & Tiwari, S. (2008). Omni-optimizer: A generic evolutionary algorithm for single and multi-objective optimization. *European Journal of Operational Research, 185*(3), 1062–1087.

Deb, K., & Deb, D. (2014). Analysing mutation schemes for real-parameter genetic algorithms. *International Journal of Artificial Intelligence and Soft Computing, 4*(1), 1–28.

Deb, K., & Jain, H. (2014). An evolutionary many-objective optimization algorithm using reference-point-based nondominated sorting approach, part I: Solving problems with box constraints. *Transactions on Evolutionary Computation, 18*(4), 577–601.

Deb, K., Pratap, A., Agarwal, S., & Meyarivan, T. (2002a). A fast and elitist multiobjective genetic algorithm: NSGA-II. *IEEE Transactions on Evolutionary Computation, 6*(2), 182–197.
Deb, K., Thiele, L., Laumanns, M., & Zitzler, E. (2002b). Scalable multi-objective optimization test problems. In *2002 IEEE Congress on Evolutionary Computation (CEC)* (pp. 825–830). IEEE.
Dellnitz, M., & Hohmann, A. (1997). A subdivision algorithm for the computation of unstable manifolds and global attractors. *Numerische Mathematik, 75*(3), 293–317.
Dellnitz, M., & Junge, O. (1999). On the approximation of complicated dynamical behavior. *SIAM Journal on Numerical Analysis, 36*(2), 491–515.
Dellnitz, M., Schütze, O., & Sertl, S. (2002a). Finding zeros by multilevel subdivision techniques. *IMA Journal of Numerical Analysis, 22*(2), 167–185.
Dellnitz, M., Schütze, O., & Zheng, Q. (2002b). Locating all the zeros of an analytic function in one complex variable. *Journal of Computational and Applied Mathematics, 138*(2), 325–333.
Dellnitz, M., Schütze, O., & Hestermeyer, T. (2005). Covering Pareto sets by multilevel subdivision techniques. *Journal of Optimization Theory and Applications, 124*(1), 113–155.
Denysiuk, R., Costa, L., & Santo, I. E. (2013). A new hybrid evolutionary multiobjective algorithm guided by descent directions. *Journal of Mathematical Modelling and Algorithms in Operations Research, 12*(3), 233–251.
Dilettoso, E., Rizzo, S. A., & Salerno, N. (2017). A weakly pareto compliant quality indicator. *Mathematical and Computational Applications, 22*(1).
Ehrgott, M. (2005). *Multicriteria optimization*. Springer.
Eiben, A. E., & Rudolph, G. (1999). Theory of evolutionary algorithms: A bird's eye view. *Theoretical Computer Science, 1*, 3–9.
Eichfelder, G. (2008). *Adaptive scalarization methods in multiobjective optimization*. Springer.
Emmerich, M., & Deutz, A. (2007). Test problems based on Lamé superspheres. In S. Obayashi, K. Deb, C. Poloni, T. Hiroyasu, & T. Murata (Eds.), *Evolutionary Multi-Criterion Optimization: 4th International Conference, EMO 2007, Matsushima, Japan, March 5–8, 2007. Proceedings* (pp. 922–936). Berlin, Heidelberg: Springer.
Everson, R. M., Fieldsend, J. E., & Singh, S. (2002). Full elite sets for multi-objective optimisation. In *Adaptive computing in design and manufacture V* (pp. 343–354). London: Springer.
Fernández, S., Alvarez, S., Díaz, D., Iglesias, M., & Ena, B. (2014). Scheduling a galvanizing line by ant colony optimization. In *Proceedings of the Swarm Intelligence: 9th International Conference, ANTS 2014* (pp. 146–157).
Fernández, J., Schütze, O., Hernández, C., Sun, J.-Q., & Xiong, F.-R. (2016). Parallel simple cell mapping for multi-objective optimization. *Engineering Optimization, 48*, 1845–1868.
Fieldsend, J. (2020). Data structures for non-dominated sets: implementations and empirical assessment of two decades of advances. In *Proceedings of the 2020 Genetic and Evolutionary Computation Conference* (pp. 489–497).
Fieldsend, J. E., Everson, R. M., & Singh, S. (2003). Using unconstrained elite archives for multiobjective optimization. *IEEE Transactions on Evolutionary Computation, 7*, 305–323.
Fliege, J. (2004a). Gap-free computation of Pareto-points by quadratic scalarizations. *Mathematical Methods of Operations Research, 59*, 69–89.
Fliege, J. (2004b). Gap-free computation of Pareto-points by quadratic scalarizations. *Mathematical Methods of Operations Research, 59*, 69–89.
Fliege, J., & Svaiter, B. F. (2000). Steepest descent methods for multicriteria optimization. *Mathematical Methods of Operations Research, 51*(3), 479–494.
Fonseca, C. M., & Fleming, P. J. (1993). *Genetic algorithms for multiobjective optimization: formulation, discussion, and generalization* (pp. 416–423).
Fonseca, C. M., & Fleming, P. J. (1995). An overview of evolutionary algorithms in multiobjective optimization. *Evolutionary Computation, 3*, 1–16.
Forster, O. (1984). *Analysis 3*. Vieweg.
Fu, Y., Wang, H., Tian, C., Li, Z., & Hu, H. (2019). Two-agent stochastic flow shop deteriorating scheduling via a hybrid multi-objective evolutionary algorithm. *Journal of Intelligent Manufacturing, 30*, 2257–2272.

Gass, S., & Saaty, T. (1955). The computational algorithm for the parametric objective function. *Naval Research Logistics Quarterly*, *2*(1), 39–45.

Gebken, B., Peitz, S., & Dellnitz, M. (2019a). A descent method for equality and inequality constrained multiobjective optimization problems. In *Numerical and Evolutionary Optimization (NEO 2017)* (pp. 29–61.) Springer.

Gebken, B., Peitz, S., & Dellnitz, M. (2019b). On the hierarchical structure of Pareto critical sets. *Journal of Global Optimization*, *73*, 891–913.

Gembicki, F. W., & Haimes, Y. Y. (1975). Approach to performance and multiobjective sensivity optimization: The goal attainment method. *IEEE Transactions on Automatic Control*, *20*, 769–771.

Goldberg, D. E. (1989). *Genetic algorithms for search, optimization, and machine learning*. Reading, Massachusetts: Addison-Wesley.

Gomes da Silva, C., Clímaco, J., & Figueira, J. (2004a). *Geometrical configuration of the Pareto frontier of bi-criteria* {0, 1}*-knapsack problems*. Technical report, University of Coimbra.

Gomes da Silva, C., Clímaco, J., & Figueira, J. (2004b). A scatter search method for the bi-criteria multi-dimensional {0, 1}-knapsack problem using surrogate relaxation. *Journal of Mathematical Modelling and Algorithms*, *3*, 183–208.

Habenicht, W. (1983). *Quad trees, a data structure for discrete vector optimization problems* (pp. 136–145). Berlin, Heidelberg: Springer.

Hanne, T. (1999). On the convergence of multiobjective evolutionary algorithms. *European Journal of Operational Research*, *117*, 553–564.

Hanne, T. (2001). Global multiobjective optimization with evolutionary algorithms: Selection mechanisms and mutation control. *Evolutionary Multi-Criterion Optimization, First International Conference, EMO 2001, Zurich, Switzerland* (pp. 197–212). Berlin: Springer.

Hanne, T. (2007a). A multiobjective evolutionary algorithm for approximating the efficient set. *European Journal of Operational Research*, *176*, 1723–1734.

Hanne, T. (2007b). A primal-dual multiobjective evolutionary algorithm for approximating the efficient set. In *2007 IEEE Congress on Evolutionary Computation (CEC)* (pp. 3127–3134). IEEE Press.

Hansen, M. P., & Jaszkiewicz, A. (1998). *Evaluating the quality of approximations of the non-dominated set*. IMM Technical Report IMM-REP-1998-7, Institute of Mathematical Modeling, Technical University of Denmark.

Harada, K., Sakuma, J., & Kobayashi, S. (2006). Local search for multiobjective function optimization: Pareto descent method. In *Proceedings of the 8th Annual Conference on Genetic and Evolutionary Computation* (pp. 659–666). ACM.

Hernández, C., Naranjani, Y., Sardahi, Y., Liang, W., Schütze, O., & Sun, J.-Q. (2013a). Simple cell mapping method for multi-objective optimal feedback control design. *International Journal of Dynamics and Control*, *1*(3), 231–238.

Hernández, C., Sun, J. Q., & Schütze, O. (2013b). Computing the set of approximate solutions of a multi-objective optimization problem by means of cell mapping techniques. In M. Emmerich et al. (Eds.), *EVOLVE—A Bridge Between Probability, Set Oriented Numerics and Evolutionary Computation IV* (pp. 171–188). Springer.

Hernández, C. I., Schütze, O., Sun, J.-Q., & Ober-Blöbaum, S. (2020). Non-epsilon dominated evolutionary algorithm for the set of approximate solutions. *Mathematical and Computational Applications*, *25*(1).

Hestermeyer, T., & Oberschelp, O. (2003). Selbstoptimierende Fahrzeugregelung - Verhaltensbasierte Adaption. In *Intelligente mechatronische Systeme* (Vol. 122). HNI-Verlagsschriftenreihe. Heinz Nixdorf Institut.

Hillermeier, C. (2001). *Nonlinear multiobjective optimization: A generalized homotopy approach* (Vol. 135). Springer Science & Business Media.

Himmelblau, D. M. (1972). *Applied nonlinear programming*. McGraw-Hill.

Holland, J. H. (1975). *Adaptation in natural and artificial systems*. MIT Press.

Horn, J., Nafpliotis, N., & Goldberg, D. E. (1994). A niched Pareto genetic algorithm for multiobjective optimization. In *Proceedings of the First IEEE Conference on Evolutionary Computation, IEEE World Congress on Computational Computation* (pp. 82–87). IEEE Press.

Huband, S., Hingston, P., Barone, L., & While, L. (2006). A review of multiobjective test problems and a scalable test problem toolkit. *IEEE Transactions on Evolutionary Comuputation*, *10*(5), 477–506.

Huttenlocher, D., Klanderman, G., & Rucklidge, W. (1993). Comparing images using the Hausdorff distance. *IEEE Transactions on Pattern Analysis and Machine Intelligence*, *15*(9), 850–863.

Ikeda, K., Kita, H., & Kobayashi, S. (2001). Failure of Pareto-based MOEAs: Does non-dominated really mean near to optimal? In *2001 IEEE Congress on Evolutionary Computation (CEC)* (pp. 957–962). IEEE.

Ishibuchi, H., Matsumoto, T., Masuyama, N., & Nojima, Y. (2020a). Effects of dominance resistant solutions on the performance of evolutionary multi-objective and many-objective algorithms. In *GECCO '20: Proceedings of the 22th annual Conference on Genetic and Evolutionary Computation* (pp. 507–515).

Ishibuchi, H., Pang, L. M., & Shang, K. (2020b). A new framework of evolutionary multi-objective algorithms with an unbounded external archive.

Ishibuchi, H., Pang, L. M., & Shang, K. (2020c). Solution subset selection for final decision making in evolutionary multi-objective optimization, arXiv:2006.08156.

Jahn, J. (2006). Multiobjective search algorithm with subdivision technique. *Computational Optimization and Applications*, *35*(2), 161–175.

Jain, H., & Deb, K. (2014). An evolutionary many-objective optimization algorithm using reference-point based nondominated sorting approach, part II: Handling constraints and extending to an adaptive approach. *IEEE Transactions on Evolutionary Computation*, *18*(4), 602–622.

Jaszkiewicz, A., & Lust, T. (2018). ND-tree-based update: A fast algorithm for the dynamic nondominance problem. *IEEE Transactions on Evolutionary Computation*, *22*, 778–791.

Jian, Y. (1992). Essential weak efficient solution in multiobjective optimization problems. *Journal of Mathematical Analysis and Applications*, *166*, 230–235.

Kacprzyk, J., & W. Pedrycz (Eds.). (2015). *Springer handbook of computational intelligence*. Springer.

Karush, W. (1939). *Minima of functions of several variables with inequalities as side constraints*. Ph.D. thesis, Master's thesis, Department of Mathematics, University of Chicago.

Kettner, L. J., & Deng, S. (2012). On well-posedness and Hausdorff convergence of solution sets of vector optimization problems. *Journal of Optimization Theory and Applications*, *153*(3), 619–632.

Kirchberg, K. J., Jesorsky, O., & Frischholz, R. W. (2002). Genetic model optimization for Hausdorff distance-based face localization. In M. Tistarelli, J. Bigun, A. K. Jain (Eds.), *Biometric Authentication. BioAW 2002* (Vol. 2359, pp. 103–111). Lecture notes in computer science. Berlin, Heidelberg: Springer.

Klamroth, K., Tind, J., & Wiecek, M. (2002). Unbiased approximation in multicriteria optimization. *Mathematical Methods of Operations Research*, *56*, 413–437.

Knowles, J. D., & Corne, D. W. (2000a). Approximating the nondominated front using the Pareto archived evolution strategy. *Evolutionary Computation*, *8*(2), 149–172.

Knowles, J. D., & Corne, D. W. (2000b). M-PAES: a memetic algorithm for multiobjective optimization. In *2000 IEEE Congress on Evolutionary Computation (CEC)* (pp. 325–332).

Knowles, J. D., & Corne, D. W. (2003). Properties of an adaptive archiving algorithm for storing nondominated vectors. *IEEE Transactions on Evolutionary Computation*, *7*, 100–116.

Knowles, J. D., & Corne, D. W. (2004). Bounded Pareto archiving: Theory and practice. In *Metaheuristics for Multiobjective Optimisation* (pp. 39–64). Springer.

Knowles, J. D., Corne, D. W., & Fleischer, M. (2003). Bounded archiving using the lebesgue measure. In *Proceedings of the IEEE Congress on Evolutionary Computation* (pp. 2490–2497). IEEE Press.

Köppen, M., & Yoshida, K. (2007). Many-objective particle swarm optimization by gradual leader selection. In B. Beliczynski, A. Dzielinski, M. Iwanowski, & B. Ribeiro (Eds.), *Adaptive and Natural Computing Algorithms. ICANNGA 2007* (Vol. 4431, pp. 323–331). Lecture Notes in Computer Science. Berlin, Heidelberg: Springer.

Kuhn, H., & Tucker, A. (1951). Nonlinear programming. In J. Neumann (Ed.), *Proceeding of the 2nd Berkeley Symposium on Mathematical Statistics and Probability* (pp. 481–492).

Lara, A., Sanchez, G., Coello Coello, C. A., & Schütze, O. (2009). HCS: A new local search strategy for memetic multiobjective evolutionary algorithms. *IEEE Transactions on Evolutionary Computation, 14*(1), 112–132.

Laumanns, M., & Zenklusen, R. (2011). Stochastic convergence of random search methods to fixed size Pareto front approximations. *European Journal of Operational Research, 213*, 414–421.

Laumanns, M., Zitzler, E., & Thiele, L. (2001). On the effects of archiving, elitism, and density based selection in evolutionary multi-objective optimization. In *International Conference on Evolutionary Multi-Criterion Optimization* (pp. 181–196). Berlin, Heidelberg: Springer.

Laumanns, M., Thiele, L., Deb, K., & Zitzler, E. (2002). Combining convergence and diversity in evolutionary multiobjective optimization. *Evolutionary Computation, 10*, 263–282.

Liu, Y., Ishibuchi, H., Yen, G. G., Nojima, Y., & Masuyama, N. (2020). Handling imbalance between convergence and diversity in the decision space in evolutionary multimodal multiobjective optimization. *IEEE Transactions on Evolutionary Computation, 24*(3), 551–565.

Loomis, L. H., & Sternberg, S. (1990). *Advanced calculus*. Jones and Bartlett Publishers.

López-Ibáñez, M., Knowles, J. D., & Laumanns, M. (2011). On sequential online archiving of objective vectors. In *Evolutionary Multi-Criterion Optimization (EMO 2011)* (pp. 46–60). Berlin, Heidelberg: Springer.

Loridan, P. (1984). ϵ-solutions in vector minimization problems. *Journal of Optimization, Theory and Application, 42*, 265–276.

Luong, H. N., & Bosman, P. A. N. (2012). Elitist archiving for multi-objective evolutionary algorithms: To adapt or not to adapt. In *Parallel Problem Solving from Nature—PPSN XII* (pp. 72–81). Berlin, Heidelberg: Springer.

Martín, A., & Schütze, O. (2018). Pareto tracer: A predictor-corrector method for multi-objective optimization problems. *Engineering Optimization, 50*(3), 516–536.

Martin, B., Goldsztejn, A., Granvilliers, L., & Jermann, C. (2013). Certified parallelotope continuation for one-manifolds. *SIAM Journal on Numerical Analysis, 51*(6), 3373–3401.

Martin, B., Goldsztejn, A., Granvilliers, L., & Jermann, C. (2014). On continuation methods for nonlinear bi-objective optimization: towards a certified interval-based approach. *Journal of Global Optimization*, 1–14.

Martínez, S. Z., & Coello, C. A. C. (2011). A multi-objective particle swarm optimizer based on decomposition. In *Genetic and Evolutionary Computation Conference (GECCO-2011)* (pp. 69–76).

Miettinen, K. (2012). *Nonlinear multiobjective optimization* (Vol. 12). Springer Science & Business Media.

Moghadam, M. E., Falaghi, H., & Farhadi, M. (2020). A novel method of optimal capacitor placement in the presence of harmonics for power distribution network using NSGA-II multi-objective genetic optimization algorithm. *Mathematical and Computational Applications, 25*(1).

Mostaghim, S., & Teich, J. (2005). Quad-trees: A data structure for storing pareto sets in multiobjective evolutionary algorithms with elitism. In *Evolutionary multiobjective optimization* (pp. 81–104). Advanced Information and Knowledge Processing. Berlin, Heidelberg: Springer.

Moubayed, N. A., Petrovski, A., & McCall, J. (2014). (DMOPSO)-M-2: MOPSO based on decomposition and dominance with archiving using crowding distance in objective and solution spaces. *Evolutionary Computation, 22*(1).

Naranjani, Y., Hernández, C., Xiong, F. R., Schütze, O., & Sun, J. Q. (2017). A hybrid method of evolutionary algorithm and simple cell mapping for multi-objective optimization problems. *International Journal of Dynamics and Control, 5*, 570–582.

Nocedal, J., Wright, S. (2006). *Numerical optimization*. Springer Science & Business Media.

Okabe, T., Jin, Y., Olhofer, M., & Sendhoff, B. (2004). On test functions for evolutionary multi-objective optimization. In et al., X. Y., editor, *Parallel Problem Solving from Nature—PPSN VIII: 8th International Conference, Birmingham, UK, September 18–22, 2004. Proceedings* (pp. 792–802).

Pang, L. M., Ishibuchi, H., & Shang, K. (2020). Algorithm configurations of MOEA/d with an unbounded external archive, arXiv:2007.13352.

Pareto, V. (1971). *Manual of Political Economy (original edition in French in 1927)*. The MacMillan Press.

Pascoletti, A., & Serafini, P. (1984). Scalarizing vector optimization problems. *Journal of Optimization Theory and Applications*, *42*, 499–524.

Patil, M. B. (2020). Improved performance in multi-objective optimization using external archive. Sādhanā 45.

Peitz, S., & Dellnitz, M. (2018a). Gradient-based multiobjective optimization with uncertainties. *Numerical and Evolutionary Optimization (NEO 2016)* (pp. 159–182). Cham: Springer.

Peitz, S., & Dellnitz, M. (2018b). A survey of recent trends in multiobjective optimal control - surrogate models, feedback control and objective reduction. *Mathematical and Computational Applications*, *23*(2).

Pereyra, V., Saunders, M., & Castillo, J. (2013). Equispaced Pareto front construction for constrained bi-objective optimization. *Mathematical and Computer Modelling*, *57*(9–10), 2122–2131.

Peschel, M., & Riedel, C. (1977). Use of vector optimization in multiobjective decision making. In D. E. Bell, R. L. Keeney, & H. Raiffa (Eds.), *Conflicting objectives in decisions* (pp. 97–121). Chichester: Wiley.

Pottharst, A., Baptist, K., Schütze, O., Böcker, J., Fröhlecke, N., & Dellnitz, M. (2004). Operating point assignment of a linear motor driven vehicle using multiobjective optimization methods. In *Proceedings of the 11th International Conference EPE-PEMC 2004, Riga, Latvia.*

Powell, M. J. D. (1973). On search directions for minimization algorithms. *Mathematical Programming*, *4*(1), 193–201.

Preuss, M. (2015). *Multimodal optimization by means of evolutionary algorithms*. Springer.

Preuss, M., Naujoks, B., & Rudolph, G. (2006). Pareto set and EMOA behavior for simple multimodal multiobjective functions. In T. P. Runarsson, H. G. Beyer, E. Burke, J. J. Merelo-Guervós, L. D. Whitley, & X. Yao (Eds.), *Parallel Problem Solving from Nature - PPSN IX: 9th International Conference, Reykjavik, Iceland, September 9–13, 2006, Proceedings* (pp. 513–522). Berlin Heidelberg: Springer.

Purshouse, R. C., & Fleming, P. J. (2007). On the evolutionary optimization of many conflicting objectives. *IEEE Transactions on Evolutionary Computation*, *11*(6), 770–784.

Qin, Z. Q., Xiong, F. R., Hernández, C., Fernandez, J., Ding, Q., Schütze, O., et al. (2017). Multi-objective optimal design of sliding mode control with parallel simple cell mapping method. *Journal of Vibration and Control*, *23*, 46–54.

Recchioni, M. C. (2003). A path following method for box-constrained multiobjective optimization with applications to goal programming problems. *Mathematical Methods of Operations Research*, *58*, 69–85.

Rechenberg, I. (1973). *Evolutionsstrategie—Optimierung technischer Systeme nach Prinzipien der biologischen Evolution*. Frommann-Holzboog.

Rigoni, E., & Poles, S. (2005). NBI and MOGA-II, two complementary algorithms for multi-objective optimizations. In J. Branke, K. Deb, K. Miettinen, & R. E. Steuer (Eds.), *Practical Approaches to Multi-Objective Optimization, number 04461 in Dagstuhl Seminar Proceedings*. Germany: Internationales Begegnungs- und Forschungszentrum (IBFI), Schloss Dagstuhl.

Ringkamp, M., Ober-Blöbaum, S., Dellnitz, M., & Schütze, O. (2012). Handling high dimensional problems with multi-objective continuation methods via successive approximation of the tangent space. *Engineering Optimization*, *44*(6), 1117–1146.

Roy, B. (1971). Problems and methods with multiple objective functions. *Mathematical Programming*, *1*, 239–266.

Rudolph, G. (1994). Convergence analysis of canonical genetic algorithms. *IEEE Transactions on Neural Networks*, *5*(1), 96–101.
Rudolph, G. (1996). Convergence of evolutionary algorithms in general search spaces. In T. Fukuda, & T. Furuhashi (Eds.), *Proceedings of 1996 IEEE International Conference on Evolutionary Computation, Nayoya University, Japan, May 20–22, 1996* (pp. 50–54). IEEE.
Rudolph, G. (1997). Local convergence rates of simple evolutionary algorithms with Cauchy mutations. *IEEE Transactions on Evolutionary Computation*, *1*(4), 249–258.
Rudolph, G. (1998a). Finite Markov chain results in evolutionary computation: A tour d'horizon. *Fundamenta Informaticae*, *35*, 67–89.
Rudolph, G. (1998b). On a multi-objective evolutionary algorithm and its convergence to the Pareto set. In *IEEE International Conference on Evolutionary Computation (ICEC 1998)* (pp. 511–516). IEEE Press.
Rudolph, G. (2001). Evolutionary search under partially ordered fitness sets. In *Proceedings of the International NAISO Congress on Information Science Innovations (ISI 2001)* (pp. 818–822). Sliedrecht, The Netherlands: ICSC Academic Press.
Rudolph, G., & Agapie, A. (2000). Convergence properties of some multi-objective evolutionary algorithms. In *2011 IEEE Congress on Evolutionary Computation (CEC)*. IEEE Press.
Rudolph, G., Naujoks, B., & Preuss, M. (2007). Capabilities of EMOA to detect and preserve equivalent Pareto subsets. In S. Obayashi, K. Deb, C. Poloni, T. Hiroyasu, & T. Murata (Eds.), *Evolutionary Multi-Criterion Optimization: 4th International Conference, EMO 2007, Matsushima, Japan, March 5–8, 2007. Proceedings* (pp. 36–50). Berlin, Heidelberg: Springer.
Rudolph, G., Schütze, O., Grimme, C., Domínguez-Medina, C., & Trautmann, H. (2016). Optimal averaged Hausdorff archives for bi-objective problems: Theoretical and numerical results. *Computational Optimization and Applications*, *64*(2), 589–618.
Sánchez, X. E., & Schütze, O. (2010). On the interplay of generator and archiver within archive based multiobjective evolutionary algorithms. In *7th International Conference on Electrical Engineering Computing Science and Automatic Control, Tuxtla Gutierrez* (pp. 375–380). IEEE.
Schaeffler, S., Schultz, R., & Weinzierl, K. (2002). Stochastic method for the solution of unconstrained vector optimization problems. *Journal of Optimization Theory and Applications*, *114*(1), 209–222.
Schaffer, J. D. (1984). *Multiple objective optimization with vector evaluated genetic algorithms*. Ph.D. thesis, Vanderbilt University.
Schaffer, J. D. (1985). Multiple objective optimization with vector evaluated genetic algorithms. In *Genetic Algorithms and their Applications: Proceedings of the First International Conference on Genetic Algorithms* (pp. 93–100).
Schandl, B., Klamroth, K., & Wiecek, M. (2002). Introducing oblique norms into multiple criteria programming. *Journal of Global Optimization*, *23*, 925–942.
Schütze, O. (2003). A new data structure for the nondominance problem in multi-objective optimization. *Evolutionary Multi-Criterion Optimization (EMO 2003)* (pp. 509–518). Berlin, Heidelberg: Springer.
Schütze, O. (2004). *Set oriented methods for global optimization*. Ph.D. thesis, Department of Mathematics, University of Paderborn.
Schütze, O., Mostaghim, S., Dellnitz, M., and Teich, J. (2003). Covering Pareto sets by multilevel evolutionary subdivision techniques. In C. M. Fonseca, P. J. Fleming, E. Zitzler, K. Deb, & L. Thiele (Eds.), *Evolutionary Multi-Criterion Optimization (EMO 2003)* (pp. 118–132). Springer.
Schütze, O., Dell'Aere, A., & Dellnitz, M. (2005). On continuation methods for the numerical treatment of multi-objective optimization problems. In J. Branke, K. Deb, K. Miettinen, & R. E. Steuer (Eds.), *Practical Approaches to Multi-Objective Optimization*, number 04461 in Dagstuhl Seminar Proceedings. Internationales Begegnungs- und Forschungszentrum (IBFI), Schloss Dagstuhl, Germany, http://drops.dagstuhl.de/opus/volltexte/2005/349.
Schütze, O., Coello, C. A. C., & Talbi, E.-G. (2007a). Approximating the ϵ-efficient set of an MOP with stochastic search algorithms. In A. Gelbukh, & A. F. K. Morales (Eds.), *Mexican International Conference on Artificial Intelligence (MICAI-2007)* (pp. 128–138).

Schütze, O., Laumanns, M., Tantar, E., Coello, C. A. C., Talbi, E.-G. (2007b). Convergence of stochastic search algorithms to gap-free Pareto front approximations. In *Genetic and Evolutionary Computation Conference (GECCO-2007)* (pp. 892–901).

Schütze, O., Coello, C. A. C., Mostaghim, S., Dellnitz, M., & Talbi, E.-G. (2008a). Hybridizing evolutionary strategies with continuation methods for solving multi-objective problems. *IEEE Transactions on Evolutionary Computation*, *19*(7), 762–769.

Schütze, O., Coello, C. A. C., Tantar, E., & Talbi, E.-G. (2008b). Computing finite size representations of the set of approximate solutions of an MOP with stochastic search algorithms. In *GECCO '08: Proceedings of the 10th Annual Conference on Genetic and Evolutionary Computation* (pp. 713–720). ACM: New York, NY, USA.

Schütze, O., Laumanns, M., & Coello, C. A. C. (2008c). Approximating the knee of an MOP with stochastic search algorithms. In *Proceedings of the International Conference on Parallel Problem Solving from Nature (PPSN 2008)* (pp. 795–805). Berlin, Heidelberg: Springer.

Schütze, O., Laumanns, M., Coello, C. A. C., Dellnitz, M., & Talbi, E.-G. (2008d). Convergence of stochastic search algorithms to finite size Pareto set approximations. *Journal of Global Optimization*, *41*(4), 559–577.

Schütze, O., Vasile, M., & Coello, C. A. C. (2008e). Approximate solutions in space mission design. In G. Rudolph, T. Jansen, S. M. Lucas, C. Poloni, & N. Beume (Eds.), *Proceedings of the International Conference on Parallel Problem Solving from Nature (PPSN 2008)* (pp. 805–814). Berlin, Heidelberg: Springer.

Schütze, O., Coello Coello, C. A., Mostaghim, S., Talbi, E.-G., & Dellnitz, M. (2008f). Hybridizing evolutionary strategies with continuation methods for solving multi-objective problems. *Engineering Optimization*, *40*(5), 383–402.

Schütze, O., Vasile, M., Junge, O., Dellnitz, M., & Izzo, D. (2009). Designing optimal low thrust gravity assist trajectories using space pruning and a multi-objective approach. *Engineering Optimization*, *41*(2), 155–181.

Schütze, O., Lara, A., Coello, C. A. C., & Vasile, M. (2010a). *Computing approximate solutions of scalar optimization problems and applications in space mission design* (pp. 1–8). IEEE.

Schütze, O., Laumanns, M., Tantar, E., Coello, C. A. C., & Talbi, E.-G. (2010b). Computing gap free Pareto front approximations with stochastic search algorithms. *Evolutionary Computation*, *18*(1), 65–96.

Schütze, O., Lara, A., & Coello, C. A. C. (2011a). On the influence of the number of objectives on the hardness of a multi-objective optimization problem. *IEEE Transactions on Evolutionary Computation*, *15*(4), 444–455.

Schütze, O., Lara, A., Coello, C. A. C., & Vasile, M. (2011b). On the detection of nearly optimal solutions in the context of single-objective space mission design problems. *Journal of Aerospace Engineering*, *225*(11), 1229–1242.

Schütze, O., Vasile, M., & Coello, C. A. C. (2011c). Computing the set of epsilon-efficient solutions in multiobjective space mission design. *Journal of Aerospace Computing, Information, and Communication*, *8*(3), 53–70.

Schütze, O., Esquivel, X., Lara, A., & Coello, C. A. C. (2012a). Using the averaged Hausdorff distance as a performance measure in evolutionary multi-objective optimization. *IEEE Transactions on Evolutionary Computation*, *16*(4), 504–522.

Schütze, O., Esquivel, X., Lara, A., & Coello, C. A. C. (2012b). Using the averaged Hausdorff distance as a performance measure in evolutionary multiobjective optimization. *IEEE Transactions on Evolutionary Computation*, *16*(4), 504–522.

Schütze, O., Alvarado, S., Segura, C., & Landa, R. (2016a). Gradient subspace approximation: A direct search method for memetic computing. *Soft Computing*, *21*, 6331–6350.

Schütze, O., Domínguez-Medina, C., Cruz-Cortés, N., de la Fraga, L. G., Sun, J.-Q., Toscano, G., et al. (2016b). A scalar optimization approach for averaged Hausdorff approximations of the Pareto front. *Engineering Optimization*, *48*(9), 1593–1617.

Schütze, O., Martín, A., Lara, A., Alvarado, S., Salinas, E., & Coello, C. A. C. (2016c). The directed search method for multi-objective memetic algorithms. *Computational Optimization and Applications*, *63*(2), 305–332.

Schütze, O., Hernandez, C., Talbi, E.-G., Sun, J. Q., Naranjani, Y., & Xiong, F. R. (2019). Archivers for the representation of the set of approximate solutions for mops. *Journal of Heuristics*, *5*(1), 71–105.

Schütze, O., Cuate, O., Martín, A., Peitz, S., & Dellnitz, M. (2020). Pareto explorer: A global/local exploration tool for many-objective optimization problems. *Engineering Optimization*, *52*(5), 832–855.

Schwefel, H. P. (1977). *Numerische Optimierung von Computer-Modellen mittels der Evolutionsstrategie*. Birkhäuser.

Sebag, M., Tarrisson, N., Teytaud, O., & Lefevre, J. (2005). A multi-objective multi-modal optimization approach for mining stable spatio-temporal patterns. *IJCAI International Joint Conference on Artificial Intelligence* (pp. 859–864). Berlin, Heidelberg: Springer.

Segura, C., Hernández Aguirre, A., Ivvan Valdez Peña, S., & Botello Rionda, S. (2015). The importance of proper diversity management in evolutionary algorithms for combinatorial optimization. In O. Schütze, L. Trujillo, P. Legrand, Y. Maldonado (Eds.), *NEO 2015—Results of the Numerical and Evolutionary Optimization Workshop NEO 2015 held at September 23–25 2015 in Tijuna, Mexico* (pp. 121–148). Springer.

Segura, C., Coello, C. A. C., Segredo, E., & Aguirre, A. H. (2016). A novel diversity-based replacement strategy for evolutionary algorithms. *IEEE Transactions on Cybernetics*, *46*(12), 3233–3246.

Segura, C., Hernández-Aguirre, A., Luna, F., & Alba, E. (2017). Improving diversity in evolutionary algorithms: New best solutions for frequency assignment. *IEEE Transactions on Evolutionary Computation*, *21*(4), 539–553.

Shim, V. A., Tan, K. C., & Cheong, C. Y. (2012). A hybrid estimation of distribution algorithm with decomposition for solving the multiobjective multiple traveling salesman problem. *IEEE Transactions on Systems, Man, and Cybernetics, Part C*, *42*, 5.

Shir, O., Preuss, M., Naujoks, B., & Emmerich, M. (2009). Enhancing decision space diversity in evolutionary multiobjective algorithms. In M. Ehrgott, C. M. Fonseca, X. Gandibleux, J.-K. Hao, & M. Sevaux (Eds.), *International Conference on Evolutionary Multi-Criterion Optimization* (pp. 95–109). Berlin, Heidelberg: Springer.

Shukla, P. K. (2007). Gradient based stochastic mutation operators in evolutionary multi-objective optimization. In *International Conference on Adaptive and Natural Computing Algorithms* (pp. 58–66). Springer.

Sindhya, K., Ruuska, S., Haanpää, T., & Miettinen, K. (2011). A new hybrid mutation operator for multiobjective optimization with differential evolution. *Soft Computing*, *15*, 2041–2055.

Sindhya, K., Miettinen, K., & Deb, K. (2012). A hybrid framework for evolutionary multi-objective optimization. *IEEE Transactions on Evolutionary Computation*, *17*, 495–511.

Sosa-Hernández, V. A., Schütze, O., Wang, H., Deutz, A., & Emmerich, M. (2020). The set-based hypervolume newton method for bi-objective optimization. *IEEE Transactions on Cybernetics*, *50*(5), 2186–2196.

Srinivas, N., & Deb, K. (1994). Multiobjective optimization using nondominated sorting in genetic algorithms. *Evolutionary Computation*, *2*, 221–248.

Stewart, T., Bandte, O., Braun, H., Chakraborti, N., Ehrgott, M., Göbelt, M., Jin, Y., & Nakayama, H. (2008). Real-world applications of multiobjective optimization. In Slowinski, R., editor, *Multiobjective optimization* (Vol. 5252, pp. 285–327). Lecture Notes in Computer Science. Berlin, Heidelberg: Springer.

Stoean, C., Preuss, M., Stoean, R., & Dumitrescu, D. (2010). Multimodal optimization by means of a topological species conservation algorithm. *IEEE Transactions on Evolutionary Computation*, *14*(6), 842–864.

Sun, M., & Steuer, R. E. (1996a). InterQuad: An interactive quad tree based procedure for solving the discrete alternative multiple criteria problem. *European Journal of Operational Research, 89*, 462–472.

Sun, M., & Steuer, R. E. (1996b). Quad trees and linear list for identifying nondominated criterion vectors. *INFORMS Journal on Computing, 8*, 367–375.

Sun, J.-Q., Xiong, F.-R., Schütze, O., & Hernández, C. (2019). *Cell mapping methods—Algorithmic approaches and applications*. Springer.

Tanabe, R., & Ishibuchi, H. (2018). An analysis of control parameters of MOEA/D under two different optimization scenarios. *Applied Soft Computing, 70*, 22–40.

Tantar, E., Schütze, O., Figueira, J. R., Coello, C. A. C., & Talbi, E.-G. (2008). Computing and selecting ϵ-efficient solutions of $\{0, 1\}$-knapsack problems. In *Proceedings of the 19th International Conference on Multiple Criteria Decision Making (MCDM-2008)* (pp. 379–389).

Teytaud, O. (2007). On the hardness of offline multi-objective optimization. *Evolutionary Computation, 15*(4), 475–491.

Tian, Y., Cheng, R., Zhang, X., & Jin, Y. (2017). PlatEMO: A matlab platform for evolutionary multi-objective optimization. *IEEE Computational Intelligence Magazine, 3*(4), 73–87.

Uribe, L., Lara, A., & Schütze, O. (2020). On the efficient computation and use of multi-objective descent directions within constrained MOEAs. *Swarm and Evolutionary Computation, 52*.

Valledor Pellicer, P., Iglesias Escudero, M., Fernández Alzueta, S., & Deb, K. (2020). Gap finding and validation in evolutionary multi- and many-objective optimization. In *Genetic and Evolutionary Computation Conference (GECCO-2020)* (pp. 1–8).

Van Veldhuizen, D. A. (1999). *Multiobjective evolutionary algorithms: Classifications, analyses, and new innovations*. Technical report, Air Force Institute of Technology.

Wagner, T., & Trautmann, H. (2010). Integration of preferences in hypervolume-based multiobjective evolutionary algorithms by means of desirability functions. *IEEE Transactions on Evolutionary Computation, 14*(5), 688–701.

Wang, R., Zhou, Z., Ishibuchi, H., Liao, T., & Zhang, T. (2016). Localized weighted sum method for many-objective optimization. *IEEE Transactions on Evolutionary Computation, 22*(1), 3–18.

Wang, F. H., Zhang, Y. L., Zhao, Y., & Rao, Q. (2018). External archive matching strategy for MOEA/D. *Soft Computing, 22*, 7833–7846.

Wessing, S., & Preuss, M. (2016). On multiobjective selection for multimodal optimization. *Computational Optimization and Applicatons, 63*(3), 875–902.

White, D. J. (1986). Epsilon efficiency. *Journal of Optimization Theory and Applications, 49*(2), 319–337.

Witting, K. (2012). *Numerical algorithms for the treatment of parametric multiobjective optimization problems and applications*. Ph.D. thesis, PhD thesis, Deptartment of Mathematics, University of Paderborn.

Xiong, F.-R., Qin, Z.-C., Xue, Y., Schütze, O., Ding, Q., & Sun, J.-Q. (2014). Multi-objective optimal design of feedback controls for dynamical systems with hybrid simple cell mapping algorithm. *Communication in Nonlinear Science and Numerical Simulation, 19*(5), 1465–1473.

Zapotecas Martínez, S., & Coello Coello, C. A. (2008). A proposal to hybridize multi-objective evolutionary algorithms with non-gradient mathematical programming techniques. In *PPSN '08: Proceedings of the 10th International Conference on Parallel Problem Solving From Nature* (pp. 837–846).

Zapotecas Martínez, S., & Coello Coello, C. A. (2010). An archiving strategy based on the convex hull of individual minima for MOEAs. In *2010 IEEE Congress on Evolutionary Computation (CEC)* (pp. 1–8). IEEE Press.

Zapotecas Martínez, S., & Coello Coello, C. A. (2012). *A direct local search mechanism for decomposition-based multi-objective evolutionary algorithms* (pp. 3431–3438).

Zapotecas Martínez, S., Montaño, A. A., & Coello Coello, C. A. (2011). A nonlinear simplex search approach for multi-objective optimization. In *2011 IEEE Congress on Evolutionary Computation (CEC)*, (pp. 2367–2374).

Zapotecas-Martínez, S., López-Jaimes, A., & García-Nájera, A. (2019). Libea: A Lebesgue indicator-based evolutionary algorithm for multi-objective optimization. *Swarm and Evolutionary Computation*, *44*, 404–419.

Zhang, Q., & Li, H. (2007). MOEA/D: A multi-objective evolutionary algorithm based on decomposition. *IEEE Transactions on Evolutionary Computation*, *11*(6), 712–731.

Zhou, A., Qu, B.-Y., Li, H., Zhao, S.-Z., Suganthan, P. N., & Zhang, Q. (2011). Multiobjective evolutionary algorithms: A survey of the state of the art. *Swarm and Evolutionary Computation*, *1*, 32–49.

Zitzler, E., & Thiele, L. (1999). Multiobjective evolutionary algorithms: A comparative case study and the strength Pareto approach. *IEEE Transactions on Evolutionary Computation*, *3*, 257–271.

Zitzler, E., Laumanns, M., & Thiele, L. (2002). SPEA2: Improving the strength pareto evolutionary algorithm for multiobjective optimization. In K. Giannakoglou et al. (Eds.), *Evolutionary Methods for Design, Optimisation and Control with Application to Industrial Problems (EUROGEN 2001)* (pp. 95–100). International Center for Numerical Methods in Engineering (CIMNE).

Zitzler, E., Thiele, L., Laumanns, M., Fonseca, C. M., & Da Fonseca, V. G. (2003). Performance assessment of multiobjective optimizers: An analysis and review. *IEEE Transactions on Evolutionary Computation*, *7*(2), 117–132.

Zitzler, E., Thiele, L., & Bader, J. (2008). SPAM: Set preference algorithm for multiobjective optimization. *Parallel Problem Solving From Nature PPSN X*, 847–858.

Zuiani, F., & Vasile, M. (2013). Multi agent collaborative search based on tchebycheff decomposition. *Computational Optimization and Applications*, *56*(1), 189–208.

The manufacturer's authorised representative in the EU is Springer Nature Customer Service Centre GmbH, Europaplatz 3, 69115 Heidelberg, Germany. If you have any concerns regarding our products, please contact ProductSafety@springernature.com

Printed and bound by CPI Group (UK) Ltd, Croydon, CR0 4YY
28/09/2026
02224073-0005